Advances in
Physical Organic Chemistry

Advances in Physical Organic Chemistry

Volume 14

Edited by

V. Gold

Department of Chemistry
King's College
University of London

Associate Editor

D. Bethell

The Robert Robinson Laboratories
University of Liverpool
England

1977

Academic Press
London · New York · San Francisco

A Subsidiary of Harcourt Brace Jovanovich, Publishers

ACADEMIC PRESS INC. (LONDON) LTD
24–28 Oval Road, London NW1

United States Edition published by
ACADEMIC PRESS INC.
111 Fifth Avenue, New York, New York 10003

Copyright © 1977 By Academic Press Inc. (London) Ltd

All Rights Reserved

No part of this book may be reproduced in any form by
photostat, microfilm, or any other means
without written permission from the publishers

Library of Congress Catalog Card Number: 62–22125
ISBN 0-12-033514–X

PRINTED IN GREAT BRITAIN AT
THE SPOTTISWOODE BALLANTYNE PRESS
BY WILLIAM CLOWES & SONS LIMITED
LONDON, COLCHESTER AND BECCLES

Contributors to Volume 14

T. W. Bentley, *Department of Chemistry, University College of Swansea, Swansea SA2 8PP, Wales, United Kingdom.*

Michael J. Blandamer, *Department of Chemistry, The University, Leicester LE1 7RH, United Kingdom.*

E. Buncel, *Department of Chemistry, Queen's University, Kingston, Ontario, Canada K7L 3N6.*

A. Pross, *Department of Chemistry, Ben Gurion University of the Negev, Beer Sheva, Israel.*

P. von R. Schleyer, *Institut für Organische Chemie, Universität Erlangen-Nürnberg, D-8520 Erlangen, West Germany.*

H. Wilson, *Department of Chemistry, Queen's University, Kingston, Ontario, Canada K7L 3N6.*

Contents

Contributors to Volume 14 v

Medium Effects on the Rates and Mechanisms of Solvolytic Reactions — *T. W. Bentley and P. von R. Schleyer*

1. Introduction 2
2. Mechanisms of Solvolysis of Secondary Substrates . . 5
3. The Solvent as Ionizing Medium 32
4. The Solvent as Electrophile 43
5. The Solvent as Nucleophile 45
6. Correlation of Solvolysis Rates 51

The Reactivity-Selectivity Principle and its Mechanistic Applications — *A. Pross*

1. Introduction 69
2. Theoretical Considerations 71
3. Mechanistic Applications of the Reactivity-Selectivity Principle 82
4. Conclusion 126

Physical Organic Chemistry of Reactions in Dimethyl Sulphoxide — *E. Buncel and H. Wilson*

1. Scope 133
2. Solute-Solvent Interactions 135

3. Proton Transfer Processes 150
4. Rate Variations in DMSO as a Guide to Mechanism . . 159
5. Role of DMSO in Selected Mechanism Studies . . . 174
6. Future Developments 188
7. Conclusion 191

Kinetics of Organic Reactions in Water and Aqueous Mixtures —
Michael J. Blandamer

1. Introduction 204
2. Scope of the Problem 211
3. Water Molecules and Hydrogen Bonding 219
4. Water in the Solid State 223
5. Water 229
6. Solutes in Water 237
7. Apolar Solutes in Water 248
8. Hydrophilic Solutes 259
9. Ionic Solutions 263
10. Aqueous Mixtures 280
11. Typically Aqueous Mixtures 290
12. Typically Non-Aqueous Mixtures with G^E Negative (TNAN Mixtures) 325
13. Typically Non-Aqueous Mixtures with G^E Positive (TNAP Mixtures) 333

Medium Effects on the Rates and Mechanisms of Solvolytic Reactions

T. W. BENTLEY

Department of Chemistry, University College of Swansea, Swansea SA2 8PP, Wales, United Kingdom

and

P. von R. SCHLEYER

Institut für Organische Chemie, Universität Erlangen-Nürnberg, D-8520 Erlangen, West Germany

1.	Introduction	2
	Scope of Review	2
	Definitions	4
2.	Mechanisms of Solvolysis of Secondary Substrates	5
	The S_N2–S_N1 Mechanistic Framework	5
	Relative Rates	8
	Stereochemistry	15
	Ion-pair Intermediates	17
	α-Deuterium Kinetic Isotope Effects	22
	The Ion-pair S_N2 Mechanism	27
	Summary	29
3.	The Solvent as Ionizing Medium	32
	Y-Scale of Solvent Ionizing Power	32
	Solvent Ionizing Power for Sulphonates Y	36
	Other Scales of Solvent Polarity	40
4.	The Solvent as Electrophile	43
5.	The Solvent as Nucleophile	45

6. Correlation of Solvolysis Rates 51
   The mY Equation 51
   Four-Parameter Equations 52
   Three-Parameter Equations 56
   Limitations 58
   Summary 61
   References 62

1. INTRODUCTION

Scope of Review

The rates of nucleophilic substitution and elimination reactions are markedly solvent dependent, and much research has been directed towards interpreting the detailed role of solvents in such processes. The mechanisms of these reactions, based on the S_N1-S_N2 framework of Ingold and coworkers, were among the first to be studied in detail and form one of the cornerstones of physical organic chemistry. The firm foundation provided by investigations, carried out over many years by independent research groups, has led to the use of solvolytic reactions as testing grounds for new interpretations and new mechanistic criteria (e.g. secondary isotope effects, p. 22). This review will be restricted to solvolytic processes, although there is also current interest in reactions of charged nucleophiles and bases with organic substrates (e.g. in dipolar aprotic solvents; Parker, 1969).

The term "solvolysis" was introduced by Steigman and Hammett (1937) for kinetically first-order nucleophilic displacements by solvent present in large excess, and considerable attention has been devoted to studying such processes over the past few decades (Streitwieser, 1956; Streitwieser, 1962; Thornton, 1964; Hartshorn, 1973). Most undergraduate texts on reaction mechanisms include discussion of solvolysis mechanisms; this review will concentrate on recent developments.

Solvolytic reactions frequently are used to derive structure/reactivity relationships (e.g. acetolysis of series of tosylates) and the results are implicitly assumed to be independent of the solvent. From recent studies in weakly nucleophilic, highly ionizing media such as 1,1,1,3,3,3-hexafluoropropan-2-ol and trifluoracetic acid, it is now known that the solvent dependence of relative rates has been underestimated. The implications of these recent developments are discussed in Section 2. For solvolytic reactions thought to proceed via carbocations or ion pairs, it is of interest to compare the solvent effects with those obtained in strongly acidic media such as "magic acid" and in the gas phase. From the thermochemical cycle (Fig. 1), it can be calculated that the activation energy for formation of a t-butyl cation and a chloride anion from t-butyl chloride is 159 kcal mol^{-1} (Gold, 1972); this is much greater than the activation energy of 23 kcal mol^{-1} for the hydrolysis of t-butyl chloride, which probably proceeds via an ion pair intermediate. However, despite minor mechanistic differences, it is clear that solvation of charged intermediates must greatly reduce the energies of solvolytic reactions compared with the corresponding gas phase reaction. Considering these large solvation energies, which would correspond to over 80 powers of ten in relative rates at 25°, it is perhaps surprising that very large solvent effects are not observed—e.g. the rate constant for hydrolysis of t-butyl chloride is only 5 powers of ten greater than the

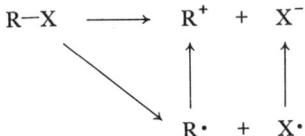

Figure 1. Thermochemical cycle for gas phase heterolysis.

corresponding ethanolysis rate constant. Also, the relative stabilities of carbenium ions appear to be surprisingly similar in strongly acidic media and in the gas phase (Fry *et al.*, 1970b).

Since solvolytic reactions involve the nucleophilic "push" and electrophilic "pull" (which, when occurring in enzymic systems, is frequently regarded as marvellous), it is possible to derive quantitative empirical measures of the role of solvent molecules as ionizing medium, electrophile and nucleophile. By comparing the sensitivity of various reactions to solvent polarity or "ionizing power" and nucleophilicity, it is possible to deduce mechanistic information which can supplement other evidence for reaction mechanisms. The basic ideas behind this approach are the qualitative proposals of Hughes and Ingold (1935); e.g. for a reaction during which charge separation increases in the transition state, an increase in solvent polarity should increase the rate of the reaction (see also Wigfield and Lem, 1975). This approach was put on a quantitative basis by Grunwald and Winstein (1948) with their development of the

Y-scale of solvent ionizing power. Further refinements have since been made and these are discussed on pp. 51–58.

To limit the scope of this review and to reduce overlap with other reviews, some related topics will only be discussed briefly. The addition of salts can affect both the rate and the course of solvolytic reactions and provides very important evidence for ion-pair intermediates. A full discussion of this topic has been published recently (Raber *et al.*, 1974); additional comments are given on pp. 27, 32. Also, we have generally excluded solvolyses known to proceed by competitive nucleophilically solvent-assisted and anchimerically assisted pathways. These solvolyses are very common (e.g. even n-propyl tosylate yields 87% of rearranged product during trifluoroacetolysis; Reich *et al.*, 1969), but a detailed account has been published recently (Harris, 1974). Recognition that solvolytic reactions could proceed by these two competitive, assisted pathways provided the key to the solution of the controversial "phenonium ion" problem (Lancelot *et al.*, 1972; Brown *et al.*, 1970), as well as inspiring the reinvestigation of the mechanisms of solvolyses of simple secondary substrates discussed in Section 2.

Definitions

The following specialized terms will be used:

Adamantyl—derivatives of structure [1];

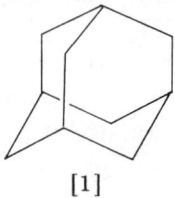

[1]

Anchimeric Assistance—a rate enhancing effect caused by electron release from a neighbouring bond or group with some movement of nuclei (purely hyperconjugative stabilization is thus excluded);

Brosylate—*p*-bromobenzenesulphonate (OBs);

Contact Ion Pair—synonymous with "tight" or "intimate" ion pair;

Internal Ion Pair Return—collapse of a contact ion pair to covalent material, sometimes referred to as "hidden return" if the covalent material is formed without scrambling of isotopic label or without partial racemization;

Nucleophilic Solvent Assistance—kinetically significant involvement of the solvent as nucleophile or base by partial bonding (as distinct from general electrostatic solvation) to any atom of the substrate (e.g. α-carbon, β-hydrogen, etc.);
Neophyl—the 2-methyl-2-phenyl-1-propyl system [2];
Norbornyl—derivatives of structure [3];
Pinacolyl—the 3,3-dimethyl-2-butyl system [4];
Tosylate—toluene-p-sulphonate (OTs).

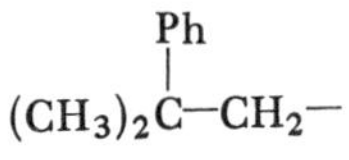

$(CH_3)_2\overset{Ph}{\underset{|}{C}}-CH_2-$ [3] $(CH_3)_3\overset{H}{\underset{|}{C}}\overset{|}{\underset{CH_3}{C}}-$

[2]    [3] 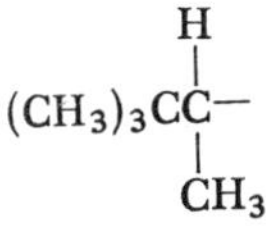   [4]

In specifying the composition of mixed solvents, the following abbreviations are used:

%EtOH, further abbreviated in Figures to %E: %v/v ethanol/water;
%(CF$_3$)$_2$CHOH: %w/w hexafluoropropan-2-ol/water;
%T: %w/w 2,2,2-trifluoroethanol/water.

2. MECHANISMS OF SOLVOLYSIS OF SECONDARY SUBSTRATES

Detailed discussions in this section will be restricted to the solvolyses of simple secondary substrates, i.e. those possessing saturated secondary alkyl groups which do not undergo neighbouring group participation. Solvolyses of other secondary substrates involving ion pair intermediates have recently been reviewed (Raber *et al.*, 1974).

The S_N2-S_N1 Mechanistic Framework

The variety of difficulties encountered when trying to fit the solvolysis of secondary substrates into the S_N2-S_N1 framework has led to much research and much discussion of the mechanisms of these reactions. Consideration has been given to "intermediate" mechanisms, possibly involving ion pairs, as well as to a spectrum of mechanisms between the S_N2 and S_N1 extremes. Since a vague theory cannot be proved wrong (Feynman, 1965), scientific progress in studies of reaction mechanisms benefits from the quantitative tests provided by studies of reactivity. Consequently, the correlations discussed later (Section 6) will be of considerable assistance in evaluating the various mechanistic proposals.

There is some confusion about the various theoretical and experimental distinctions between the $S_N 1$ and $S_N 2$ mechanisms. Since molecularity is the number of molecules necessarily undergoing covalency change during the rate determining step (Ingold, 1969), the unimolecular reaction ($S_N 1$) involves rate determining heterolysis of the R—X bond (k_1, Fig. 2) without assistance from nucleophilic attack. This definition is independent of the nature of the first intermediate, which in many cases is probably a contact ion pair rather than a free (i.e. symmetrically solvated) carbocation (see

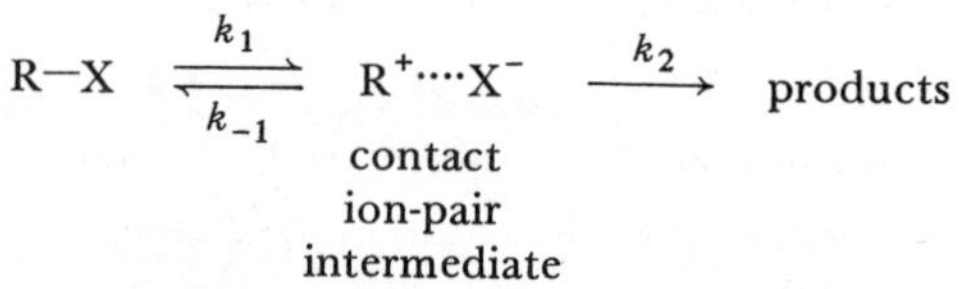

$$R-X \underset{k_{-1}}{\overset{k_1}{\rightleftarrows}} \underset{\substack{\text{contact}\\\text{ion-pair}\\\text{intermediate}}}{R^+ \cdots X^-} \xrightarrow{k_2} \text{products}$$

Figure 2. Simplified ion-pair mechanism.

Section 3, p. 36). Also, the theoretical definition does not specify the stereochemistry of the products, which are very rarely (if ever) found to be the 50/50 mixture of inverted and retained configurations predicted by classical theory (Doering and Zeiss, 1953; Okamoto et al., 1975a, b; Bone et al., 1975). We shall use the term $S_N 1$ to refer to mechanisms in which k_1 is rate determining and the term "classical $S_N 1$" to refer to the mechanism (possibly defunct) in which rate-determining formation of a free carbocation is envisaged. However, k_1 (Fig. 2) is only rate determining if $k_2 \gg k_{-1}$; when internal ion pair return is important ($k_2 < k_{-1}$) a more complex mechanistic notation is required (e.g. for solvolysis of t-butyl chloride in trifluoroethanol, p. 36).

According to the theoretical definition of molecularity, it is necessary for both RX and the nucleophile to undergo covalency change during the rate-determining step of the bimolecular reaction ($S_N 2$). Consequently, an $S_N 2$ reaction should be first order in nucleophile, but frequently this is not easy to establish experimentally because solvents are present in large excess and rates of reactions involving charged nucleophiles show complex salt effects (Section 3). The theoretical definition does not specify the stereochemistry of the $S_N 2$ reaction nor any other experimentally accessible parameter such as the entropy of activation (cf. Pritt and Whiting, 1975; see also Sinnott and Whiting, 1975).

Several mechanisms satisfy the theoretical requirements of the $S_N 2$ reaction (Fig. 3). We shall use the term "classical $S_N 2$" to refer to the reaction in which nucleophilic attack on covalent substrate leads directly to product. Two simple ion pair mechanisms will also be considered: the $S_N 2$ (intermediate) is similar to the classical $S_N 2$ mechanism except that the pentavalent species [5] is an intermediate rather than a transition state; the ion-pair $S_N 2$ mechanism involves

classical $S_N 2$:

$$R-X \xrightarrow{Nuc} [Nuc \cdots R \cdots X]^{\ddagger} \longrightarrow Nuc-R$$

$S_N 2$ (intermediate):

$$R-X \xrightarrow[\text{(slow)}]{Nuc} \underset{[5]}{[Nuc \cdots R \cdots X]} \longrightarrow Nuc-R$$

ion-pair $S_N 2$:

$$R-X \rightleftharpoons R^+ \cdots X^- \xrightarrow[\text{(slow)}]{Nuc} Nuc-R$$

Figure 3. Possible $S_N 2$ mechanisms.

rate-limiting nucleophilic attack on a preformed ion pair intermediate $R^+ X^-$. The latter mechanism corresponds to that shown in Fig. 2 with $k_2 [\text{Nuc}] \ll k_{-1}$, where k_{-1} is the rate constant for internal ion pair return and $k_2 [\text{Nuc}]$ is the rate constant for nucleophilic attack multiplied by the concentration of nucleophile; the overall rate of reaction of covalent RX is then given by eqn (1), which satisfies the theoretical definition of an $S_N 2$ reaction (Sneen, 1973).

$$\frac{-d[RX]}{dt} = k_t [RX] = \frac{k_1 k_2 [\text{Nuc}] [RX]}{k_{-1}} \quad (1)$$

It follows that, whatever detailed mechanisms might be associated with the terms $S_N 2$ and $S_N 1$, the vital difference between an $S_N 2$ and $S_N 1$ reaction is not dependent on the extent to which the bond to the leaving group is broken but *depends on whether or not nucleophilic attack occurs before the transition state in the rate determining step*. Since there is a clear-cut difference between whether one molecule does or does not attack, there is *in principle* a clear-cut difference between the two mechanisms (Gold, 1956). However, just as there is a whole range of covalent bond strengths of molecules in their normal states, there can be a whole range of extents to which the nucleophile can assist heterolytic cleavage by partial formation of a covalent bond at the rearside (e.g. in [5]). Even formation of a very weak, partial, covalent bond (as opposed to an electrostatic interaction) would constitute the covalency change required by Ingold's definition of the term $S_N 2$. Thus, in practice, it seems preferable to imagine that there is a spectrum of $S_N 2$ mechanisms involving varying amounts of nucleophilic attack and that the $S_N 1$ mechanism is the limiting case where nucleophilic attack does not occur before the transition state of the rate determining step (Thornton, 1964a). This interpretation corresponds to the $N - Lim$ spectrum of Winstein et al. (1951) (see also Swain and Mosely, 1955).

Relative Rates

In order to fit solvolytic reactions into the above framework it is important to establish the magnitude of nucleophilic solvent assistance, the extent to which nucleophilic attack by solvent assists heterolytic cleavage of the bond to the leaving group. 2-Adamantyl tosylate [6] was selected as a model for $S_N 1$ behaviour because non-bonded interactions [7] would reduce the possibility of nucleophilic attack (Fry et al., 1970a); extensive experimental evidence

[6]   [7]

consistent with the absence of kinetically significant nucleophilic attack in solvolyses of 2-adamantyl tosylate is summarized below:

(i) solvolyses of 2-adamantyl tosylate are more sensitive to solvent ionizing power than solvolyses of other secondary tosylates (Fry et al., 1970a; Bentley and Schleyer, 1976);

(ii) the ratio $[k_{EW}/k_{AcOH}]_Y$[1] is lower than for solvolyses of other secondary tosylates, suggesting that the other solvolyses are accelerated because mixtures of ethanol and water are

[1]Defined and further discussed on p. 44.

more nucleophilic than acetic acid (see Table 7 and also Streitwieser, 1956a);

(iii) the relative rate of 2-adamantyl tosylate and bromide (OTs/Br ratio) is similar to that of tertiary substrates (Fry et al., 1970a); this criterion of mechanism was proposed by Hoffmann (1965), but it now appears that OTs/Br ratios in secondary and tertiary systems are determined largely by steric effects (Slutsky et al., 1974);

(iv) 2-adamantyl tosylate does not undergo an $S_N 2$ reaction with azide ion in 80% ethanol/water (Raber et al., 1971);

(v) the α-deuterium isotope effect is higher than for most secondary solvolyses and is independent of the solvent (Harris et al., 1971; Shiner and Fisher, 1971)—see p. 22 for discussion;

(vi) there is a good correlation between rates of solvolyses of 2-adamantyl tosylate and of 1-bicyclo[2,2,2]octyl tosylate (Fig. 15)—see also pp. 36–38.

The above evidence firmly establishes that 2-adamantyl tosylate behaves very much like a tertiary substrate in its lack of response to changes in solvent nucleophilicity. As shown by correlations using eqn (9)—Section 6, Table 9—and by the relative rates in Table 1, other secondary tosylates are more responsive to solvent nucleophilicity. Schleyer et al. (1970) proposed that the magnitude of

TABLE 1

Relative Rates of Secondary Solvolyses[a]

Solvent Tosylate	EtOH	CH_3CO_2H	HCO_2H	97 wt % $(CF_3)_2CHOH$	CF_3CO_2H
2-Adamantyl [6]	1·0[b]	1·0[b]	1·0[b]	1·0[b]	1·0[b]
exo-2-Norbornyl	10,400	4,140	1,940	~1300[f]	520[c]
endo-2-Norbornyl	34	14	1·14	~0·8[f]	0·46[c]
Cyclohexyl	108	8·3	1·5	0·19[d]	0·30[e]
Cyclopentyl	6,250	280	27	3·1[g]	2·7[c]
3-Pentyl	1,560	40	5·3	0·51	0·85[e]
2-Propyl	910	13	0·9	0·016[d]	0·028[c]

[a] Titrimetric or conductimetric rate constants (k_t) at 25°; kinetic data summarised by Schadt et al., 1976, and Bentley and Schleyer (1976).
[b] Defined as 1·0 for each solvent.
[c] Nordlander et al., 1974.
[d] Schadt et al., 1974.
[e] Peterson et al., 1965.
[f] Lenoir et al., 1976 (estimated assuming OBs/OTs rate ratio = 3).
[g] Unpublished results from three independent laboratories.

nucleophilic solvent assistance in any solvent could be calculated by eqn (2), where k_t is the titrimetric rate constant. Thus, in acetic acid,

$$\begin{bmatrix} \text{Nucleophilic solvent} \\ \text{assistance} \end{bmatrix} = \frac{[k_t(\text{ROTs})/k_t(\text{2-AdOTs})]_{\text{ANY SOLVENT}}}{[k_t(\text{ROTs})/k_t(\text{2-AdOTs})]_{\text{CF}_3\text{CO}_2\text{H}}} \quad (2)$$

2-propyl tosylate appears to be nucleophilically assisted by a factor of $13/0\cdot028 = 470$ whereas the more hindered cyclopentyl tosylate is assisted by a factor of $280/2\cdot7 = 105$. An extensive tabulation is given in Table 2. As Schadt et al. (1976) have noted, the above argument and calculations are based on the following assumptions.

(a) Rates of solvolysis of 2-adamantyl tosylate are not nucleophilically assisted by the solvent. (Considering the extensive evidence discussed above, it is difficult to see how this assumption could be in error by more than a small rate factor).

(b) Rates of solvolysis of all secondary tosylates in trifluoroacetic acid are not nucleophilically assisted by solvent. Evidence supporting this assumption comes from Fig. 6 (discussed later), from the high α-deuterium isotope effect for trifluoroacetolysis of 2-propyl tosylate (Streitwieser and Daffom, 1969), and from the similarity of relative rates in 97% w/w hexafluoroisopropyl alcohol/water and in trifluoroacetic acid (Table 1), suggesting that a limiting value is being closely approached (Schadt et al., 1974).

(c) The variation in relative rates (Table 1) is caused solely by nucleophilic solvent assistance in solvolyses of tosylates less hindered than 2-adamantyl in media more nucleophilic than trifluoroacetic acid. As discussed later (Section 6), the precision of the correlations using eqn (9) suggests that this is a satisfactory assumption because the worst deviations from the correlations are less than a rate factor of five (Fig. 20).

The most likely direction of errors in these three assumptions are such that the calculations of Schleyer et al. (1970) will underestimate the magnitude of nucleophilic solvent assistance.[2] This can readily be seen for assumptions (a) and (b), but for assumption (c) it will be shown later that the major deviations from Fig. 20 are such that the calculated rate constants are too small. If there is an

---

[2]Since the original work, some of the experimental rate constants have been revised (Bentley and Schleyer, 1976).

TABLE 2

Minimum Estimates of Nucleophilic Solvent Assistance (Equation 2)[a]

Substrate	CF_3CO_2H	97 wt % $(CF_3)_2CHOH$	97 wt % CF_3CH_2OH	HCO_2H	70 wt % CF_3CH_2OH	H_2O	AcOH	50% EtOH	80% EtOH	EtOH
2-Adamantyl	$1 \cdot 0^c$	$1 \cdot 0^b$	$1 \cdot 0^b$	$1 \cdot 0^b$	$1 \cdot 0^b$	$1 \cdot 0^b$	$1 \cdot 0^b$	$1 \cdot 0^b$	$1 \cdot 0^b$	$1 \cdot 0^f$
Pinacolyl	$1 \cdot 0^c$	—	$3 \cdot 6$	$2 \cdot 4$	$3 \cdot 2$	$2 \cdot 2^d$	$8 \cdot 6$	$9 \cdot 5$	12	—
Cyclohexyl	$1 \cdot 0^c$	$0 \cdot 62$	$3 \cdot 2$	$5 \cdot 0$	$6 \cdot 6$	10^d	28	61	104	256^f
Cyclopentyl	$1 \cdot 0^c$	$1 \cdot 0$	—	10	—	15^d	105	160^e	455	1680^f
4-Heptyl	$1 \cdot 0^c$	$0 \cdot 87$	—	$3 \cdot 9$	—	—	28	46	146	—
3-Pentyl	$1 \cdot 0^c$	$0 \cdot 60$	—	$6 \cdot 2$	—	12^d	46	103	310	1310^f
2-Pentyl	$1 \cdot 0^c$	—	—	$9 \cdot 6$	—	—	88	195	616	—
2-Butyl	$1 \cdot 0^c$	—	$6 \cdot 3$	13	22	22^d	140	292	979	—
2-Propyl	$1 \cdot 0^c$	$0 \cdot 57$	$15 \cdot 3$	32	57	57^d	472	1130	4430	$23{,}500^f$

[a] Schadt *et al.*, 1976.
[b] From assumption (a); see text.
[c] From assumption (b); see text.
[d] Using kinetic data for the hydrolysis of mesylates (Bentley, 1974; Bowen, 1976).
[e] Bowen, 1976.
[f] Based on an extrapolated rate constant for 2-adamantyl tosylate in ethanol at 25°C.

additional effect (such as internal ion-pair return) causing deviations from eqn. (9), then the rate constants in the absence of this effect would be greater; it follows that the calculated magnitude of nucleophilic solvent assistance would also be greater.

An independent method for estimating nucleophilic solvent assistance has been applied to solvolyses of bromides with similar results (Fry et al., 1970b). In this approach, secondary solvolyses are compared with solvolyses of the corresponding tertiary substrates (e.g. 2-propyl compared with t-butyl). Again the reference substrate was 2-adamantyl (and 2-methyl-2-adamantyl [8]). It was found that

[8]

the tertiary substrate ([8], X = Br) solvolysed 10^8 times faster than 2-adamantyl bromide. Since this corresponds to an energy difference between tertiary and secondary systems of 11 kcal mol^{-1}, which is close to the value expected for free carbenium ions, it was argued that this energy or rate difference should be found in other systems in which both the secondary and the corresponding tertiary substrate solvolyse by $S_N 1$ mechanisms. Deviations from this value were taken to be a measure of nucleophilic solvent assistance in the solvolysis of the secondary substrate, e.g. in acetic acid, t-butyl-bromide solvolyses only $10^{4.2}$ times faster than 2-propyl bromide, suggesting that acetolysis of 2-propyl bromide is accelerated by nucleophilic solvent assistance corresponding to a rate factor of up to $10^{3.8}$ (i.e. $10^8/10^{4.2}$). Implicit in this argument is the assumption that the replacement of a hydrogen by a methyl group has the same steric and electronic effects in both 2-propyl and 2-adamantyl systems. Whilst it may be true in the above case (Fry et al., 1972), the assumption should be used cautiously (Bentley, 1974a). To test this approach, it would be useful to compare solvolyses of 2-propyl and t-butyl systems in a very weakly nucleophilic solvent, in which nucleophilic solvent assistance does not occur even for the secondary substrate. Unfortunately, in these solvents it appears that t-butyl halides solvolyse anomalously slowly (pp. 34–36).

In order to measure the large rate differences (up to 10^8) discussed above, it is necessary to carry out extrapolations of rate

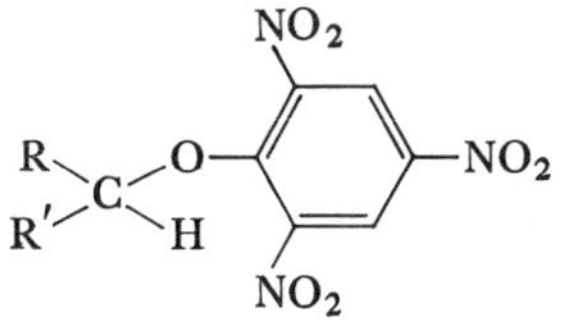

$$[9] \qquad [10]$$

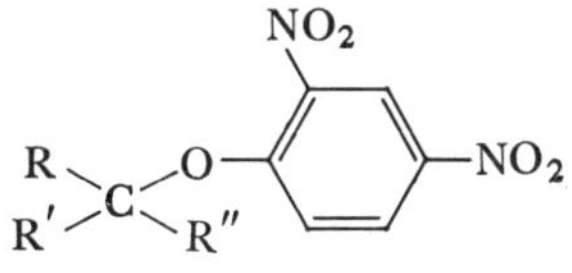

constants from high temperatures (>150°C) to 25°C. By comparing secondary alkyl picrates [9] with tertiary alkyl 2,4-dinitrophenyl ethers [10], Pritt and Whiting (1975) avoided errors due to temperature extrapolations, but introduced uncertainties due to the comparison of different leaving groups. The subtle problems that can arise when different leaving groups are compared can be illustrated by the *solvent* dependence of OBs/OTs ratios, referring to leaving groups which differ only in a remote *para*-substituent; in ethanol/water mixtures OBs/OTs ratios of five are observed, whereas in more acidic solvents values closer to three are usually found (Roberts, 1972).[3] However, a major advantage of Pritt and Whiting's approach is that the two methods for estimating nucleophilic solvent assistance (i.e. using solvolyses in trifluoroacetic acid or solvolyses of the 2-methyl-2-adamantyl system as reference point) can be based on the same leaving group (picrate); for acetolysis of 2-propyl picrate a rate factor of 400 (3·5 ± 0·1 kcal mol^{-1}) can be obtained using eqn (2), which is comparable to, but somewhat larger than the value of about 20 (1·7 ± 1 kcal mol^{-1}) obtained by using the tertiary system ([8], X = 2,4-dinitrophenolate) as reference. Thus, to a first approximation the results are similar, but it was suggested that the difference between the two estimates may be due to appreciable anchimeric assistance to trifluoroacetolysis of 2-adamantyl picrate, which is lower in magnitude for acetolysis. Consequently, values calculated by the equivalent of eqn (2) for picrates would be too large. At first sight this suggestion is attractive because anchimeric assistance in 2-adamantyl solvolyses would also explain a variety of other results:

(*i*) the preference for substitution with retention of configuration (Bone *et al.*, 1975), which is also observed to a lesser extent, however, in solvolyses of the corresponding tertiary system ([8], X = OTs), where anchimeric assistance should be absent.

---

[3] Whilst this small effect is well established, the relative rates of sulphonates are in general remarkably independent of substrate, mechanism and solvent (Su *et al.*, 1969; Kevill *et al.*, 1973).

[6] → [11] ← [12]

(ii) the formation of 0·4–0·5% of the rearranged isomer, the acetate of *exo*-protoadamantan-4-ol [11] during acetolysis of 2-adamantyl tosylate [6] (Lenoir et al., 1974a); rearrangement is greater during solvolysis of the 1-methyl-2-adamantyl tosylate [13] in 60% acetone, which yields the corresponding

[13] →(60% acetone/water, 75°) [14] +

protoadamantyl derivative [14] in 28% yield (Lenoir et al., 1974b) and during deamination of 2-aminoadamantane by acetolysis of the triazene [12] to the acetate [11] in 5·8% yield (Storesund and Whiting, 1975).

Unfortunately the relationship between rate and product determining steps is not clear, and anchimeric assistance should be determined from *kinetic* rather than product data (for an appropriate kinetic analysis of the 1-methyl system [13], see Lenoir et al., 1974b, c). The possibility of weak anchimeric assistance amounting to less than a factor of 10 in rate (at 25°C) has been noted (Lenoir et al., 1974a). However, the suggestion of Pritt and Whiting (1975) that its magnitude might be solvent dependent is novel, though not consistent with comparisons of 2-adamantyl tosylate with 1-bicyclo[2,2,2]octyl tosylate (see discussion of Fig. 15), nor with quantitative calculations of the solvent dependence of 2-phenylethyl tosylate in which anchimerically and solvent assisted pathways compete (Schadt and Schleyer, 1973).

The minimum estimates of nucleophilic solvent assistance for a variety of secondary tosylates (Table 2) show a marked dependence on the solvent and the structure of the tosylate. Lower values are obtained as solvent nucleophilicity decreases, as solvent ionizing power increases, and as the tosylate becomes more hindered towards rearside nucleophilic attack. For over twenty years, the rates of acetolyses of tosylates of many substrates have been assumed to be a measure of the carbocation stability. However, it now appears that the rates of many acetolyses are assisted by nucleophilic attack, so that these studies are unreliable guides to S_N1 reactivity. Both acetic and formic acids appear to be relatively nucleophilic compared with trifluoroacetic acid. Consequently, comparisons of relative carbocation reactivity should be based on rates of trifluoroacetolysis. Since kinetic studies using trifluoroacetic acid can be both unpleasant and unreproducible, Schadt et al. (1974) pointed out that hexafluoroisopropyl alcohol may be a preferable alternative. It is a weakly nucleophilic, highly ionizing solvent, which has the additional advantage that rate constants can be determined easily and accurately from conductance measurements.

These results provide strong evidence for the role of the solvent as a kinetically significant nucleophile in many secondary solvolyses, and hence suggest that they should be regarded as S_N2 reactions.

Stereochemistry

Although stereochemistry is not part of Ingold's theoretical definition of S_N2 and S_N1 mechanisms, it has nevertheless been used as an important guide to reaction mechanism. In the past, a major argument against putting solvolytic reactions such as acetolyses into the S_N2 category would have been that substitution occurred to give partially racemized product. Over the past ten years considerable evidence has been obtained showing that the substitution process occurs with essentially complete inversion of configuration and that the partial racemization is caused by competing side reactions.

Weiner and Sneen (1965a) showed that the hydrolysis of 2-octyl mesylate in pure water occurred with complete inversion of configuration, whereas some alcohol of retained configuration was formed from 2-octyl brosylate when "inert" cosolvents such as dioxan and acetone were present. Because addition of azide ion suppressed formation of the alcohol of retained configuration a double inversion process was implicated, in which the mesylate was first attacked nucleophilically by the cosolvent to give an intermediate [15] which was then attacked by water or by azide. As

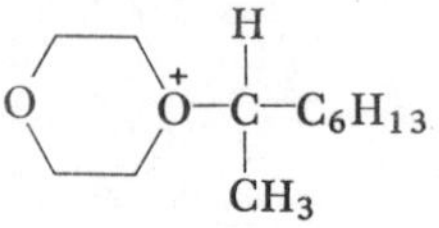

[15]

might be expected ethanolysis of 2-octyl tosylate and 2-butyl brosylate also proceed with essentially complete inversion of configuration (Streitwieser and Waiss, 1962).

Similar results have been obtained for acetolysis and formolysis. Streitwieser *et al.* (1965) showed that the amount of racemized product formed during acetolysis of 2-octyl tosylate at 75° depends on the reaction time. Presumably 2-octyl acetate is racemized by the liberated toluene-*p*-sulphonic acid. Attempts to remove the acid by addition of lithium acetate led to even more racemization (see also Streitwieser and Walsh, 1965). Using isotopically labelled acetic acid, they showed that some of the racemized 2-octyl acetate must arise by addition of acetic acid to octenes formed during the reaction (see also Campbell *et al.*, 1968). The unreacted 2-octyl tosylate recovered after 1·4 half lives was more highly racemized than the acetate product, indicating that the alkyl tosylate is racemized by liberated tosylate ion and/or toluene-*p*-sulphonic acid. Enabled by these control experiments to correct quantitatively for the various side reactions, Streitwieser *et al.* showed that the solvolytic displacement reaction itself proceeds with essentially complete inversion of configuration. Although the major product of acetolysis of cyclohexyl tosylates are alkenes, and the reaction is complicated by hydride shifts, Nordlander and McCrary (1972) observed complete inversion

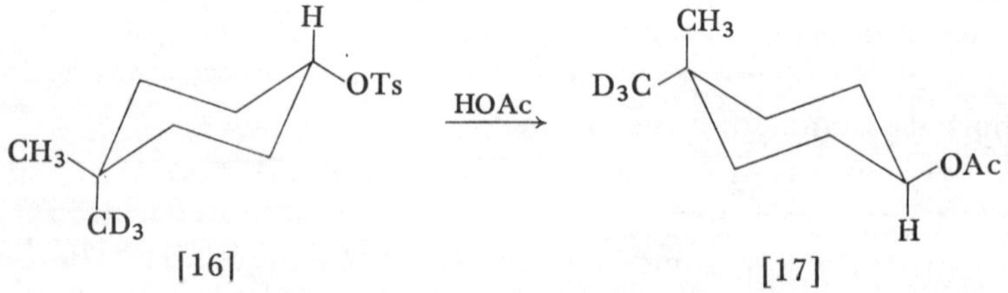

[16]   [17]

of configuration (±2% by nmr) in the non-rearranging displacement of [16] to give the acetate [17]. Similar results were obtained for acetolysis and formolysis of the isotopically labelled cyclohexyl tosylate [18] (Lambert *et al.*, 1972; Lambert and Putz, 1973). In solvolyses not complicated by hydride shifts, Humski *et al.* (1973) showed that the *p*-bromobenzenesulphonate [19] was substituted

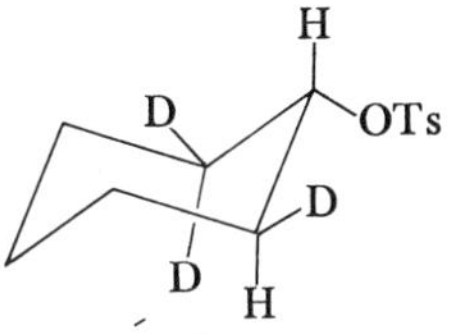

[18] 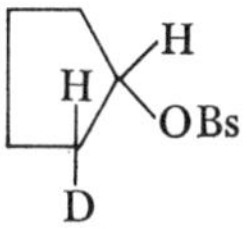   [19]

with >95% (from nmr) inversion of configuration at the reacting carbon in 80% v/v ethanol/water and 70% w/w trifluoroethanol/water.

Thus, as well as involving the kinetically significant nucleophilic role of solvent required by the definition of the term $S_N 2$, *these reactions show the essentially complete inversion of configuration characteristic of the classical $S_N 2$ mechanism.* However solvolyses of more hindered secondary tosylates such as menthyl tosylate (Hiršl-Starčević *et al.*, 1974) and 2-adamantyl tosylate (Bone *et al.*, 1975) appear to solvolyse with excess retention of configuration.

Ion Pair Intermediates

It is accepted, with varying amounts of caution by most researchers in the field, that the Winstein ion-pair mechanism for sequential ionization (Fig. 4) provides a reasonable general frame-

$$R-X \underset{k_{-1}}{\overset{k_1}{\rightleftarrows}} R^+ \cdots X^- \underset{k_{-2}}{\overset{k_2}{\rightleftarrows}} R^+ \| X^- \underset{k_{-3}}{\overset{k_3}{\rightleftarrows}} R^+ + X^-$$
[20]

Figure 4. Winstein's general ion-pair mechanism for sequential ionization.

work for discussion of solvolytic reactions. Since there are two recent reviews of this work (Harris, 1974; Raber *et al.*, 1974), detailed discussion here is not warranted, but several additional points may need emphasis.

(i) The scheme was deduced by fitting together evidence from a variety of solvolyses, usually involving substrates leading to cationic intermediates which could be stabilized by electron-releasing substituents. There are no examples involving *one* thoroughly studied system in which all three intermediates have been clearly implicated.

(ii) Winstein did not intend the representation R^+X^- to imply that formation of the contact ion pair occurred without nucleophilic solvent assistance (e.g. Winstein, 1965; Winstein and Takahashi, 1958). Unfortunately, nucleophilic solvation of R^+X^- has been explicitly excluded in interpretations of more recent work (Shiner, 1970; Sneen, 1973).

(iii) The ion pair $R^+ \parallel X^-$ is usually referred to as a solvent-separated ion pair and separation by at least one solvent molecule is implied. A rather different terminology has sometimes been used in the chemistry of carbanions for which there is spectroscopic evidence for the presence of "solvent-shared" ion pairs having one solvent molecule between the ions, and "solvent-separated" ion pairs having two solvent molecules between the ions (Griffiths and Symons, 1960; Symons, 1967; see, however, Hogen-Esch and Smid, 1965; Szwarc, 1968; Smid, 1972). This indicates a possible area of confusion as well as the possibility that the subtleties of solvation could give rise to a wide variety of ion pairs.

Based on Winstein's extended ion pair scheme (Fig. 4), Shiner and coworkers have provided a comprehensive interpretation of the rates and products of solvolytic reactions. Whereas we have based our discussions on 2-adamantyl tosylate [6], Shiner selected 3,3-dimethyl-2-butyl (pinacolyl) brosylate [21], which solvolyses with essentially complete rearrangement. Because addition of *p*-bromobenzenesulphonic acid to t-butylethylene [23] in trifluoroacetic acid produces the rearranged trifluoroacetate [25], Shiner *et al.* (1969b) argued that rearrangement in the contact ion pair [22] must be more rapid than internal return to pinacolyl brosylate [21]. Assuming that the same ion pair [22] would be produced from both the addition and solvolytic reactions, they proposed that, during solvolytic reactions internal ion pair return ($k_2 < k_{-1}$, Fig. 4) was prevented by a rapid methyl shift to the rearranged ion pair [24]. In contrast, addition of *p*-bromobenzenesulphonic acid to propene in

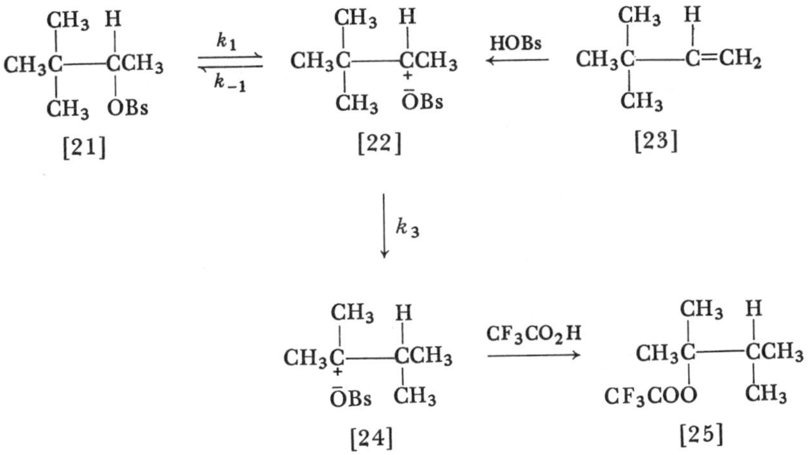

$$
\begin{array}{c}
\underset{[21]}{\overset{\text{CH}_3\ \ \text{H}}{\text{CH}_3\text{C}-\text{CCH}_3}}\ \underset{\substack{| \ \ \ |\\ \text{CH}_3\ \text{OBs}}}{\ \ } \quad \underset{k_{-1}}{\overset{k_1}{\rightleftarrows}} \quad \underset{[22]}{\overset{\text{CH}_3\ \ \text{H}}{\text{CH}_3\overset{+}{\text{C}}-\text{CCH}_3}}\ \underset{\substack{| \ \ \ |\\ \text{CH}_3\ \overline{\text{O}}\text{Bs}}}{\ \ } \quad \xleftarrow{\text{HOBs}} \quad \underset{[23]}{\overset{\text{CH}_3\ \ \text{H}}{\text{CH}_3\text{C}-\text{C}=\text{CH}_2}}\ \underset{\substack{|\\ \text{CH}_3}}{\ \ }
\end{array}
$$

$\downarrow k_3$

$$
\underset{[24]}{\overset{\text{CH}_3\ \ \text{H}}{\text{CH}_3\text{C}-\text{CCH}_3}}\ \underset{\substack{| \ \ \ |\\ \overline{\text{O}}\text{Bs}\ \ \text{CH}_3}}{\ \ } \quad \xrightarrow{\text{CF}_3\text{CO}_2\text{H}} \quad \underset{[25]}{\overset{\text{CH}_3\ \ \text{H}}{\text{CH}_3\text{C}-\text{CCH}_3}}\ \underset{\substack{| \ \ \ \ \ \ |\\ \text{CF}_3\text{COO}\ \ \text{CH}_3}}{\ \ }
$$

Figure 5. Shiner's ion-pair mechanism for pinacolyl solvolyses.

trifluoroacetic acid produced 2-propyl brosylate, so that collapse of the contact ion pair (k_{-1}, Fig. 4) appears to be more rapid than further dissociation (k_2) to produce the corresponding trifluoroacetate (Shiner and Dowd, 1969). Thus variations in k_2 relative to k_{-1} could account for the solvent effects on relative rates (Table 1), but the assumption that the same contact ion pair is produced from both solvolytic and addition reactions is highly questionable (Raber et al., 1974a). The mechanisms of electrophilic addition reactions are complex, but it appears that the ion pair produced by addition is solvated differently and collapses more readily than the ion pair produced in solvolysis (e.g. Brown and Liu, 1975).

More direct experimental evidence against Shiner's proposal was obtained by studying the 1-adamantylmethyl carbinyl system [26], which is structurally very similar to pinacolyl [21] except that the increased ring strain by rearrangement should make this step less favourable (Bentley et al., 1974). Although some rearranged product [28] was observed, the major products were the unrearranged acetate [27] and the alkene [29]. According to Shiner's interpretation this lack of rearrangement indicates that ion pair return (k_{-1}) in [26] should increase and therefore its rate of solvolysis should be slower than that of a pinacolyl brosylate [21]. The experimental results did not support this prediction. In agreement with classical theories of electron release from alkyl groups, [26] solvolysed slightly faster than [21] when allowance was made for the well-established OBs/OTs rate factor of three.

[Scheme showing structures [26], [27], [28], [29] with interconversions via k_1, k_{-1}, and HOAc]

Furthermore Shiner et al. (1969b) estimated a rate constant for trifluoroacetolysis of pinacolyl brosylate which appears to be about ten times too fast (Nordlander et al., 1973; Harper, 1968). This error means that calculations based on the incorrect rate constant will overestimate the importance of internal ion pair return in trifluoroacetic acid; Shiner et al. (1969b) estimated a rate factor of ten for the rate acceleration of pinacolyl brosylate over 2-propyl brosylate due to inductive effects. Using the revised experimental pinacolyl/2-propyl rate ratio of 160, we calculate a factor of 16 (i.e. 160/10) due apparently to reduction of the rate constant for trifluoroacetolysis of 2-propyl brosylate by internal ion pair return. However, even this revised value is probably too high because inductive/hyperconjugative effects in trifluoroacetolysis are greater than in formolysis and acetolysis (Peterson et al., 1965). The linear free energy relationship (Fig. 6) suggests that trifluoroacetolysis of simple alkyl tosylates must proceed by similar mechanisms (Bentley et al., 1974), presumably rate-determining formation of intimate ion pairs. Also the results for pinacolyl tosylate (Fig. 6) support the evidence from γ-deuterium isotope effects (Shiner et al., 1969b; Schubert and Le Fevre, 1969) which suggest that methyl migration does not occur in the rate determining step. The anomalous behaviour of 3-methyl-2-butyl tosylate (Pross and Koren, 1974) is probably due to anchimeric assistance by the neighbouring hydrogen atom (Winstein and Takahashi, 1958; Shiner, 1970a).

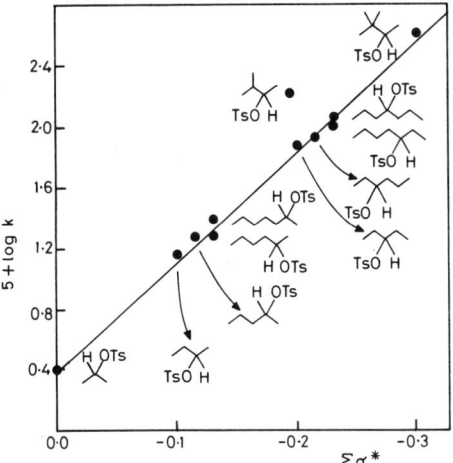

Figure 6. Correlation of trifluoroacetolysis rate constants for secondary tosylates at 25° and Taft σ^* (data from Bentley *et al.*, 1974, and Pross and Koren, 1974).

Whilst these results cast considerable doubt on Shiner's evidence for appreciable internal ion pair return, it should be emphasized that Shiner's interpretation is also based on α-deuterium isotope effects, which are discussed and criticized below.

There is also relatively direct evidence for internal ion pair return from ^{18}O-scrambling studies [eqn (3)] during solvolyses of simple

$$R-O-\overset{^{18}O}{\underset{O}{S}}-Ar \;\rightleftharpoons\; R^+\cdots{}^-O-\overset{^{18}O}{\underset{O}{S}}-Ar \;\rightleftharpoons\; R^+\cdots{}^{18}-O-\overset{O}{\underset{O}{S}}-Ar$$

[30]   [31]   [32]

$$R-{}^{18}O-\overset{O}{\underset{O}{S}}-Ar \quad (3)$$

[33]

secondary tosylates (Diaz *et al.*, 1968a). Although the extent of ^{18}O-scrambling in "unreacted" material was small (up to 20%, but usually much less), it could reasonably be argued that the ion pair intermediate [31] is much more likely to collapse to covalent starting material [30] than to isomerize with ^{18}O-exchange to [32].

Assuming that the ion pairs undergoing ^{18}O-*scrambling are the same as those undergoing solvolysis*, it would follow that such studies underestimate the magnitude of internal ion pair return in solvolysis. However, it may be that there are several contact ion pairs, all represented simply by [31] but differing in their solvation (Hammett, 1970a). Perhaps (for the sake of argument) ion pairs which are more strongly solvated electrophilically (e.g. by hydrogen bonding) are more likely to undergo solvolysis than ^{18}O-scrambling. Whilst this is only conjecture, it is generally agreed that the importance of internal ion pair return in solvolysis is difficult to quantify.

Other evidence for ion pair return is less relevant to solvolyses of simple secondary substrates. It is known that during solvolyses of certain optically active substrates racemization occurs more rapidly than solvolysis, but it is not known whether the ion-pair intermediates undergoing racemization are the same as those undergoing solvolysis (Hammett, 1970a). Such behaviour is usually observed in solvolyses where neighbouring group participation occurs and the intermediates are probably more stable than those from simple secondary solvolyses. As the stabilities of the intermediates increase, there appears to be a general trend towards formation of more dissociated species,[4] and thus the relevance of these results is questionable.

α-Deuterium Kinetic Isotope Effects

There can be no doubt that secondary kinetic isotope effects provide one of the most subtle probes of reaction mechanism currently available. The perturbation of the system under study is small in comparison with many other methods of probing reaction mechanisms such as changing solvent, replacing one substituent by another, or adding extra reagents to the reaction medium. Progress in this field is continuing and developments depend on a theoretical framework which can be thoroughly tested in a well-established area of mechanistic organic chemistry. The solvolysis reaction has been selected by several groups for studies of various secondary isotope

---

[4] For example, anchimerically assisted solvolyses may show the "special salt effect", implicating solvent-separated ion pairs, and solvolyses of highly stabilized substrates (e.g. diphenylmethyl compounds) show the "common ion effect", implicating kinetically free cations (Raber *et al.*, 1974). Also, Hammett (1970a) has drawn attention to an inconsistency in the widely accepted interpretation of the special salt effect.

effects (Thornton and Thornton, 1970; Williams and Taylor, 1974), but discussion here will be restricted to α-deuterium isotope effects. As detailed reviews (Shiner, 1970; Sunko and Borčić, 1970) and critical comments (Harris, 1974; Raber et al., 1974b) have been published recently, only more recent developments and additional comments will be included here. It will become apparent that there is no general agreement about the mechanistic interpretation of α-deuterium isotope effects. Since interpretation of more remote deuterium isotope effects appears to be even more complex (see for example, Fry and Badger, 1975), further discussion of more remote effects would be premature.

Isotope effects in the range 0·97-1·26 have been observed (Table 3) but according to Shiner (1970) an α-deuterium isotope effect close to unity (range 0·97-1·06) indicates a classical $S_N 2$ reaction. The isotope effect is theoretically related to the molecular vibrational frequencies of the initial and transition states for the protium and deuterium compounds. Thus, nucleophilic attack must reduce the change in vibrational frequencies.

Presumably less nucleophilically assisted solvolyses could show higher α-deuterium isotope effects, and there is a linear relationship between the magnitude of nucleophilic solvent assistance (Table 2) and the α-deuterium isotope effect for solvolyses of 2-propyl sulphonates (Fig. 7). Another measure of nucleophilic assistance is the ratio $k_2(OH^-)/k_1$, where k_2 is the second-order rate constant for nucleophilic attack by OH^- and k_1 is the first-order rate constant for reaction with the solvent water, and a linear correlation was obtained by plotting the ratio versus the experimentally observed isotope effects for methyl and trideuteriomethyl sulphonates, chlorides, bromides and iodides (Hartman and Robertson, 1960). Using fractionation factors the latter correlation may also be explained by a leaving group effect on initial state vibrational frequencies (Hartshorn and Shiner, 1972), but there seems to be no sound evidence to support the view that S_N2 reactions must give α-deuterium isotope effects of 1·06 or less.

In a study of the effect of azide ion on the solvolysis of 2-octyl brosylate in aqueous acetone, Raaen et al. (1974) showed that the α-deuterium isotope effects for both the alcohol and azide produced were almost identical (1·097 ± 0·007 and 1·106 ± 0·007 respectively). These results imply that the rate determining step(s) for formation of alcohol and azide are either identical or very similar. A likely explanation is that the two products are formed in parallel

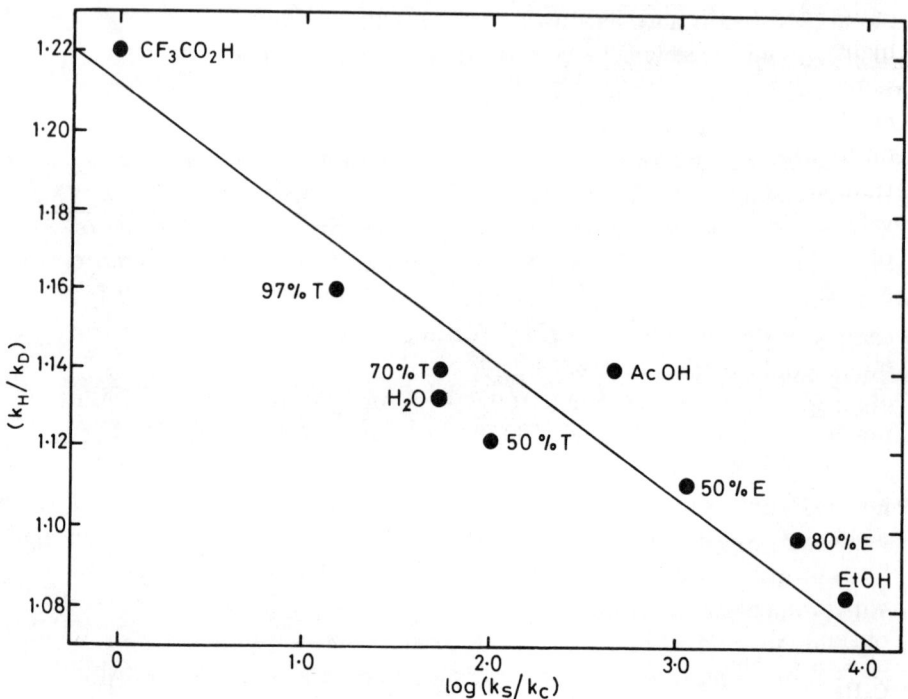

Figure 7. Correlation of α-deuterium isotope effect (k_H/k_D) and magnitude of nucleophilic solvent assistance for 2-propyl sulphonates (data from Schadt et al., 1976, and Table 3). Abbreviations for solvents: T = wt % trifluoroethanol/water; E = vol % ethanol/water.

S_N2 reactions for which the magnitudes of the nucleophilic assistance (and hence the α-deuterium isotope effect) are similar. Alternatively, rate-determining formation of a nucleophilically solvated intermediate ion pair may be followed by more rapid product determining steps (Fig. 8), although this mechanism may not be consistent with rate-product correlations (pp. 27, 28). Other interpretations seem even less probable. Surprisingly, Raaen et al. (1974) suggested that the results were consistent with classical competing S_N2 and S_N1 processes; even if one neglects the incorrect impli-

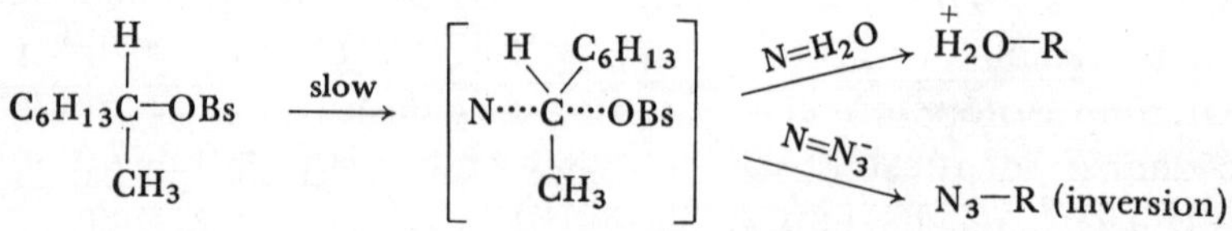

Figure 8. Possible S_N2 (intermediate) mechanism for hydrolysis in the presence of azide ion.

cation that free cations may be present in these reactions, it seems highly unlikely that nucleophilic attack by azide will occur in the same proportion of $S_N 2$ and $S_N 1$ reactions as attack by water. Criticisms of another alternative mechanism involving nucleophilic attack on a contact ion pair intermediate are discussed in the following section. Since both tenable mechanisms involve $S_N 2$ attack on the covalent substrate, these results suggest that α-deuterium isotope effects of 1·10 are consistent with an $S_N 2$ mechanism. Furthermore, similar experiments on cyclopentyl methanesulphonate in aqueous diglyme gave an α-deuterium isotope effect of 1·16, which was only marginally reduced (to 1·15) in the presence of sodium azide (Kovacĕvić et al., 1972). Consequently, for solvolyses of simple secondary sulphonates, α-deuterium isotope effects of up to 1·15 appear to be consistent with weakly nucleophilically assisted displacement on covalent substrate. Further, after considering several possible ion pair mechanisms, Westaway (1975) proposed the $S_N 2$ mechanism for the nucleophilic substitution (4) of the benzyl ammonium salt [35], for which an α-deuterium isotope effect of 1·09 per deuterium atom was observed.

$$C_6H_5S^- + C_6H_5CH_2\overset{+}{N}(CH_3)_2C_6H_5 \xrightarrow[0°C, NaNO_3]{DMF} C_6H_5SCH_2C_6H_5 + (CH_3)_2NC_6H_5$$
[35] (4)

Unfortunately, the prevailing view expressed in the literature is not in agreement with the above conclusion; the view that "no $S_N 2$ reactions are known which show isotope effects larger than about 1·04" (Humski et al., 1974) conflicts with the observations of Raaen et al. and Westaway (discussed above) which suggest to us that some $S_N 2$ reactions show α-deuterium isotope effects of at least 1·10. Observation of an α-deuterium isotope effect of 1·114 ± 0·003 led Vitullo and Wilgis (1975) to discount the classical $S_N 2$-type mechanism for the *intramolecular* substitution of the chloride [36] to give [37].

[36]   [37]

At least three distinct additional mechanisms have been proposed to account for the higher α-deuterium isotope effects (Shiner, 1970). One of these mechanisms is illustrated by the behaviour of pinacolyl brosylate. Because the isotope effects for pinacolyl brosylate were less than the maximum values observed (Table 3), some nucleophilic attachment of the leaving group in the transition state was proposed (Shiner et al., 1969b). This suggests that the rate determining step is formation of a contact ion pair, consistent with earlier discussion (p. 18). Also, very weak nucleophilic assistance (if any) is suggested

TABLE 3

α-Deuterium Isotope Effects in Solvolysis Reactions

Substrate[a]	Solvent	k_H/k_D[b] at 25°	Reference
Methyl tosylate	H_2O	0·985[c]	Llewellyn et al., 1960
Ethyl tosylate	H_2O	1·020[c]	Leffek et al., 1960
2-Propyl tosylate	H_2O	1·13	Leffek et al., 1960
2-Propyl brosylate	50% ethanol/water	1·114	Shiner et al., 1969b
	90% ethanol/water	1·083	Shiner et al., 1969b
	50% trifluoro-ethanol/water	1·122	Shiner et al., 1969b
	70% trifluoro-ethanol/water	1·140	Shiner et al., 1969b
	97% trifluoro-ethanol/water	1·16	Shiner et al., 1969b
	trifluoroacetic acid	1·22	Streitwieser and Dafforn, 1969
Pinacolyl brosylate	97% trifluoro-ethanol/water	1·153	Shiner et al., 1969b
	70% trifluoro-ethanol/water	1·152	Shiner et al., 1969b
	50% ethanol/water	1·159	Shiner et al., 1969b
2-Adamantyl tosylate	as for pinacolyl	1·23	Harris et al., 1971
Cyclopentyl brosylate	97·5% trifluoro-ethanol/water	1·25[d]	Humski et al., 1973

[a] Isotope effects for sulphonates are directly comparable, but iodides, bromides, chlorides, and fluorides give inherently different isotope effects (Shiner and Dowd, 1971).
[b] Per deuterium atom.
[c] Calculated by Shiner (1970b) from experimental data at other temperatures.
[d] At 30°C.

both by our calculations (Table 2) and by the lack of dependence of isotope effect on solvent nucleophilicity (Table 3; see also Winstein and Marshall, 1952). Thus there is considerable agreement from independent studies about the mechanism of solvolysis of pinacolyl brosylate.

However, it has been suggested that α-deuterium isotope effects of 1·23 must involve a more dissociated transition state, and rate-determining formation of solvent-separated ion pair from the contact ion pair (k_2, Fig. 4) was postulated (i.e. $k_{-1} > k_2$); this proposal has already been criticized (p. 18). It is now clear that such a small change in α-deuterium isotope effect is insufficient grounds for proposing a completely different mechanism, because *isotope effects depend on the structure of the alkyl group*; thus, for example, α-deuterium isotope effects have been reported as high as 1·25 for cyclopentyl brosylate in 97·5% trifluoroethanol/water (Humski *et al.*, 1973) and as low as 1·11 for 1-adamantyl methyl carbinyl tosylate [26] (Bentley *et al.*, 1974). Some indication of the expected dependence of isotope effects on structure can be obtained from published fractionation factors (Hartshorn and Shiner, 1972), but further calculations on more complex structures are needed. In retrospect, it is remarkable that a variety of secondary substrates, giving isotope effects greater than 1·15–1·16 were all shifted from the classical heterolysis mechanism (k_1, rate determining) to Shiner's mechanism (k_2, rate determining). This left only *one* firmly established example, pinacolyl solvolyses, of the classical heterolysis mechanism and it would appear to be a good example of the tail wagging the dog!

Shiner's third mechanism is intended to explain isotope effects in the range 1·06–1·15. It requires rate-determining nucleophilic attack on a preformed contact ion pair and corresponds with Sneen's mechanism. An alternative interpretation, based on varying amounts of nucleophilic assistance, has been discussed above and criticisms of Sneen's mechanism are discussed in the following section.

The Ion Pair S_N2 *Mechanism*

This section will be brief because thorough, critical summaries of the literature on Sneen's ion-pair S_N2 mechanism have recently been published (Raber *et al.*, 1974c; McLennan, 1976). According to this mechanism, nucleophilic attack occurs as a discrete, rate determining step on a contact ion pair. Sneen's claim to have trapped the reactive intermediate with azide ion was based on a rate-product correlation, which has been criticized extensively because of complications from salt effects (Raber *et al.*, 1974c; Dais and Gregoriou, 1974; Queen and Matts, 1975). By repeating the original work of Sneen and Larsen (1966, 1969) on 2-octyl mesylate at constant total salt concentration, which was maintained at 2·0 M by addition of $NaClO_4$ to the

0–0.3 M NaN$_3$, McLennan (1974) found a rate-product correlation in agreement with the classical S$_N$2 mechanism. Similar rate/product correlations were obtained when thiourea was used as nucleophile instead of azide; thiourea has the advantage that salt effects can be ignored because, by control experiments, it was established that the non-nucleophilic urea had a negligible effect on the reaction rate (McLennan, 1975).

To these apparently overwhelming criticisms of Sneen's interpretation, two discrepancies in the original experimental work can be added. Weiner and Sneen (1965b) reported that addition of 0.03 M NaN$_3$ to 2-octyl mesylate in 25% dioxan/water *reduced* the rate of reaction, consistent with the "negative salt effect" discussed by Raber et al., (1974c). However, in later work by Sneen and Larsen (1966, 1969), salt concentrations greater than 0.05 M were used and only increased rates of reaction were reported. Furthermore, Weiner and Sneen (1965a) reported that the azide product formed from reaction at 0.03 M or 0.04 M NaN$_3$ was only 80% optically pure. Consequently, additional pathways such as ones via the dioxonium ion [15] may be important, whereas Sneen and Larsen (1969) stated that the azide was "highly inverted" and implicitly assumed that such additional pathways were insignificant.

Raaen et al. (1974) have interpreted ^{12}C/^{14}C kinetic isotope effects as evidence against the ion pair mechanism; they also dispute an alternative interpretation by Graczyk and Taylor (1974), who used ^{35}Cl/^{37}Cl isotope effects in a study of the hydrolysis of *p*-methoxybenzyl chloride.

What is the present status of Sneen's ion-pair S$_N$2 proposals? For simple, primary, and secondary substrates without stabilizing groups there is no clear evidence requiring such a mechanism. Indeed, for methyl substrates the ion-pair S$_N$2 mechanism appears to be 40 kcal mol^{-1} less favourable than the observed (presumably classical S$_N$2) mechanism (Abraham, 1973). Systems such as allyl [38] (Sneen and Bradley, 1972; Sneen and Kay, 1972; Sneen and Carter, 1972), the interesting deactivated allyl system [39] (Bordwell and Mecca, 1975; Bordwell et al., 1975), α-phenylethyl [40] (Sneen and Robbins, 1972) and benzyl [41] (Sneen et al., 1973) are more capable of stabilizing the developing charge and may undergo the ion-pair S$_N$2 mechanism. This would be in agreement with the trend of the results discussed on p. 36, in which t-butyl chloride appears to undergo the ion-pair S$_N$2 or ion-pair E2 mechanism in the less nucleophilic solvents. At the extreme of this trend are the highly stabilized systems (e.g. triphenylmethyl [42]), the carbenium ions of which are known to be attacked by nucleophiles in the rate-determining step (Ritchie, 1972). However, these systems ([38-[41]) have not yet been examined by the linear free energy

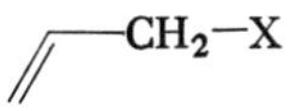

[38] CH₂=CH—CH₂—X

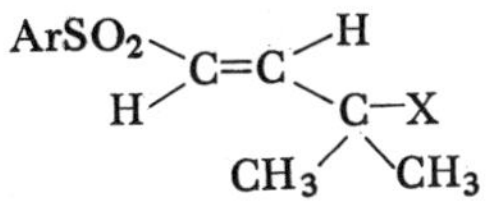

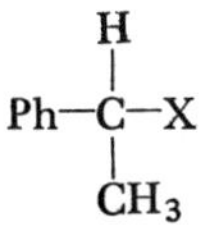

[39] ArSO₂(H)C=C(CH(CH₃)₂)—X (with CH₃ groups)

[40] Ph—CH(CH₃)—X

[41] PhCH₂—X

[42] Ph₃C—X

relationships (e.g. Figs. 6, 11, and 12) which provide a relatively direct probe for the presence of appreciable internal ion-pair return implicit in the ion-pair $S_N 2$ mechanism.

Summary

The solvent acts as a kinetically significant nucleophile in the overall solvolysis process for many simple secondary substrates, and this appears to be the major cause of the variation in relative rates with changes in solvent (Table 2, p. 11). This conclusion is supported by the quantitative correlations discussed in Section 6. The stereochemical evidence further suggests that, even when the magnitude of nucleophilic solvent assistance is less than a rate factor of 10 at 25°, solvolyses (e.g. of cylcohexyl tosylate in formic acid) can proceed with essentially complete inversion of configuration. These results are consistent with an $S_N 2$ mechanism and the evidence for ion pair intermediates can then be considered in one of two ways.

(1) It may be argued that the evidence for ion pair intermediates is too indirect, since it has not been established that the ion pairs undergoing ^{18}O-scrambling or internal return are the same as those undergoing solvolysis. Evidence for ion pairs would then be explained by side-reactions and the solvolytic reactions for which nucleophilic solvent assistance is greater

than a rate factor of 10 would be taken to proceed by the classical S_N2 mechanism.

(2) Alternatively, if it is accepted that ion pairs are involved in the solvolysis process and that nucleophilic solvent assistance is also significant, then the reaction might proceed via a nucleophilically-solvated ion-pair intermediate (e.g. [34], Fig. 8).

The latter mechanism has been termed "S_N2 (intermediate)" (Bentley and Schleyer, 1976), and the intermediate [34] has been referred to as an "ion-sandwich" (Bordwell et al., 1975; see also, Doering and Zeiss, 1953). The former term emphasizes the S_N2 character of the reaction, whereas the latter emphasizes the ionic

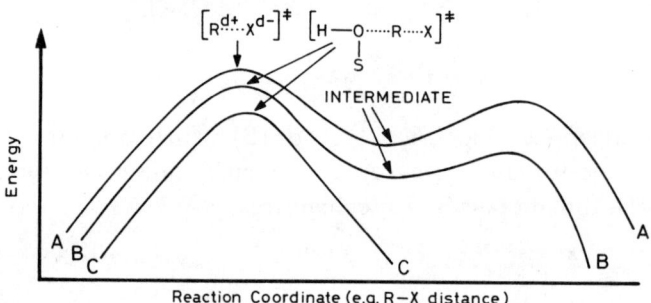

Figure 9. Schematic representation of upper portion of potential energy surface for merging of substitution mechanisms. A: S_N1 mechanism. No nucleophilic solvation in transition state; ion pair intermediate (possibly nucleophilically solvated); B: S_N2 (intermediate). Transition state is nucleophilically solvated by solvent (SOH); intermediate is a nucleophilically solvated ion pair (see Fig. 8); C: Classical S_N2. No energy minimum. In curves A and B, the second transition state may be of higher energy than the first in cases where internal return is important.

character of the intermediate. Possibly similar intermediates (e.g. RBr_2^-) have been postulated to explain some recent observations of gas phase reactions (Dougherty, 1974; Dougherty and Roberts, 1974), although calculations suggest that such species are very unsymmetrical (Dedieu and Veillard, 1972).

A change of mechanism from S_N1 to S_N2 can be envisaged (Fig. 9) to involve an increase in the magnitude of nucleophilic solvent assistance from zero (S_N1). We suggest that solvolyses for which nucleophilic solvent assistance is greater than a rate factor of 10 (Table 2) should be classified as S_N2.[5] This includes many solvolyses

---

[5] This criterion should be applied cautiously because misleading estimates of nucleophilic solvent assistance could be obtained for solvolyses of substrates which are insensitive to solvent nucleophilicity but are less sensitive than 2-adamantyl to solvent ionizing power.

which were formerly classified as $S_N 1$, borderline, or mixed $S_N 1$ and $S_N 2$. Intuitively, it seems likely that a substrate would take advantage of the amount of nucleophilic solvent assistance at its disposal and so competitive $S_N 1$ and $S_N 2$ mechanisms would only take place where nucleophilic solvent assistance was very small (<10). Then it would be necessary also to take account of a possible Boltzmann distribution of transition states. Thornton (1964a) has discussed the consequences of weak nucleophilic solvent assistance and arrived at similar conclusions; he also pointed out that an intermediate formed with nucleophilic solvent assistance $<RT$ (i.e. <0.6 kcal mol^{-1} at room temperature) could not be detected because of thermal agitation of the molecules. Also, it seems likely that these $S_N 2$ reactions which lie towards the $S_N 1$ end of the spectrum will show appreciable carbenium ion character.

The above refinements of the interpretation of earlier work have been brought about by two important developments, recognition from studies in trifluoroacetic acid that acetic and formic acids are nucleophilic (Peterson et al., 1965) and the use of 2-adamantyl tosylate as a reference point for $S_N 1$ reactivity (Schleyer et al., 1970). These developments have led to sensitive ways of detecting nucleophilic assistance and have shown that solvolyses of secondary substrates in acetic and formic acids may be significantly nucleophilically assisted (Table 2). Previous arguments were based on less sensitive tests such as the rate-accelerating effect of low concentrations of added nucleophiles, which must be appropriately oriented and desolvated before they are able nucleophilically to assist in the displacement reaction (Peterson et al., 1967).

All the currently available evidence for attack by nucleophile on an ion-pair intermediate in a discrete step can be explained by alternative mechanisms, at least for solvolyses of simple primary and secondary sulphonates. However, such ion pair mechanisms might be favoured by alkyl or aryl groups which strongly stabilize positive charges, and by weakly nucleophilic solvents. Moreover, other leaving groups (e.g. chloride or bromide) may be more prone to internal return; e.g. in the common ion effect, the mass law constants (α) depend on leaving-group nucleophilicity (Hine, 1962).

In the following Sections (3 and 6) more quantitative interpretations of the kinetic data will be discussed.

3. THE SOLVENT AS IONIZING MEDIUM

The background to the studies described below has been discussed in earlier reviews (Reichardt, 1965; Reichardt and Dimroth, 1968; Kosower, 1968; Dack, 1970), and also in the text by Bethell and Gold (1967).

Y-Scale of Solvent Ionizing Power

The empirical "ionizing power" parameter, Y, was introduced by Grunwald and Winstein (1948) and was defined by eqn (5) with

$$\log (k/k_0)_{RX} = mY \tag{5}$$

$m = 1$ for t-butyl chloride at $25°$; k is the rate constant for solvolysis in any solvent and k_0 that for solvolysis in 80% v/v ethanol/water. Thus all Y values are defined relative to 80% v/v ethanol/water, which is assigned a Y value of 0·0. A more complete tabulation of Y values was later reported by Fainberg and Winstein (1956). This parameter has proved to be of considerable utility in correlating solvolysis rates (Section 6).

Implicit in the choice of t-butyl chloride as the model compound is the proposal that the mechanism of solvolysis of t-butyl chloride involves rate-limiting formation of a carbocation or ion pair without nucleophilic solvent assistance. Initially, this suggestion was criticized using arguments which were later shown to be unsound. Thus, from a comparison of rates of solvolysis, the behaviour of t-butyl chloride was concluded to be intermediate between that of triphenylmethyl fluoride [42, X = F] and n-butyl bromide (Swain and Mosely, 1955), i.e. dependent on solvent nucleophilicity. However, Winstein et al. (1957) showed that this was probably due to a leaving group effect, the more electronegative fluorine being more susceptible than chlorine to solvent electrophilicity. Various studies (Le Roux and Swart, 1955; Hughes et al., 1955; de la Mare, 1955) alleging that an S_N2 reaction occurs between t-butyl bromide and lithium bromide or lithium chloride in anhydrous acetone did not adequately account for salt effects (Winstein et al., 1959a, 1959b), and did not even establish that the products were derived from a substitution reaction.

Since the solvent is present in large excess in a solvolytic reaction, it is not possible to determine from the kinetic order of the reaction whether the solvent acts as nucleophile. As mentioned above, and on p. 27, addition of nucleophilic salts leads to complications from salt effects, which are not yet fully understood (Raber et al., 1974; Grunwald and Effio, 1974; Dais and Gregoriou, 1974; Queen and Matts, 1975). Consequently it is not possible to justify directly the suitability of t-butyl chloride as a model compound. One indirect approach is to compare solvolysis rates for t-butyl chloride with those for another model compound. For this purpose Winstein et al., (1951) selected *trans*-2-bromocyclohexyl brosylate [43] and their

[43]

results are shown in Fig. 10; the good linear correlation appeared to give strong support to choice of *t*-butyl chloride. It is not clear, however, why the point for acetic acid should fit the correlation line, because it is now known that sulphonates in acidic solvents do not

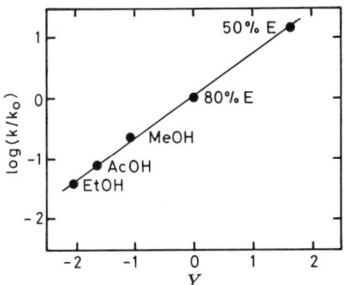

Figure 10. Correlation of solvolytic rate constants for *trans*-2-bromocyclohexyl *p*-bromobenzenesulphonate [43] at 50° and Y (data from Winstein et al., 1951).

generally correlate well (Smith et al., 1961; Kevill et al., 1970); as noted above for fluorides, differential electrophilic solvation effects become important. Two bridgehead caged systems 1-bicyclo-[2,2,2]octyl bromide [44] and 1-adamantyl bromide [45, X = Br]

[44]   [45]

have also been examined; [44] was studied in only five solvents (Finkelstein, 1955), but an extensive range of solvents was investigated for [45, X = OTs) (Kevill et al., 1970), and for [45, X = Br] (see Fig. 11; Raber et al., 1970). Again the correlation is somewhat better than would have been expected because the leaving groups are not the same. Closer inspection shows that there is a good linear correlation for 40%, 50%, 60%, 70%, and 80% aqueous ethanol, from which the points for 90% ethanol and acetic acid deviate. It is to be expected (Fainberg and Winstein, 1957b, c) that the more electronegative chlorine is more electrophilically solvated in acetic acid,

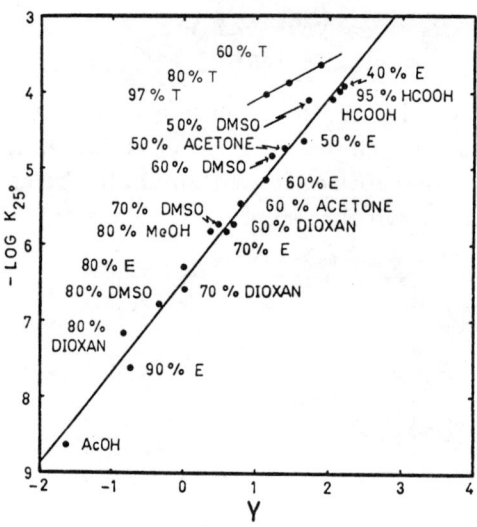

Figure 11. Correlation of solvolytic rate constants for 1-adamantyl bromide (45, X = Br) at 25° and Y. (From Raber et al., 1970, and reproduced by permission of the American Chemical Society.)

which explains why it would fall significantly below the correlation line. The point for 90% ethanol lies off the correlation line because of experimental errors exaggerated by errors of extrapolation to 25° (Bentley, 1971). However, it is surprising that the point for formic acid correlates well. Thus, the precision of the correlation in Fig. 11 appears to be somewhat fortuitous.

Having noted that two correlations (Figs. 10 and 11), which appeared at first sight to support fully the choice of t-butyl chloride as a model compound, cannot be completely explained, it is not surprising that rates of a solvolysis of 1-adamantyl chloride do not correlate well with Y as shown in Fig. 12 (Morten, 1975). Similar

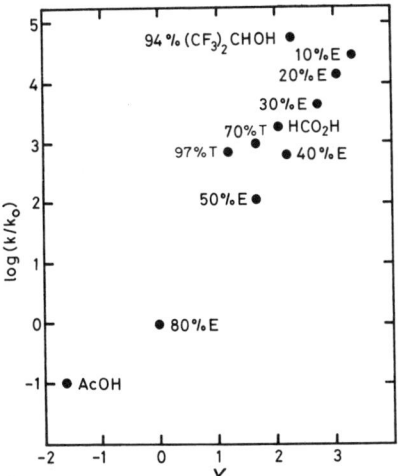

Figure 12. Correlation of solvolytic rate constants for 1-adamantyl chloride [45, X = Cl] at 25° and Y (data from Bentley, 1971).

results for ethanol/water mixtures and acetic and formic acids were obtained for neophyl chloride [46, X = Cl] (Fainberg and Winstein, 1957c.) Relative to the correlation line for aqueous ethanol, all the more acidic solvents give rate constants which are either too large for 1-adamantyl chloride or too small for t-butyl chloride. It is not clear how the solvolysis rates of 1-adamantyl chloride in acidic solvents could be anomalously fast, but as discussed below there is independent evidence from deuterium kinetic isotope effects (Table 4) that (titrimetric) solvolysis rates for t-butyl chloride in trifluoroethanol are reduced by ion-pair return. (Solvolysis rate constants are often determined from the rate of production of acid and may conveniently be referred to as titrimetric rate constants, although many are actually determined from conductance measurements.)

TABLE 4

Isotope Effects on Solvolysis of t-Butyl-d_9 Chloride

Solvent	k_H/k_{d_9}	Olefin fraction
97 wt % trifluoroethanol/water[a]	2·616	0·30
70 wt % trifluoroethanol/water[a]	2·493	0·17
54 vol % ethanol/water[b]	2·35 ± 0·03	0·083
29 vol % acetic acid/formic acid[b]	2·30 ± 0·05	0·118

[a] Shiner et al., 1969a.
[b] Frisone and Thornton, 1968.

Since both the deuterium isotope effects (Table 4) and the olefin fractions were larger in trifluoroethanol than in ethanol, Shiner et al. (1969a) argued that elimination from an ion pair may be rate-limiting. Thus, in the mechanism shown in Fig. 2, k_{-1} is greater than k_2 (leading to substitution and elimination), and the rate constant for ionization (k_1) is reduced by the fraction $k_2/(k_{-1} + k_2)$. Formation of the intermediate ion pair (R^+X^-) is consistent with the product studies of Cocivera and Winstein (1963) and Darwish and Tourigny (1972); in contrast to the predictions of the classical E1 mechanism, the products of solvolyses of t-butyl halides are dependent on the leaving group and, because of uncertainties in the interpretation of salt effects, there is no clear evidence for a rate depression when a common ion is added to the solution (common ion or mass law effect). A classical E2 mechanism also seems unlikely because trifluoroethanol is less basic than ethanol; if the E2 mechanism operated less olefin (not more) would be expected.

As shown in Table 4, the extent of rate-limiting elimination from the t-butyl chloride contact ion pair (R^+X^-) is greater in 97% trifluoroethanol than in 70% trifluoroethanol. This is entirely consistent with the deviations of the points for these solvents from the correlation lines of Fig. 11 and Fig. 12. (Elimination from 1-adamantyl halides is very unlikely because of the highly strained bridgehead olefin that would be produced.) Unfortunately, kinetic isotope effect data for trifluoroacetic acid, hexafluoroisopropanol, formic acid and acetic acid are not available, but the results in Table 4 for a mixture of acetic and formic acids show that the amount of elimination is small. However, the internal return from a contact ion pair probably occurs during substitution processes, causing a lower kinetic isotope effect (Shiner et al., 1969a); this explanation implies that internal return from the contact ion pair would be more important for t-butyl chloride than for 1-adamantyl chloride even if elimination did not occur. Further work is needed, but it appears that 1-adamantyl chloride would be a better model compound than t-butyl chloride as a basis for an empirical scale of solvent ionizing power for chlorides (see also Harris et al., 1974; Rappoport and Kaspi, 1974a, b; Sunko et al., 1972; Sunko and Szele, 1972).

Solvent Ionizing Power for Sulphonates (Y_{OTs})

The possibility that internal return from a contact ion pair might reduce titrimetric rate constants was considered by Fainberg and

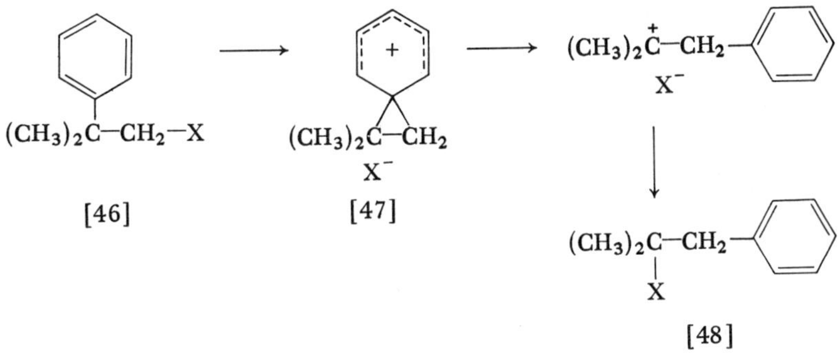

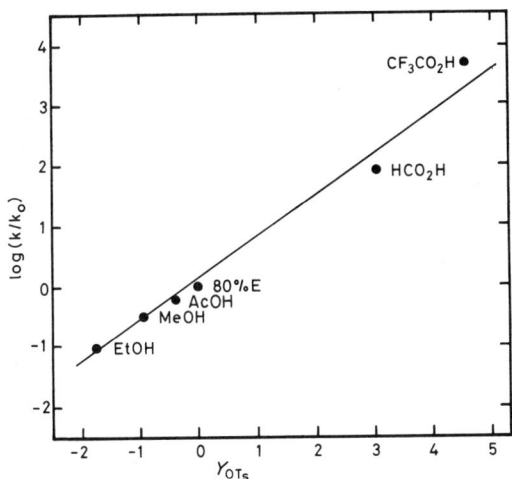

Figure 13. Winstein's mechanism for neophyl solvolyses.

Winstein (1957c). They argued that during solvolysis of neophyl substrates [46, X = Cl, Br] internal return from the proposed phenonium ion-pair intermediate [47] would yield the tertiary derivative [48] (Fig. 13) and this would solvolyse rapidly. Thus, the titrimetric rate constants should correspond to the ionization rate constant k_1 (Fig. 2). Later neophyl tosylate [46, X = OTs] and its p-methoxy derivative were used (Smith et al., 1961; Diaz et al., 1968b; Yamataka et al., 1973), and recently Schadt et al. (1976) defined as a scale of solvent ionizing power for tosylates, designated $Y_{\text{2-AdOTs}}$ or Y_{OTs}, using eqn (5) but based on 2-adamantyl tosylate [6] instead of t-butyl chloride. A correlation of the rates of solvolysis of neophyl tosylate with Y_{OTs} (Fig. 14) is satisfactory (correlation coefficient

Figure 14. Correlation of solvolytic rate constants for neophyl tosylate [46] at 50° and Y_{OTs} (data from Smith et al., 1961, and Diaz et al., 1968b).

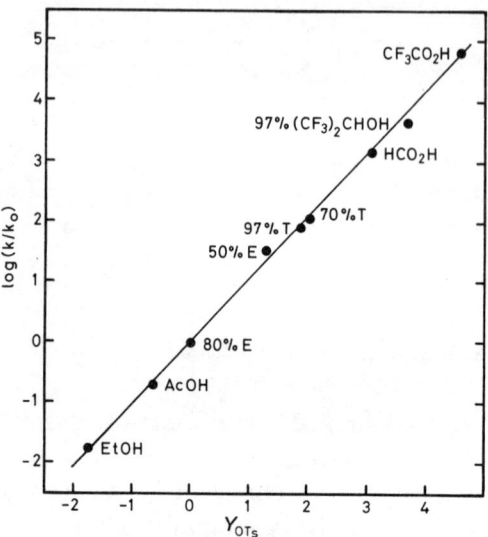

Figure 15. Correlation of solvolytic rate constants for bicyclo[2,2,2]octyl tosylate [49] at 25° and Y_{OTs} (data from Morten, 1975).

0·993) and a similar good correlation for 1-bicyclo[2,2,2]octyl tosylate [49] (Fig. 15) has a correlation coefficient of 0·999 (Morten, 1975). Remarkably, the latter correlation is very close to unit slope (1·04 ± 0·02). Of particular importance in both correlations is the absence of significant "dispersion" into solvent types frequently observed using standard Y values (e.g. Fig. 12).

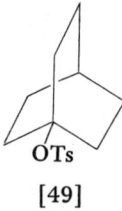

[49]

Consequently there appears to be a sound empirical basis for the use of the Y_{OTs} scale of solvent ionizing power. Its use should be restricted to sulphonates, however, because of the differential effects of electrophilic solvation in acidic solvents (see Section 4). The importance of these effects can be seen by comparing the Y and Y_{OTs} values for carboxylic acids (Table 5); it appears that, relative to 80% ethanol/water, a carboxylic acid ionizes a tosylate about ten times more rapidly than a chloride.

TABLE 5
Empirical Scales of Solvent Polarity

Solvent	Y^a	$Y_{OTs}{}^b$	$d_2{}^c$	Z^d	E_T [51]^e	E_T [52]	Dielectric Constant (25°)^g
Trifluoroacetic acid	1·84^h	4·57	—	—	—	—	8·4^i
Formic acid	2·05	3·04	6·53	—	—	—	57·9
Acetic acid	−1·64	−0·61	3·12	79·2	—	—	6·19
97% w/w Hexafluoropropan-2-ol/water	2·46^j	3·61	—	—	—	—	(16·7)^p
2,2,2-Trifluoroethanol	1·045^k	1·80	—	—	—	73·0^f	26·7^l
Water	3·49	4·0^m	4·01	94·6^n	63·1	69·5^f, 69·4^e	78·5
Methanol	−1·09	−0·92	−0·73	83·6	55·5	64·6^e	32·6
Ethanol	−2·03	−1·75	−1·03	79·6	51·9	61·2^e	24·3
Propan-2-ol	−2·73	—	—	76·3	48·6	57·2^e	18·3
t-Butyl alcohol	−3·26	—	—	71·3	43·9°	51·1^e	12·2

^{a} Fainberg and Winstein, 1956, except where otherwise noted.
^{b} Schadt et al., 1976 and references there cited, except where noted otherwise.
^{c} Swain et al., 1955a [equation (13)]. For discussion see Section 6.
^{d} kcal mol^{-1} at 25° Kosower 1968.
^{e} kcal mol^{-1} at 25° (Dimroth et al., 1963; see also Tamura and Imoto, 1975).
^{f} kcal mol^{-1} at 25° (Rappoport and Kaspi, 1974b).
^{g} Reichardt, 1965 and references there cited, except where otherwise noted.
^{h} Haake and Ossip, 1971.
^{i} Harriss and Milne, 1971.
^{j} Sunko and Szele, 1972.
^{k} Shiner et al., 1969a; da Roza et al., (1973) found $Y = 1 \cdot 147$.
^{l} Murto and Heino 1966; Eckstrom et al., 1960.
^{m} Estimate based on solvolysis of 2-adamantyl methanesulphonate in 5% v/v acetone/water (Bentley, 1974b).
^{n} Estimated by extrapolation (see text).
^{o} 30°.
^{p} Value for the pure solvent quoted by Matesich et al., 1973.

Other evidence to support the selection of 2-adamantyl tosylate as model compound has been discussed in Section 2 (p. 8), and the main conclusion can be summarized as follows. Independent evidence and the results in Fig. 15 firmly establish that nucleophilic solvent assistance cannot be appreciable for solvolyses of 2-adamantyl tosylate, since its response to solvents of widely varying nucleophilicities is almost exactly the same as 1-bicyclo[2,2,2]octyl tosylate [49] for which rearside attack is impossible; if solvolyses of 2-adamantyl tosylate are anchimerically assisted, the extent of anchimeric assistance does not appear to be dependent on solvent (cf. Pritt and Whiting, 1975). In addition to these conclusions, the results in Fig. 14 suggest that internal return from intermediate contact ion pairs does not occur to a detectable extent for solvolyses of 2-adamantyl tosylate (see also Bentley and Schleyer, 1976).

Other Scales of Solvent Polarity

The scales of solvent polarity discussed above are based on rate processes (6) in which a polar transition state is formed from a covalent initial state. There are also several scales based on an electronic transition in which an electron is transferred from one species to another or from one end of a molecule to another [eqn (7); Reichardt, 1965; Reichardt and Dimroth, 1968; Kosower, 1968]. All these scales represent aspects of the microscopic behav-

$$D, A \rightarrow D^+A^- \qquad (6)$$

$$D^+A^- \rightarrow D, A \qquad (7)$$

iour of the solvent, whereas the dielectric constant of the solvent, another measure of polarity, is a macroscopic property.

As might be expected the macroscopic property, dielectric constant, appears to be an unsuitable parameter to compare with microscopic rate processes. One of the most outstanding anomalies can be seen by comparing acetic acid with trifluoroacetic acid (Table 5). Despite the fact that the dielectric constant of trifluoroacetic acid is only slightly greater than that of acetic acid, the ionizing power of trifluoroacetic acid on the Y scale is considerably greater; it is even higher on the Y_{OTs} scale than formic acid. It appears that the ability of a solvent to assist formation of an ion pair ("ionizing power") bears no direct relation to its ability to separate those charges further ("dissociating power"). Presumably, dielectric constant is more of a guide to the "dissociating power" of a solvent than to its "ionizing power", but in certain limited ranges of solvents, electrostatic theories of solvation based on dielectric constant have achieved modest success (Laidler, 1960; Tommila, 1967; see also Fainberg and Winstein, 1956).

Since the parameters based on electronic transitions represent microscopic properties, these should be appropriate for comparison with rate processes. Unfortunately, experimental difficulties limit their usefulness for our purposes. The Z parameter (Kosower, 1958), based on the solvent dependence of charge transfer transitions in pyridinium iodides (e.g. [50]), is limited to solvents of $Y < 1$

$$\underset{[50]}{\underset{\overset{|}{CH_2 CH_3}}{\overset{CO_2 CH_3}{\bigodot}} I^-}$$

because the longest wavelength charge transfer band moves to such short wavelengths in highly polar solvents that it cannot be distinguished from the much stronger $\pi \rightarrow \pi^*$ transition of the pyridinium ion. However, a Z value for pure water was obtained by extrapolation of a graph of Y versus Z for methanol/water, ethanol/water, and acetone/water mixtures. Because the Z value of acetic acid can be measured directly, a similar extrapolation could lead to estimates of the Z values of formic and trifluoroacetic acids, but it has been pointed out that such procedures should be used with caution (Kosower, 1968).

By constructing betaines (e.g. [51], [52]) in which the solvent-sensitive absorption band is displaced to longer wavelengths, Dimroth et al. (1963) were able to obtain directly a solvent polarity scale (E_T), including more polar solvents. Unfortunately, acidic solvents cannot be studied because the oxygen atom of the indicator is protonated by these solvents. Thus in two of the solvents (formic acid and trifluoroacetic acid), independent measures of solvent polarity, which would have been particularly helpful in analysing the results of rate correlations (e.g. Fig. 12), are not available.

A summary of the empirical scales of solvent polarity for the pure solvents frequently used for solvolytic studies is shown in Table 5.

Because of the experimental difficulties discussed above, there is limited opportunity for direct comparison between the different scales. There is a satisfactory qualitative trend in all the scales (including dielectric constant) for water, methanol, ethanol, propan-2-ol, and t-butyl alcohol. However, it has been noted that a graph of Z versus E_T for pure alcohols and water is markedly curved (Fig. 16) and that 2,2,3,3-tetrafluoropropan-1-ol does not fit the curve (Dimroth *et al.*, 1963). The results for both Y and Z scales suggest that the E_T parameter for water is surprisingly low. Consequently,

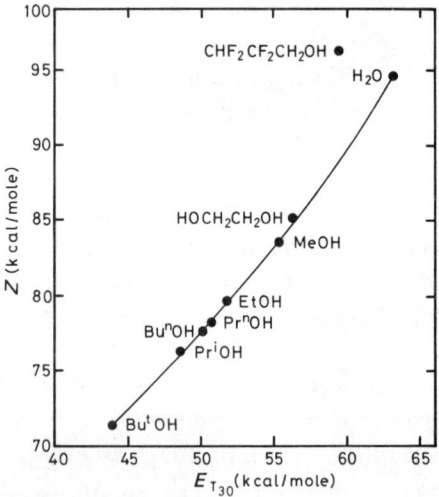

Figure 16. Comparison of Z and E_T scales of solvent polarity (data from Dimroth *et al.*, 1963).

the recent suggestion (Rappoport and Kaspi, 1974b) based on E_T values that the ionizing power of 2,2,2-trifluoroethanol may be greater than that of water is of doubtful significance. Both Y and E_T scales indicate that 2,2,2-trifluoroethanol is a much better ionizing medium than ethanol, although they have similar dielectric constants. Conductance measurements of electrolytes in 2,2,2-trifluoroethanol show that mobility and association behaviour are consistent with enhanced solvation of anions (e.g. by hydrogen bonding; see also Sep et al., 1975) and diminished solvation of cations relative to ethanol (Evans et al., 1971). Similar results have been obtained for 1,1,1,3,3,3-hexafluoro-2-propanol (Matesich et al., 1973).

Because of the different effects of electrophilic solvation of the various negative charges (i.e. Cl^- for Y, ^-OTs for Y_{OTs}, I^- for Z, O^- for E_T), direct comparisons between the various scales should be done cautiously. A wide variety of correlations giving clear indications of trends, has been reported by Reichardt and Dimroth (1968), but the significance of a recent general survey of scales of solvent polarity is doubtful, because of the many parameters used in the correlations (Fowler et al., 1971). The multi-parameter approach has also been adopted and reviewed by Koppel and Palm (1972).

4. THE SOLVENT AS ELECTROPHILE

The idea that electrophilic attachment to the leaving group assists heterolytic cleavage is well established (Streitwieser, 1956b). In the case of metal salts such as those of silver and mercury, the importance of the metal ion in the transition state is apparent from the rate law (Hartshorn, 1973a; Bach and Willis, 1975). For solvolytic reactions carried out in the absence of metal salts, evidence for electrophilic solvation is less direct but extensive. Farinacci and Hammett (1937) showed that addition of small amounts of water increased the rate of alcoholysis of diphenylmethyl chloride, whilst the product remained almost entirely the corresponding ethyl ether. Thus it appears that the water may hydrogen bond to the developing chloride ion to give a transition state [53], and presumably in the absence of water,

$$\left[\begin{matrix} Ph \\ Ph \end{matrix} \!\!>\!\! CH \cdots Cl \cdots H\!\!-\!\!OH \right]^{\ddagger}$$

[53]

ethanol performs the same function but does it less effectively (Hammett, 1970b). Similar catalytic effects have been observed for the addition of hydroxylic solvents to aprotic solvents (Hartshorn, 1973a; Swain, 1948). Brown and Hudson (1953) showed that, for several $S_N 1$ solvolyses in various binary solvent mixtures, the logarithms of the rate constants were proportional to log $[H_2O]$ v/v for solutions of low water content. According to their simple theory, the slopes of these graphs (4 ± 0·5) correspond to the number of water molecules specifically solvating the $S_N 1$ transition state (see also Böhme and Schürhoff, 1951; Tommila, 1955).

Inevitably this specific electrophilic solvation must be included in the Y, Z and E_T scales of solvent ionizing power. Ideally some correction for the different extents of solvation of the anions Cl^-, OTs^-, I^-) should be made, and then the various scales of solvent polarity may be more comparable.

One possible method of correction would be to use the heats of solution of salts; from the heats of solution of KI, KCl, NaI and NaCl in water and in methanol (Slansky, 1940) it appears that, compared with I^-, Cl^- is about 2 kcal mol^{-1} more strongly solvated in water than in methanol. This effect would complicate comparisons of Y and Z scales because the former is based on solvation of chloride whereas the latter is based on solvation of iodide. A correction factor of 2 kcal mol^{-1} (increase in Z) would improve a correlation of Y versus Z for water, methanol, ethanol, propan-2-ol and t-butyl alcohol (see Table 5), but significant progress in this approach would require heats of solution in all five solvents. One might have predicted that the oxygen anions [51] and [52], on which the E_T is based, would show behaviour more like Cl^- than I^-. Consequently, the upward curvature of Fig. 16 is surprising, because it suggests that, compared with [51] and [52] the iodide anion in [50] is more solvated in water than in methanol (Dimroth et al., 1963). Such discrepancies may indicate the importance of specific solvation effects stabilizing other parts of the molecules [50], [51] and [52]. Alternatively, they may arise because the iodide anion in [50] would be relatively free in solution, whereas the negatively charged oxygen in [51] and [52] is part of a complex molecule.

Quantitative indications of the relative magnitudes of solvation of the anions during solvolytic reactions can be obtained using eqn (5). By comparing the rate constants for solvolysis in acetic or formic acids with the rate constant for the ethanol/water mixture having the same Y value, a parameter $[k_{EW}/k_{RCO_2H}]_Y$ can be calculated. As

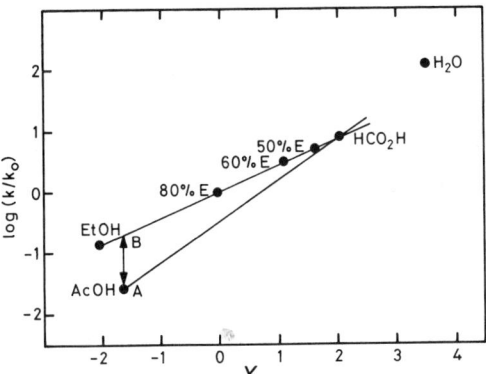

Figure 17. Correlation of solvolytic rate constants at 25° for 2-propyl tosylate and Y (data from Bentley and Schleyer, 1976, and Leffek et al., 1960). Log $[k_{EW}/k_{AcOH}]$ = B − A.

Y is conventionally the abscissa of a correlation, this parameter corresponds to the "vertical gap" or "dispersion" between a correlation line for mixtures of ethanol and water and a correlation line for acetic and formic acids (see Fig. 17). Although most of the ratios in Table 6 are complicated by the effect of ion pair return on titrimetric rate constants, the values of $[k_{EW}/k_{AcOH}]_Y$ vary with leaving group in the order Br > Cl > OTs ≫ F. This can be explained by hydrogen bonding to the leaving groups, which is greater in acetic acid than in ethanol/water mixtures and much greater for fluorides than tosylates; similarly hydrogen bonding is greater for tosylates than for chlorides (Winstein et al., 1957; Streitwieser, 1956b).

5. THE SOLVENT AS NUCLEOPHILE

Various scales of nucleophilic reactivity for anions have been in use for several years (Swain and Scott, 1953; Edwards, 1956; Pearson et al., 1968; Ritchie, 1972; see also Bunnett, 1963; Ibne-Rasa, 1967; Hartshorn, 1973b), but quantitative scales of solvent nucleophilicity have only recently been developed. Peterson and Waller (1972) derived a scale of solvent nucleophilicity (N_{PW}) based on the

$$\underset{Cl^+}{\bigcirc} + RCO_2H \longrightarrow H\overset{+}{-}O\underset{\underset{R}{C=O}}{\diagdown}\diagup Cl \qquad (8)$$

TABLE 6

Leaving Group Effects on Relative S_N1 Solvolysis Rates

Substrate	$[k_{EW}/k_{AcOH}]Y$	$[k_{EW}/k_{HCO_2H}]Y$	Temp (°C)
t-Butyl bromide	3.3^a		25°
t-Butyl chloride	1.0^b	1.0^b	25°
Neophyl tosylate [46, X = OTs]	0.37^c	0.07^c	75°
Neophyl bromide [46, X = Br]	$1.0^a (2.8)^d$	—	50°
Neophyl chloride [46, X = Cl]	$0.5^a (1.0)^{b,d}$	—	50°
1-Adamantyl tosylate [45, X = OTs]	0.18^e	—	25°
1-Adamantyl bromide [45, X = Br]	3^f	1^f	25°
1-Adamantyl chloride [45, X = Cl]	$<0.5^g$	0.5^g	25°
Bicyclo[2,2,2] octyl tosylate [49]	0.2^h	0.07^h	25°
Bicyclo[2,2,2] octyl bromide [44]	0.7^a	—	100°
2-Adamantyl tosylate [6]	0.2^i	0.04	25°
Diphenylmethyl fluoride	$0.0002^a (0.00001)^d$	—	25°
Diphenylmethyl bromide	$90^a (8.5)^d$	—	25°
Diphenylmethyl chloride	$20^a (1.0)^{b,d}$	—	25°
α-Phenylethyl bromide [40, X = Br]	$30^a (6.3)^d$	—	25°
α-Phenylethyl chloride [40, X = Cl]	$5^a (1.0)^{b,d}$	—	25°
Triphenylmethyl fluoride [42, X = F]	0.0008^a	—	25°

[a] Winstein et al., 1957.
[b] By definition.
[c] Smith et al., 1961.
[d] Y value based on the rate constant for solvolysis of the corresponding chloride, Winstein et al., 1957.
[e] Kevill et al., 1970.
[f] See discussion of Fig. 11, Section 3.
[g] Fig. 12.
[h] Morten, 1975.
[i] Bentley and Schleyer, 1976.

TABLE 7

Empirical Scales of Solvent Nucleophilicity

Solvent	$N_{PW}{}^a$	N^b	$N_{OTs}{}^b$	$(d_1{}^c)$
Trifluoroacetic acid	−5·33	−4·74	−5·56	−
Dichloroacetic acid	−3·91	−	−	−
Chloroacetic acid	−2·65	−	−	−
Formic acid	−1·52	−2·05	−2·35	(−4·40)
Acetic acid	−1·66	−2·05	−2·35	(−4·82)
80% v/v Ethanol/water	$0·00^d$	$0·00^d$	$0·00^d$	$(0·00^d)$
Ethanol	−	0·09	0·00	(−0·53)
Methanol	−	0·01	−0·04	(−0·05)
Water	−	−0·26	$−0·41^f$	(−0·44)
2,2,2-Trifluoroethanol	−	$−2·78^e$	−3·0	−
97% w/w Hexafluoro-propan-2-ol/water	−	−3·93	−4·27	−

^{a} Peterson and Waller, 1972.
^{b} From equation (11); for a more complete tabulation see Schadt et al., 1976.
^{c} Swain et al., 1955a [equation (13)]; for discussion see Section 6.
^{d} By definition.
^{e} See also Raber et al., 1974d, and Bentley and Lacadie, 1971.
^{f} Bentley, 1974b.

second-order rates of displacement by solvent on tetramethylene halonium ions in liquid sulphur dioxide using eqn (8). To date, this method has only been applied to carboxylic acid solvents, but the results suggest that acetic and formic acids are almost equally nucleophilic (Table 7).

The four-parameter, linear free energy relationship (9), a more generally applicable equation than eqn (5), had been proposed by

$$\log (k/k_0)_{RX} = mY + lN \qquad (9)$$

Winstein et al. (1957). In (9), k, k_0 and Y have the same meaning as for eqn (5), m is the sensitivity of the solvolysis of RX to solvent ionizing power, Y, and l is the sensitivity of the solvolysis of RX to solvent nucleophilicity, N. However, this equation was not evaluated until recently, when Bentley et al. (1972) made the following substitutions in the rearranged form (10). Methyl tosylate was

$$N = [\log (k/k_0)_{RX} - mY]/l \qquad (10)$$

selected as the source of relative rate data and its sensitivity to solvent nucleophilicity l was *defined* as equal to unity, because in solvolysis methyl tosylate is the most sensitive of all simple alkyl tosylates to changes in solvent nucleophilicity (Fig. 18). The necessity of a definition such as $l = 1$ for methyl tosylate is similar to other linear free energy relationships, e.g. $m = 1$ for t-butyl chloride [eqn (5)], or $\rho = 1$ for pK_a of substituted benzoic acids to define the Hammett σ constants. Finally the relative rate data k/k_0 must be corrected for the sensitivity of methyl solvolyses to solvent ionizing power. This was done using the slope ($m = 0\cdot3$) of the correlation line for acetic and formic acids (Fig. 18), i.e. it is assumed that these two acids are equally nucleophilic, which is indicated to be a good approximation by the data of Peterson and Waller (1972). Both Y or Y_{OTs} values can be used with the same value of m ($0\cdot3$) because, fortuitously, the difference ($Y_{HCO_2H} - Y_{CH_3CO_2H}$) is the same on

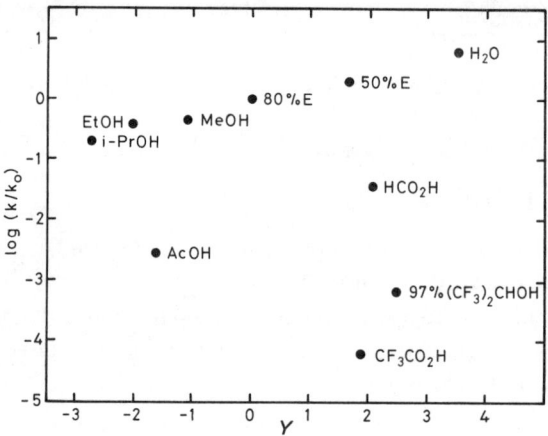

Figure 18. Correlation of solvolytic rate constants for methyl tosylate at 50° and Y (data from Schadt *et al.*, 1976).

both Y and Y_{OTs} scales. Thus N values can be calculated using eqn (11) and a selection of results is given in Table 7.

$$N = \log (k/k_0)_{CH_3OTs} - 0\cdot3Y \tag{11}$$

In general, agreement between the three scales is satisfactory. It is particularly encouraging that the very low value for N_{OTs} ($-5\cdot56$), arising from the remarkably slow rate of solvolysis of methyl tosylate in trifluoroacetic acid (Fig. 18), is in good agreement with the N_{PW} value ($-5\cdot33$). However, there does not appear to be a simple explanation of the low N values of trifluoroacetic acid. Whilst it is the most acidic solvent (as might be expected, the more acidic solvents tend to be less nucleophilic), the difference in N value between trifluoroacetic acid and formic acid is greater than between formic acid and water! The fluorinated alcohols, 2,2,2-trifluoroethanol and 1,1,1,3,3,3-hexafluoropropan-2-ol also have low N values (Table 7). Further comparisons between N_{PW} and N_{OTs} data await a more extensive collection of N_{PW} values for alcohols.

In a theoretical investigation of the stabilization of carbenium ions by trifluoroacetic acid, Dannenberg (1975) has calculated by the INDO method that the interaction between a CF_3 group and a carbenium ion centre is favoured at distances greater than 3 Å; only at shorter distances does stabilization by the more nucleophilic carboxylic acid group become overwhelmingly important. In contrast, on the basis of similar calculations it was suggested that, for acetic acid, interaction with the carboxylic acid group is always the more favourable. Whilst this might explain certain peculiarities for solvolyses in trifluoroacetic acid, and possibly in other fluorinated solvents, it should be emphasized that the preference for stabilization by CF_3 appears to be about 1 kcal mol^{-1}, which is well within the expected error of such a simple calculation.

Dougherty (1975) has pointed out that the first ionization potential (IP) of a solvent may reflect its nucleophilicity. Earlier we had considered this possibility, but were disappointed to find a lack of correspondence between the two parameters (Table 8). The IP of water is over 2 eV higher than that of ethanol, whereas water and ethanol have similar N values. The IP of acetic acid is also less than that of the more nucleophilic methanol. However, in agreement with qualitative expectations, the IP of 2,2,2-trifluoroethanol is higher than the other, more nucleophilic alcohols, and the IP of 1,1,1,3,3,3-hexafluoropropan-2-ol is even higher than that of 2,2,2-trifluoroethanol. As the original data of Watanabe (1957) are frequently quoted, revised experimental values for the IPs of acetic and formic acids should also be noted (Table 8); the measurements are complicated by formation of carboxylic acid dimers of lower IP than the monomers (Thomas, 1972). An unusually large range of results for the IP of trifluoroacetic acid has resulted from three recent, independent measurements.

The full equation used by Dougherty to correlate solvolyses of methyl tosylate and t-butyl chloride utilizes both ionization potentials (IP) and electron affinities (EA) of the solvents, and the solvent-ion interaction energy (∂E_{solv}), deduced from PMO theory, is given by eqn (12).

$$\partial E_{solv} \cong C_1(\text{IP} + \text{EA}) + C_2(\text{IP}) + C_3(\text{IP})^2 + C_4 \qquad (12)$$

TABLE 8

Comparison of First Ionization Potentials with the Solvent Nucleophilicity Parameter N

Parameter	Solvent						
	H_2O	CH_3OH	CH_3CH_2OH	HCO_2H	CH_3CO_2H	CF_3CH_2OH	$(CF_3)_2CHOH$	CF_3CO_2H
N^a	−0·26	0·01	0·09	−2·05	−2·05	−2·78	~ −4·0	−4·74
Ionization Potential (eV)	12·61[b]	10·83[c]	10·46[c]	11·33[b,d]	10·70[e]	11·7[f]		11·46[e]
(1 eV ≡ 23 kcal mol^{-1})				11·35[g]	10·69[g]			12·0[g]
				(11·05)[h]	(10·35)[h]	11·75[i]	12·35[i]	11·7[i]

[a] From Table 4.
[b] Turner et al., 1970.
[c] Dewar and Worley, 1969, and references there cited.
[d] Brundle et al., 1969.
[e] Sweigart and Turner, 1972.
[f] Determined by Mrs. S. A. Cowling (University of Liverpool), using a Perkin-Elmer PS 16 photoelectron spectrometer.
[g] Thomas, 1972.
[h] Watanabe, 1957.
[i] We are grateful to Miss C. Bowen, Mrs. P. Bradley and Dr. D. Betteridge (University College of Swansea) for these results.

Unfortunately, only five solvents appear to have been studied, so it is not surprising that the four parameters (C_1, C_2, C_3 and C_4) could be optimized to give a good fit to the solvolytic data. As would be expected, the nucleophilicity term (C_2) was found to be small for solvolyses of t-butyl chloride, but, surprisingly, the value of C_1 (associated with solvent ionizing power) was found to be similar for solvolyses of both t-butyl chloride and methyl tosylate.

6. CORRELATION OF SOLVOLYSIS RATES

Equation (12) is the most recent of several equations used to correlate solvolysis rates. Whilst in principle the equation is a theoretical one, in practice it is highly empirical because four constants have to be evaluated for each set of data correlated. Other correlations utilize experimental data to define one or more solvent parameters, and then correlate sets of experimental data using only one or at most two adjustable parameters.

The mY *equation*

The simplest two-parameter equation is the mY eqn (5), which has been tested extensively by Fainberg and Winstein (1956, 1957a, b, c), and by Winstein *et al.* (1951, 1957). Using t-butyl chloride as the source of Y values, the mY equation correlates a surprisingly wide variety of solvolyses. As discussed in Section 3 (p. 32), the equation was designed to correlate $S_N 1$ solvolyses. However, solvolyses of methyl tosylate in alcohols and mixtures of water and alcohols give a good correlation (Fig. 18). This is surprising because solvolyses of methyl tosylate are known to be sensitive to solvent nucleophilicity which is not included in eqn (5); it now appears that these solvents are of similar nucleophilicities (Table 7). In other cases, for example solvolyses of 1-adamantyl chloride [45, X = Cl] (Fig. 12), the correlation is surprisingly poor considering that both t-butyl chloride and 1-adamantyl chloride should be in the $S_N 1$ category. A comparison of rates of solvolysis of t-butyl chloride, 1-adamantyl chloride, neophyl chloride [46, X = Cl] and α-phenylethyl chloride [40, X = Cl] suggests that the last three behave similarly and the behaviour of t-butyl chloride may be anomalous (Bentley, 1971). This may be because of internal ion-pair return in solvolyses of t-butyl chloride (Section 3) or because of effects of ground state solvation (Section 6, p. 60).

An important characteristic of correlations based on eqn (5) is the tendency of different pairs or groups of solvents to form separate correlation lines (Fig. 19). Whilst some of this behaviour may be attributed to the choice of t-butyl chloride as model compound, there are probably a variety of more subtle solvation effects causing slight differences in behaviour even for solvents as similar as

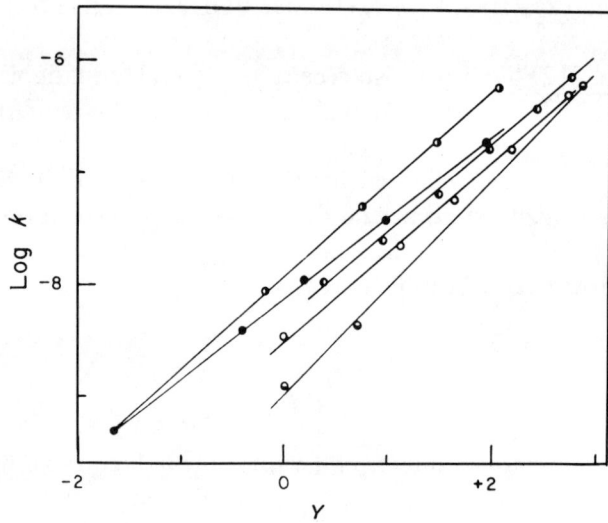

Figure 19. Correlation of solvolysis rate constants for neophyl chloride (46, X = Cl) at 50° and Y. ○, EtOH/H_2O; ◐, MeOH/H_2O; ◓, dioxane/H_2O; ●, AcOH/H_2O; ◑, AcOH/HCO_2H. (From Fainberg and Winstein, 1957 and reproduced by permission of the American Chemical Society.)

methanol/water and ethanol/water; it appears, for example, that alkyl and aryl groups make different contributions to heats of hydration (Krishnan and Friedman, 1971). Because of specific solvation of the leaving group, correlations of rates of solvolysis of substrates other than chlorides with Y (i.e. t-butyl chloride) are still less satisfactory (Fainberg and Winstein, 1957b).

Four-Parameter Equations

The success of four-parameter correlations is highly dependent on an appropriate selection of reference systems. Using the complete

Grunwald–Winstein eqn (9) with Y_{OTs} values derived from 2-adamantyl tosylate (Section 3, p. 36) and with N_{OTs} values based on methyl tosylate (Section 5), Schadt et al. (1976) have correlated a variety of solvolyses of other tosylates. Whilst their work is more extensive in that $S_N 2$ type solvolyses were studied (Table 9), only tosylates have been examined and solvation effects caused by different leaving groups are thus minimized. However, the subtlety of solvation effects is emphasized by the dependence of the relative rates of p-bromobenzenesulphonates and tosylates on solvent electro-

TABLE 9

Equations for the Correlation of Rates of Solvolysis of Tosylates

	Equations		
	$lN_{OTs} + mY_{OTs}$ (9)		Equation (17)
Tosylate	l	m	Q
2-Adamantyl [6]	$0·0^a$	$1·00^a$	$1·0^a$
Bicyclo[2,2,2] octyl [49][b]	$0·02 \pm 0·02$	$1·05 \pm 0·02$	$1·04 \pm 0·02$
Cyclohexyl[c]	0·23	0·75	0·72
Cyclopentyl[c]	0·26	0·71	0·67
3-Pentyl[c]	0·26	0·72	0·69
2-Propyl[c,d]	0·40	0·58	0·52
Ethyl[c]	0·83	0·41	0·16
Methyl	$1·0^a$	$0·3^a$	$0·0^a$
Benzyl[c]	0·75	0·64	0·51

[a] By definition.
[b] Morten, 1975.
[c] Schadt et al., (1976); calculated from data for 5–9 solvents, the values for l, m and Q depending slightly on the range of solvents calculated.
[d] See also Fig. 20.

philicity (Section 2, p. 13). In general the correlations are satisfactory (e.g. Fig. 20) and the results are consistent with a spectrum of reaction mechanisms, with the reactions differing in sensitivity to solvent ionizing power and nucleophilicity.

A similar eqn (13), developed twenty years ago, now appears to be based on inappropriate choices of model substrates and arbitrary assumptions (Swain et al., 1955a). For eqn (13), c_1 and c_2 are

$$\log (k/k_0)_{RX} = c_1 d_1 + c_2 d_2 \qquad (13)$$

constants depending only on the compound undergoing solvolysis and d_1 and d_2 are constants depending only on the solvent. In

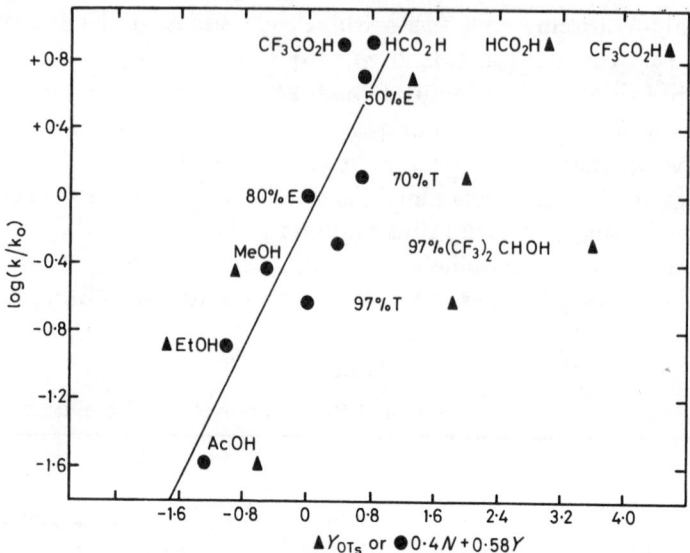

Figure 20. Correlation of solvolysis rate constants at 25° for 2-propyl tosylate and Y_{OTs} or $(0.4N_{OTs} + 0.58Y_{OTs})$. (From Schadt et al., 1976, and reproduced with permission of the American Chemical Society.)

contrast to a mechanism based on the roles of solvent as nucleophile, electrophile and ionizing medium, eqn (13) is based on Swain's push–pull mechanism, in which d_1 is intended to measure solvent nucleophilicity and d_2 measures solvent electrophilicity; as before k_0 refers to 80% v/v ethanol/water for which $d_1 = d_2 = 0$, but the following arbitrary "normalization conditions" were chosen.

$$c_1 = 3.00c_2 \text{ for methyl bromide}$$

$$c_1 = c_2 = 1.00 \text{ for t-butyl chloride}$$

$$3.00c_1 = c_2 \text{ for triphenylmethyl fluoride [42, X = F]}$$

This procedure was necessary because unique solutions were not obtained when 25 values of c_1 and c_2 as well as 17 values of d_1 and d_2 were derived by an iterative computer-procedure. Swain's own doubts about the results of this treatment have been emphasised by Streitwieser (1956a), who pointed out that values of c_1 do not reflect sensitivity to solvent nucleophilicity (Swain used c_1/c_2 to characterize reactivity).

There is also considerable doubt about the significance of d_1 and d_2, because a variety of leaving groups were studied (F, OTs, Cl, Br, SCN and OAc). As might be expected for a solvent nucleophilicity parameter, many aqueous or alcoholic solvents have $d_1 \sim 0$; acetic and formic acids give negative values. Whilst the absolute magnitudes are different, the relative magnitudes of d_1 are in moderate agreement with other measures of solvent nucleophilicity (Table 7), but it is a little surprising that values of d_1 for ethanol and water are slightly less than for 80% v/v ethanol/water. For reasons which were not stated Swain *et al.* (1955a) used $(d_1 - d_2)$ as a measure of solvent reactivity, which was claimed to parallel solvent nucleophilicity. The parameter d_2 appears to increase as the ionizing power of the solvent increases (Table 5), although the important contribution of electrophilic solvation probably causes the high value (3·12) for acetic acid. Using the alternative assumptions that, for t-butyl chloride, $c_1 = 0·00$ and $c_2 = 1·00$, it can be shown that $(d_1 + d_2)$ should correspond to Y values (Peterson and Waller, 1972).

It now seems clear that the ability of eqn (13) to correlate experimental data is due mainly to the use of four adjustable parameters. Whilst the precision of the correlations is good, the values of c_1 and c_2 are of limited utility in predicting other solvolysis rates and in mechanistic interpretations. In comparing the above applications of eqns (9) and (13), which appear to be very similar, it should be noted that the solvent parameters N and Y are constants and were calculated directly from experimental data, whereas d_1 and d_2 only became constant after they had been adjusted by an iterative computation to give the best fit to the experimental results. Whilst it could be argued that this adjustment is a correction for possible experimental errors, there are usually more chemical means of achieving this. It is preferable, in our opinion, to use raw experimental data and to accept a less precise correlation rather than to optimize a parameter which is subsequently regarded as a constant. Using the former approach the significance of the correlation is clearer and thus ways of making improvements are easier to see, and refinements to existing data as well as inclusion of new experimental data can be done readily. If the latter approach is adopted, it is usually not possible to make any changes without a complete reparameterization. Regrettably, a consequence of the ready availability of computers is an increasing tendency to develop computer programs containing "constants" which are optimized experimental values, and the term "pseudo-empirical constant" has been suggested to describe such values (Bentley, 1976).

Three-Parameter Equations

As if anticipating some of the above criticisms, Swain *et al.* (1955b) introduced a simpler eqn (14), containing only two adjustable

$$\log (k/k_0)_{RX} - \log (k/k_0)_{R'X} = ab \qquad (14)$$

parameters. In this equation k and k_0 have their usual meanings, a is an adjustable parameter depending on the substrate and b is an adjustable parameter depending on the solvent; thus $b = 0.00$ for 80% v/v ethanol/water. To make unique solutions for values of a, the conditions that $a = 1.00$ for t-butyl chloride and that $a = 0.00$ for methyl bromide were imposed. Consequently $R'X$ in eqn (14) corresponds to methyl bromide and correlations for any substrate RX are made using eqn (15). When RX is t-butyl chloride, $a = 1.00$,

$$\log (k/k_0)_{RX} = \log (k/k_0)_{CH_3Br} + ab \qquad (15)$$

$$b = \log (k/k_0)_{t\text{-BuCl}} - \log (k/k_0)_{CH_3Br} \qquad (16)$$

so that "crude" b values can be obtained from eqn (16). Thus b should be an indication of the extent to which a solvent favours an $S_N 1$ reaction more than does the reference solvent, 80% v/v ethanol/water. To obtain the best fit to the experimental data both a and b were varied. Since the correlations were restricted to chlorides and bromides, the effect of solvation of the leaving groups should largely cancel, and the precision of the correlations was encouraging. The parameter a measures the extent to which the solvolysis of a substrate resembles that of t-butyl chloride or methyl bromide. In agreement with qualitative expectations primary halides have low a values and secondary halides have larger a values (Table 10).

TABLE 10

Comparison of Equations (9), (15) and (17)

Parameter	Substrate			
	CH_3X	CH_3CH_2X	$(CH_3)_2CHX$	$PhCH_2X$
a	0.00^a	0.15^c	0.42^c	0.19^c
$(m-0.3)/0.7^b$	0.00	0.16	0.40	0.49
$1-l^b$	0.00	0.17	0.60	0.25
Q^b	0.00	0.16	0.52	0.51

[a] By definition.
[b] Values of m, l and Q from Table 9.
[c] Swain *et al.*, 1955b.

Quantitative agreement with other equations *for tosylates but not for halides* is also illustrated in Table 10. If a measures the extent to which a solvolysis is more like t-butyl chloride than methyl bromide in terms of response to solvent ionizing power then it should correspond to $(m - 0.3)/0.7$, or if a measures the extent to which a solvolysis is more like methyl bromide than t-butyl chloride in terms of response to solvent nucleophilicity then a should correspond to $1 - l$ (Schadt *et al.*, 1976). Fortunately, for simple aliphatic systems, it appears that an increasing response to solvent nucleophilicity causes a decreasing response to solvent ionizing power (see eqn 20) so that only one parameter is required to quantify the appropriate blend of behaviour. Benzyl chloride appears to show a more complex response to solvent nucleophilicity and ionizing power and agreement between the various parameters is less satisfactory (Table 10).

Equation (17) has recently been developed (Bentley *et al.*, 1972; Schadt *et al.*, 1976). This is similar to eqn (15) in that two reference

$$\log (k/k_0)_{\text{ROTs}} + (1 - Q) \log (k/k_0)_{\text{CH}_3\text{OTs}} + Q \log (k/k_0)_{\text{2-AdOTs}} \tag{17}$$

substrates are used, but is different in that there is only one adjustable parameter. Although eqn (17) is designed to correlate solvolysis rates of tosylates, a more general form of the eqation could be used to express (or to test for) a gradation of solvent dependent reactivity (and hence mechanism) between any two reference substrates. The form of eqn (17) was originally conceived to test the possibility that mole fractions of substrate $(1 - Q)$ and Q were proceeding by S_N2 and S_N1 mechanisms respectively. Although a different value of Q might be expected for each solvent if solvolysis occurred by competing S_N2 and S_N1 mechanisms, the idea was tested. In practice it turns out that one value of Q gives correlations of a precision comparable with eqn (9) (see Fig. 20 and Table 9) so that eqn (17) is a very useful way to quantify the merging of behaviour between extremes represented by methyl tosylate (S_N2) and 2-adamantyl tosylate (S_N1).[6] Since Q is not markedly dependent on solvent, it seems highly unlikely that the merging of behaviour is due to competing S_N2 and S_N1 mechanisms.

When eqn (9) utilizes values of N_{OTs} [based on methyl tosylate; eqn (11)] and Y_{OTs} (based on 2-adamantyl tosylate), there is an

---

[6] In Winstein's terminology eqn (17) applies to the k_s–k_c spectrum and should not be used for k_Δ solvolyses.

exact mathematical relationship between (9) and (17). It has been shown by Schadt et al. (1976) that:

$$l = 1 - Q \qquad (18)$$

$$m = 0\cdot 3 + 0\cdot 7Q \qquad (19)$$

$$l = (1 - m)/0\cdot 7 \qquad (20)$$

There is also an approximate correspondence between values of a [eqn (15)] and Q [eqn (17)], as illustrated in Table 10.

Limitations

Temperature dependence of relative rates

A perennial problem in interpreting reaction rates is the well established fact that relative rates (e.g. log k/k_0 values) will in general be temperature dependent, except in the very unlikely circumstances where either the reactions to which k and k_0 refer have the same enthalpy of activation ($\Delta H^{\ddagger}$) and the same temperature dependence of enthalpy of activation ($\Delta C_P^{\ddagger}$, the heat capacity of activation), or changes in $\Delta H^{\ddagger}$ are compensated exactly by changes in $\Delta S^{\ddagger}$. At the present level of sophistication in interpreting rate data, it seems generally agreed that relative rates (i.e. free energy differences) are the most appropriate parameters (Dewar, 1969a; Streitwieser, 1956c), and these have been used in the correlations discussed above. Dewar (1969b) has gone so far as to suggest, "... the measurement of activation energies, and heats of activation, has become a fetish among physical organic chemists." In contrast, Arnett et al. (1969) have pointed out that because little is known about the temperature coefficients of activation parameters ($\Delta G^{\ddagger}$, $\Delta H^{\ddagger}$, $\Delta S^{\ddagger}$), "... there is no experimental justification for the common practice of interpreting small reactivity differences within a series of related compounds in solution at ordinary temperatures strictly in terms of stereoelectronic factors whose theoretical development applies to the gas phase at absolute zero." Streitwieser (1956c) has calculated that a 3 kcal mol^{-1} change in activation energy will cause a three-fold change in relative rate between 25 and 100°C. Consequently, it is very difficult to interpret properly small differences in relative rates.

However, the temperature dependence of relative rates may be reduced considerably if changes in $\Delta H^{\ddagger}$ are compensated by changes in $\Delta S^{\ddagger}$. As part of a comprehensive analysis of the limitations of the mY eqn (5), Winstein and Fainberg (1957) noted that the simplest temperature dependence of m would arise if $\Delta S^{\ddagger}$ and $\Delta H^{\ddagger}$ were linearly related (see also Leffler and Grunwald, 1963). Unfortunately, as shown in Table 11, such a simple relationship is not found experimentally either for t-butyl chloride (the model for the Y scale) or 2-adamantyl tosylate (the model for the Y_{OTs} scale). The results for 80% ethanol/water illustrate another difficulty, because very accurate measurements of rate constants and temperatures are

TABLE 11

Activation Parameters for Solvolysis[a]

Substrate Solvent	t-Butyl chloride[b]		2-Adamantyl tosylate[c]	
	$\Delta H^{\ddagger}$	$\Delta S^{\ddagger}$	$\Delta H^{\ddagger}$	$\Delta S^{\ddagger}$
Ethanol	26·13	−3·2	29·6[d]	−2·2[d]
Methanol	24·88	−3·1	—	—
80% Ethanol/water	22·34	−6·6	26·9	−3·0
80% Ethanol/water	21·8[e]	−8·5[e]		
Water	23·22	+12·2	—	—
Acetic acid	25·80	−2·5	28·1	−2·1
Formic acid	21·0	−1·7	24·8	+3·6
97% Trifluoroethanol/water	19·7[f]	−10·0[f]	23·2	−7·1
97% Hexafluoropropan-2-ol/water	—	—	17·9	−17·0

[a] Units $\Delta H^{\ddagger}$, kcal mol^{-1}; $\Delta S^{\ddagger}$ cal deg^{-1} mol^{-1}.
[b] Winstein and Fainberg, 1957, except where noted otherwise.
[c] Bentley and Schleyer, 1976; Schadt et al. 1976, except where noted otherwise.
[d] Kevill et al., 1973.
[e] Shiner, 1954; see also Winstein and Fainberg, 1957 (Table 1, footnote g).
[f] Bentley, 1971.

required, and independent measurements of $\Delta H^{\ddagger}$ and $\Delta S^{\ddagger}$ do not always agree within the combined error limits expected.

The results in Table 11 also illustrate the dubious basis of the generalization (e.g. Pritt and Whiting, 1975) that $S_N 1$ reactions have $\Delta S^{\ddagger} \sim 0$, since t-butyl chloride in water has $\Delta S^{\ddagger} = +12·2$, and 2-adamantyl tosylate in 97% hexafluoropropan-2-ol/water has $\Delta S^{\ddagger} = -17$ cal deg^{-1} mol^{-1}. Whilst it could be argued that solvation in mixed solvents is more complex than in pure solvents, it is nevertheless clear that there is a wide range of values of $\Delta S^{\ddagger}$ for these $S_N 1$ solvolyses.

Solvation of ground states

It has long been recognized that, because the free energy of activation corresponds to the free energy change between the ground and transition states, the solvent effect on the ground state may be important (Winstein and Fainberg, 1957). However, relatively little research has been carried out on solvation of ground states and

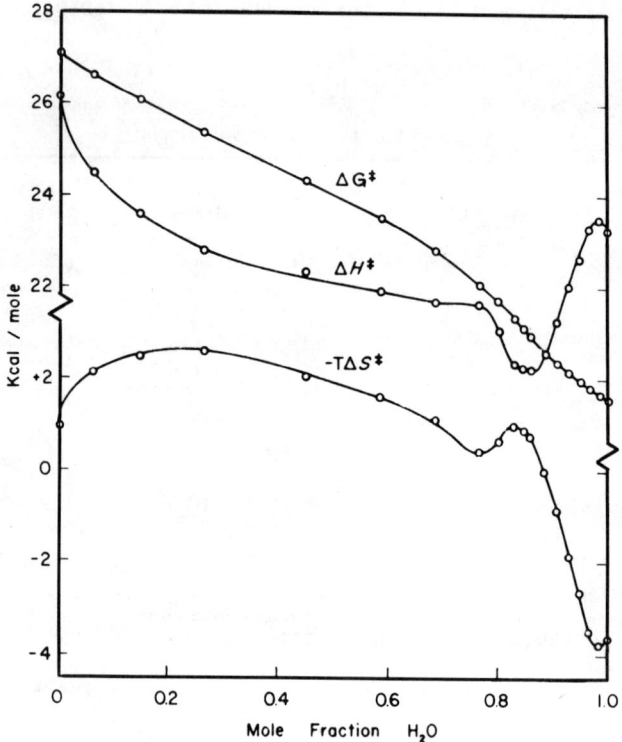

Figure 21. Plot of $\Delta H^{\ddagger}$, $\Delta G^{\ddagger}$ and $-T\Delta S^{\ddagger}$ versus mole fraction for solvolysis of t-butyl chloride in EtOH/H$_2$O at 25°. (From Winstein and Fainberg, 1957, and reproduced by permission of the American Chemical Society.)

explanations of reactivity are nearly always based on more fashionable theories which are usually applied to transition states.

Although the ground state solvation effects are quite small they have been shown to account for the Baker–Nathan order in the methanolysis of p-alkyl substituted benzyl chlorides (Clement and Naghizadeh, 1959; see also Hehre et al., 1974; Schubert and Sweeney, 1956).

Perhaps the most spectacular success of explanations based on solvation of ground states, published to date, is the dissection of activation parameters for solvolysis of t-butyl chloride in mixtures of ethanol and water, first discussed by Winstein and Fainberg (1957). The complex variation of $\Delta H^{\ddagger}$ and $\Delta S^{\ddagger}$ (Fig. 21) has been shown to be due almost entirely to ground state solvation effects, at least for the solvents ethanol—40% ethanol/water studied by Arnett et al. (1965). For 90%, 80%, 70%, 60%, 50% and 40% ethanol/water the parameter $\Delta \bar{H}^t$ for solvation of the transition state (by transfer from the gas phase) was calculated to be linearly proportional to the corresponding value of $\Delta \bar{S}^t$, as expected from the behaviour of simple salts. The point for pure ethanol did not fall on the calculated line, and this was attributed to nucleophilic solvent assistance. The variation in $\Delta G^{\ddagger}$, $\Delta H^{\ddagger}$ and $\Delta S^{\ddagger}$ (Fig. 21) can be reproduced remarkably well using ethane and the zwitterionic α-amino acid, glycine, as model compounds (Abraham et al., 1975; see also Abraham, 1974; Abraham and Abraham, 1974).

Summary

From the variety of correlations discussed above, it seems clear that considerable progress has been made in accounting quantitatively for the role of the solvent as a nucleophile. However, the most successful correlations have been restricted to a limited range of leaving groups, e.g. tosylates, chlorides and bromides. It has not yet been possible to separate the role of the solvent as an electrophile from its role as an ionizing medium, so that parameters for "ionizing power" (Y, d_2 in Table 5) contain a combination of these two effects. Research in progress in our laboratories is aimed at obtaining more extensive experimental data in order to define a consistent set of Y values for each leaving group in each solvent. It may then be possible to dissect these Y values into a term for solvent ionizing power, independent of leaving group, and a term for electrophilic solvation of the leaving group, an approach which has been outlined by Fainberg and Winstein (1957c).

Having quantitatively accounted for electrophilic effects of solvent, discrepancies between calculated and observed rates of solvolysis could indicate fruitful areas for mechanism-oriented research. However, such correlations would still not account for effects of ground state solvation or the temperature dependence of relative rates due to different activation energies. Consequently using the correlations of the above type, it will not be possible to reproduce experimental data precisely, but the correlation of relative rates within a rate factor of two or three (e.g. Fig. 20) is encouraging.

REFERENCES

Abraham, M. H. (1973). *J. C. S. Perkin II*, 1893.
Abraham, M. H. (1974). *Prog. Phys. Org. Chem.* 11, 1.
Abraham, M. H., and Abraham, R. J. (1974). *J. C. S. Perkin II*, 47.
Abraham, M. H. Buisson, D. H., and Schulz, R. A. (1975). *J. C. S. Chem. Comm.* 693.
Arnett, E. M., Bentrude, W. G., and Duggleby, P. McC. (1965). *J. Amer. Chem. Soc.* 87, 2048.
Arnett, E. M., Kover, W. B., and Carter, J. V. (1969). *J. Amer. Chem. Soc.* 91, 4028.
Bach, R. D., and Willis, C. L. (1975). *J. Amer. Chem. Soc.* 97, 3844.
Bentley, M. D., and Lacadie, J. A. (1971). *Tetrahedron Lett.* 741.
Bentley, T. W. (1971) unpublished results, Princeton University.
Bentley, T. W. (1974a). *Ann. Rep. Prog. Chem.* 71B, 119.
Bentley, T. W. (1974b). Unpublished results.
Bentley, T. W. (1976), MTP International Review of Science, Series 2, Vol. 5 (D. Ginsburg, ed.), Butterworths, p. 343.
Bentley, T. W., and Schleyer, P. von R. (1976). *J. Amer. Chem. Soc.* 98, 7658.
Bentley, T. W., Schadt, F. L., and Schleyer, P. von R. (1972). *J. Amer. Chem. Soc.* 94, 992.
Bentley, T. W., Liggero, S. H., Imhoff, M. A., and Schleyer, P. von R. (1974). *J. Amer. Chem. Soc.* 96, 1970.
Bethell, D., and Gold, V. (1967). "Carbonium Ions, an Introduction", Academic Press, London, pp. 139-148.
Böhme, H., and Schürhoff, W. (1951). *Chem. Ber.* 84, 28.
Bone, J. A., Pritt, J. R., and Whiting, M. C. (1975). *J. C. S. Perkin II*, 1447.
Bordwell, F. G., and Mecca, T. G. (1975). *J. Amer. Chem. Soc.* 97, 123, 127.
Bordwell, F. G., Wiley, P. F., and Mecca, T. G. (1975). *J. Amer. Chem. Soc.* 97, 132.
Bowen, C. (1976). Unpublished results, University College of Swansea.
Brown, D. A., and Hudson, R. F. (1953). *J. Chem. Soc.* 3352.
Brown, H. C., and Liu, K.-T. (1975). *J. Amer. Chem. Soc.* 97, 600.
Brown, H. C., Kim, C. J., Lancelot, C. J., and Schleyer, P. von R. (1970). *J. Amer. Chem. Soc.* 92, 5244.
Brundle, C. R., Turner, D. W., Robin, M. B., and Basch, H. (1969). *Chem. Phys. Lett.* 3, 292.
Bunnett, J. F. (1963). *Ann. Rev. Phys. Chem.* 14, 271.
Campbell, N. C. G., Muir, D. M., Hill, R. R., Parish, J. H., Southam, R. M., and Whiting, M. C. (1968). *J. Chem. Soc. (B)* 355.
Clement, R. A., and Naghizadeh, J. N. (1959). *J. Amer. Chem. Soc.* 81, 3154.
Cocivera, M., and Winstein, S. (1963). *J. Amer. Chem. Soc.* 85, 1702.
Dack, M. R. J. (1970). *Chem. Brit.* 6, 347.
Dais, P. J. and Gregoriou, G. A. (1974). *Tetrahedron Lett.* 3827.
Dannenberg, J. J. (1975). *Angew. Chem. Int. Edn.* 14, 641.

da Roza, D. A., Andrews, L. J., and Keefer, R. M. (1973). *J. Amer. Chem. Soc.* 95, 7003.
Darwish, D., and Tourigny, G. (1972). *J. Amer. Chem. Soc.*, 94, 2191.
Dedieu, A., and Veillard, A. (1972). *J. Amer. Chem. Soc.* 94, 6730.
de la Mare, P. B. D. (1955). *J. Chem. Soc.* 3180.
Dewar, M. J. S. (1969). "The Molecular Orbital Theory of Organic Chemistry", McGraw-Hill, New York, (a) pp. 279-283, (b) p. 283.
Dewar, M. J. S., and Worley, S. D. (1969). *J. Chem. Phys.* 50, 654.
Diaz, A. F., Lazdins, I., and Winstein, S. (1968a). *J. Amer. Chem. Soc.* 90, 1904.
Diaz, A. F., Lazdins, I., and Winstein, S. (1968b). *J. Amer. Chem. Soc.* 90, 6546.
Dimroth, K., Reichardt, C., Siepmann, T., and Bohlmann, F. (1963). *Annalen* 661, 1.
Doering, W. von E., and Zeiss, H. H. (1953). *J. Amer. Chem. Soc.* 75, 4733.
Dougherty, R. C. (1974). *Org. Mass Spectrometry* 8, 85.
Dougherty, R. C. (1975). *Tetrahedron Lett.* 385.
Dougherty, R. C., and Roberts, J. D. (1974). *Org. Mass Spectrometry* 8, 81.
Eckstrom, H. C., Berger, J. E., and Dawson, L. R. (1960). *J. Phys. Chem.* 64, 1458.
Edwards, J. O. (1956). *J. Amer. Chem. Soc.* 78, 1819.
Evans, D. F., Nadas, J. A., and Matesich, M. A. (1971). *J. Phys. Chem.* 75, 1708.
Fainberg, A. H., and Winstein, S. (1956). *J. Amer. Chem. Soc.* 78, 2770.
Fainberg, A. H., and Winstein, S. (1957a). *J. Amer. Chem. Soc.* 79, 1597.
Fainberg, A. H., and Winstein, S. (1957b). *J. Amer. Chem. Soc.* 79, 1602.
Fainberg, A. H., and Winstein, S. (1957c). *J. Amer. Chem. Soc.* 79, 1608.
Farinacci, N. T., and Hammett, L. P. (1937). *J. Amer. Chem. Soc.* 59, 2542.
Feynman, R. (1965). "The Character of Physical Law", MIT Press, Cambridge, Mass., p. 158.
Finkelstein, M. (1955). Ph.D. Thesis, Yale University, cited by Streitwieser (1962) and Fort and Schleyer (1964).
Fort, R. C., and Schleyer, P. von R. (1964). *Chem. Rev.* 64, 277.
Fowler, F. W., Katritzky, A. R., and Rutherford, R. J. D. (1971). *J. Chem. Soc. (B)* 460.
Frisone, G. J., and Thornton, E. R. (1968). *J. Amer. Chem. Soc.* 90, 1211.
Fry, J. L., and Badger, R. C. (1975). *J. Amer. Chem. Soc.* 97, 6276.
Fry, J. L., Lancelot, C. J., Lam, L. K. M., Harris, J. M., Bingham, R. C., Raber, D. J., Hall, R. E., and Schleyer, P. von R. (1970a). *J. Amer. Chem. Soc.* 92 2538.
Fry, J. L., Harris, J. M., Bingham, R. C., and Schleyer, P. von R. (1970b). *J. Amer. Chem. Soc.* 92, 2540.
Fry, J. L., Engler, E. M., and Schleyer, P. von R. (1972). *J. Amer. Chem. Soc.* 94, 4628.
Griffiths, T. R. and Symons, M. C. R. (1960). *Mol. Phys.* 3, 90, and references there cited.
Gold, V. (1956). *J. Chem. Soc.* 4633.
Gold, V. (1972). *J. C. S. Faraday I*, 1611.
Graczyk, D. G., and Taylor, J. W. (1974). *J. Amer. Chem. Soc.* 96, 3255.
Grunwald, E., and Effio, A. (1974). *J. Amer. Chem. Soc.* 96, 423.
Grunwald, E., and Winstein, S. (1948). *J. Amer. Chem. Soc.* 70, 846.
Haake, P. and Ossip, P. S. (1971). *J. Amer. Chem. Soc.* 93, 6924.
Hammett, L. P. (1970). "Physical Organic Chemistry", 2nd Edn., McGraw-Hill, New York, N.Y., (a) pp. 164-167; (b) p. 154.
Harper, J. J. (1968). Ph.D. Thesis, Princeton University.
Harris, J. M. (1974). *Prog. Phys. Org. Chem.* 11, 89.

Harris, J. M., Hall, R. E., and Schleyer P. von R. (1971). *J. Amer. Chem. Soc.* **93**, 2551.
Harris, J. M., Raber, D. J., Neal, W. C., Jr., and Dukes, M. D. (1974). *Tetrahedron Lett.* 2331.
Harriss, M. G., and Milne, J. B. (1971). *Can. J. Chem.* **49**, 1888, and references there cited.
Hartman, S., and Robertson, R. E. (1960). *Can. J. Chem.* **38**, 2033.
Hartshorn, S. R. (1973). "Aliphatic Nucleophilic Substitution", Cambridge University Press, (a) pp. 20–25; (b) pp. 48–52.
Hartshorn, S. R., and Shiner, V. J., Jr. (1972). *J. Amer. Chem. Soc.* **94**, 9003.
Hehre, W. J., McIver, R. T., Pople, J. A., and Schleyer, P. von R. (1974). *J. Amer. Chem. Soc.* **96**, 7162.
Hine, J. (1962). "Physical Organic Chemistry", 2nd Edn. McGraw-Hill, New York, p. 134.
Hiršl-Starčević, S. Majerski, Z., and Sunko, D. E. (1974). *J. Amer. Chem. Soc.* **96**, 3659.
Hoffmann, H. M. R. (1965). *J. Chem. Soc.* 6753, 6762.
Hogen-Esch, T. E., and Smid, J. (1965). *J. Amer. Chem. Soc.* **87**, 669.
Hughes, E. D., and Ingold, C. K. (1935). *J. Chem. Soc.* 244.
Hughes, E. D., Ingold, C. K., and Mackie, J. D. H. (1955). *J. Chem. Soc.* 3173.
Humski, K., Sendijarević, V., and Shiner, V. J., Jr (1973). *J. Amer. Chem. Soc.* **95**, 7722.
Humski, K., Sendijarević, V., and Shiner, V. J., Jr. (1974). *J. Amer. Chem. Soc.* **96**, 6187.
Ibne-Rasa, K. M. (1967). *J. Chem. Educ.* **44**, 89.
Ingold, C. K. (1969). "Structure and Mechanism in Organic Chemistry", 2nd Edn., Cornell University Press, Ithaca, N.Y., p. 428.
Kevill, D. N., Kolwyck, K. C., and Weitl, F. L. (1970).*J. Amer. Chem. Soc.* **92**, 7300.
Kevill, D. N., Kolwyck, K. C., Shold, D. M., and Kim, C. (1973). *J. Amer. Chem. Soc.* **95**, 6022.
Koppel, I. A., and Palm, V. A. (1972). In "Advances in Linear Free Energy Relationships" (N. B. Chapman and J. Shorter, eds.), Plenum, London, p. 208.
Kosower, E. M. (1958). *J. Amer. Chem. Soc.* **80**, 3253.
Kosower, E. M. (1968). "An Introduction to Physical Organic Chemistry", Wiley, New York, pp. 293–333.
Kovačević, D., Majerski, Z., Borčić, S., and Sunko, D. E. (1972). *Tetrahedron* **28**, 2469.
Krishnan, C. V., and Friedman, H. L. (1971). *J. Phys. Chem.* **75**, 388.
Laidler, K. J. (1960). *Suomen Kemistilehti* A, **33**, 40.
Lambert, J. B., and Putz, G. J. (1973). *J. Amer. Chem. Soc.* **95**, 6313.
Lambert, J. B., Putz, G. J., and Mixan, C. E. (1972). *J. Amer. Chem. Soc.* **94**, 5132.
Lancelot, C. J., Cram, D. J., and Schleyer, P. von R. (1972). In "Carbonium Ions" (G. A. Olah and P. von R. Schleyer eds.), Vol. 3, Wiley, New York, Chapter 27.
Leffek, K. T., Llewellyn, J. A., and Robertson, R. E. (1960), *Can. J. Chem.* **38**, 1505.
Leffler, J. E., and Grunwald, E. (1963). "Rates and Equilibria of Organic Reactions", Wiley, New York, p. 319 ff.
Lenoir, D., Hall, R. E., and Schleyer, P. von R. (1974). *J. Amer. Chem. Soc.* **96**, 2138.

Lenoir, D., Raber, D. J., and Schleyer, P. von R. (1974b). *J. Amer. Chem. Soc.* **96**, 2149.
Lenoir, D., Mison, P., Hyson, E., Schleyer, P. von R., Saunders, M., Vogel, P., and Telkowski, L. A. (1974c). *J. Amer. Chem. Soc.* **96**, 2157.
Lenoir, D., Röll, W., Weiss, E., and Wenke, G. (1976). *Tetrahedron Lett.*, 1991.
Le Roux, L. J., and Swart, E. R. (1955). *J. Chem. Soc.* 1475.
Llewellyn, J. A., Robertson, R. E., and Scott, J. M. W. (1960). *Can. J. Chem.* **38**, 222.
Matesich, M. A., Knoefel, J., Feldman, H., and Evans, D. F. (1973). *J. Phys. Chem.* **77**, 366.
McLennan, D. J. (1974). *J. C. S. Perkin II*, 481.
McLennan, D. J. (1975). *Tetrahedron Lett.* 4689.
McLennan, D. J. (1976). *Accounts Chem. Res.* **9**, 281.
Morten, D. H. (1975). M.Sc. Thesis, University College of Swansea.
Murto, J., and Heino, E. (1966). *Suomen Kemistilehti B* **39**, 263.
Nordlander, J. E., and McCrary, T. J., Jr. (1972). *J. Amer. Chem. Soc.* **94**, 5133.
Nordlander, J. E., Gruetzmacher, R. R., and Miller, F. (1973). *Tetrahedron Lett.* 927.
Nordlander, J. E., Gruetzmacher, R. R., Kelley, W. J., and Jindal, S. P. (1974). *J. Amer. Chem. Soc.* **96**, 181.
Okamoto, K., Kinoshita, T., and Osada, Y. (1975a). *J. C. S. Perkin II*, 253.
Okamoto, K., Kinoshita, T., Takemura, Y., and Yoneda, H. (1975b). *J. C. S. Perkin II*, 1427.
Parker, A. J. (1969). *Chem. Rev.* **69**, 1.
Pearson, R. G., Sobel, H., and Songstad, J. (1968). *J. Amer. Chem. Soc.* **90**, 319.
Peterson, P. E. and Waller, F. J. (1972). *J. Amer. Chem. Soc.* **94**, 991.
Peterson, P. E., Kelley, R. E., Jr., Belloli, R., and Sipp, K. A. (1965). *J. Amer. Chem. Soc.* **87**, 5169.
Peterson, P. E., Bopp, R. J., Chevli, D. M., Curran, E. L., Dillard, D. E. and Kamat, R. J. (1967). *J. Amer. Chem. Soc.* **89**, 5902.
Pritt, J. R. and Whiting, M. C. (1975). *J. C. S. Perkin II*, 1458.
Pross, A., and Koren, R. (1974). *Tetrahedron Lett.* 1949.
Queen, A., and Matts, T. C. (1975). *Tetrahedron Lett.* 1503.
Raaen, V. F., Juhlke, T., Brown, F. J., and Collins, C. J. (1974). *J. Amer. Chem. Soc.* **96**, 5928.
Raber, D. J., Bingham, R. C., Harris, J. M. Fry, J. L., and Schleyer, P. von R. (1970). *J. Amer. Chem. Soc.* **92**, 5977.
Raber, D. J., Harris, J. M., Hall, R. E., and Schleyer, P. von R. (1971). *J. Amer. Chem. Soc.* **93**, 4821.
Raber, D. J. Harris, J. M., and Schleyer, P. von R. (1974). In "Ion and Ion Pairs in Organic Reactions" (M. Szwarc, ed.), Vol. 2, Wiley, New York, pp. 247-374; (a) pp. 361-363 and references there cited; (b) pp. 297-302; (c) p. 328-345.
Raber, D. J., Dukes, M. D., and Gregory, J. (1974d). *Tetrahedron Lett.* 667.
Rappoport, Z., and Kaspi, J. (1974a). *J. Amer. Chem. Soc.* **96**, 586.
Rappoport, Z., and Kaspi, J. (1974b). *J. Amer. Chem. Soc.* **96**, 4518.
Reich, I. L., Diaz, A., and Winstein, S. (1969). *J. Amer. Chem. Soc.* **91**, 5635.
Reichardt, C. (1965). *Angew. Chem. Int. Edn.* **4**, 29.
Reichardt, C., and Dimroth, K. (1968). *Fortschr. Chem. Forsch.* **11**, 1.
Ritchie, C. D. (1972). *Accounts Chem. Res.* **5**, 348.
Roberts, D. D. (1972). *J. Org. Chem.* **37**, 1510.
Schadt, F. L., and Schleyer, P. von R. (1973). *J. Amer. Chem. Soc.* **95**, 7860.

Schadt, F. L., Schleyer, P. von R., and Bentley, T. W. (1974). *Tetrahedron Lett.* 2335.
Schadt, F. L., Bentley, T. W., and Schleyer, P. von R. (1976). *J. Amer. Chem. Soc.* 98, 7667.
Schleyer, P. von R., Fry, J. L., Lam, L. K. M., and Lancelot, C. J. (1970). *J. Amer. Chem. Soc.* 92, 2542.
Schubert, W. M., and Le Fevre, P. H. (1969). *J. Amer. Chem. Soc.* 91, 7746.
Schubert, W. M., and Sweeney, W. A. (1956). *J. Org. Chem.* 21, 119.
Sep, W. J., Verhoeven, J. W. and de Boer, Th. J. (1975). *Tetrahedron* 31, 1065.
Shiner, V. J., Jr. (1954). *J. Amer. Chem. Soc.* 76, 1603.
Shiner, V. J., Jr. (1970). In "Isotope Effects in Chemical Reactions" (C. J. Collins and N. S. Bowman, eds.). Van Nostrand Reinhold, New York, pp. 91-159; (a) pp. 131-133; (b) p. 128.
Shiner, V. J., Jr., and Fisher, R. D. (1971). *J. Am. Chem. Soc.* 93, 2553.
Shiner, V. J., Jr., and Dowd, W. (1969). *J. Amer. Chem. Soc.* 91, 6528.
Shiner, V. J., Jr., and Dowd, W. (1971). *J. Amer. Chem. Soc.* 93, 1029.
Shiner, V. J., Jr., Dowd, W., Fisher, R. D., Hartshorn, S. R., Kessick, M. A., Milakofsky, L., and Rapp, M. W., (1969a). *J. Amer. Chem. Soc.* 91, 4838.
Shiner, V. J., Jr., Fisher, R. D., and Dowd, W. (1969b). *J. Amer. Chem. Soc.* 91, 7748.
Sinnott, M. L., and Whiting, M. C. (1975). *J. C. S. Perkin II*, 1446.
Slansky, C. M. (1940). *J. Amer. Chem. Soc.* 62, 2430.
Slutsky, J., Bingham, R. C., Schleyer, P. von R., Dickason, W. C., and Brown, H. C. (1974). *J. Amer. Chem. Soc.* 96, 1969.
Smid, J. (1972). *Agnew. Chem. Int. Ed.* 11, 112,
Smith, S. G., Fainberg, A. H., and Winstein, S. (1961). *J. Amer. Chem. Soc.* 83, 618.
Sneen, R. A. (1973). *Accounts Chem. Res.* 6, 46.
Sneen, R. A., and Bradley, W. (1972). *J. Amer. Chem. Soc.* 94, 6975.
Sneen, R. A., and Carter, J. V. (1972). *J. Amer. Chem. Soc.* 94, 6990.
Sneen, R. A., and Kay, P. S. (1972). *J. Amer. Chem. Soc.* 94, 6983.
Sneen, R. A., and Larsen, J. W. (1966). *J. Amer. Chem. Soc.* 88, 2593.
Sneen, R. A., and Larsen, J. W. (1969). *J. Amer. Chem. Soc.* 91, 362.
Sneen, R. A., and Robbins, H. M. (1972). *J. Amer. Chem. Soc.* 94, 7868.
Sneen, R. A., Felt, G. R., and Dickason, W. C. (1973). *J. Amer. Chem. Soc.* 95, 638.
Steigman, J., and Hammett, L. P. (1937). *J. Amer. Chem. Soc.* 59, 2536.
Storesund, H. J., and Whiting, M. C. (1975). *J. C. S. Perkin II*, 1452.
Streitwieser, A., Jr. (1956). *Chem. Rev.* 56, (a) pp. 635-638; (b) p. 621-623; (c) p. 580.
Streitwieser, A. Jr. (1962). "Solvolytic Displacement Reactions", McGraw-Hill, New York. N.Y.
Streitwieser, A., Jr., and Dafforn, G. A. (1969). *Tetrahedron Lett.* 1263.
Streitwieser, A., Jr., and Waiss, A. C., Jr. (1962). *J. Org. Chem.* 27, 290.
Streitwieser, A., Jr., and Walsh, T. D. (1965). *J. Amer. Chem. Soc.* 87, 3686.
Streitwieser, A., Jr., Walsh, T. D., and Wolfe, J. R., Jr. (1965). *J. Amer. Chem. Soc.* 87, 3682.
Su, T. M., Sliwinski, W. F., and Schleyer, P. von R. (1969). *J. Amer. Chem. Soc.* 91, 5386.
Sunko, D. E. and Borčić, S. (1970). In "Isotope Effects in Chemical Reactions" (C. J. Collins and N. S. Bowman, eds.), Van Nostrand Reinhold, New York, pp. 160-212.
Sunko, D. E., and Szele, I. (1972). *Tetrahedron Lett.* 3617.

Sunko, D. E., Szele, I., and Tomić, M. (1972). *Tetrahedron Lett.* 1827.
Swain, C. G. (1948). *J. Amer. Chem. Soc.* 70, 1119.
Swain, C. G. and Scott, C. B. (1953). *J. Amer. Chem. Soc.* 75, 141.
Swain, C. G., and Mosely, R. B. (1955). *J. Amer. Chem. Soc.* 77, 3727.
Swain, C. G., Mosely, R. B., and Bown, D. E. (1955a). *J. Amer. Chem. Soc.* 77, 3731.
Swain, C. G., Dittmer, D. C., and Kaiser, L. E. (1955b). *J. Amer. Chem. Soc.* 77, 3737.
Sweigart, D. A., and Turner, D. W. (1972). *J. Amer. Chem. Soc.* 94, 5592.
Symons, M. C. R. (1967). *J. Phys. Chem.* 71, 172.
Szwarc, M. (1968). "Carbanions, Living Polymers and Electron Transfer Processes", Wiley-Interscience, New York.
Tamura, K., and Imoto, T. (1975). *Bull. Chem. Soc. Japan* 48, 369.
Thomas, R. K. (1972). *Proc. Roy. Soc.* A331, 249.
Thornton, E. K., and Thornton, E. R. (1970). In "Isotope Effects in Chemical Reactions" (C. J. Collins and N. S. Bowman, eds.), Van Nostrand Reinhold, New York, pp. 213-285.
Thornton, E. R. (1964). "Solvolysis Mechanisms", Ronald Press, New York, N.Y., (a) p. 96 ff.
Tommila, E. (1955). *Acta Chem. Scand.* 9, 975.
Tommila, E. (1967). *Ann. Acad. Sci. Fenn. Ser. A II* No. 139; *Chem. Abs.* (1968) 69, 90177.
Turner, D. W., Baker, C., Baker, A. D., and Brundle, C. R. (1970). "Molecular Photoelectron Spectroscopy", Wiley-Interscience, London.
Vitullo, V. P., and Wilgis, F. P. (1975). *J. Amer. Chem. Soc.* 97, 5616.
Watanabe, K. (1957). *J. Chem. Phys.* 26, 542.
Weiner, H., and Sneen, R. A. (1965a). *J. Amer. Chem. Soc.* 87, 287.
Weiner, H., and Sneen, R. A. (1965b). *J. Amer. Chem. Soc.* 87, 292.
Westaway, K. C. (1975). *Tetrahedron Lett.* 4229.
Wigfield, D. C., and Lem, B. (1975). *Tetrahedron* 31, 9.
Williams, R. C., and Taylor, J. W. (1974). *J. Amer. Chem. Soc.* 96, 3721.
Winstein, S. (1965). Chimica Teorica, VIII Corso Estivo Di Chimica, Accademia Nazionale Dei Lincei, Rome, p. 240.
Winstein, S., and Fainberg, A. H. (1957). *J. Amer. Chem. Soc.* 79, 5937.
Winstein, S., and Marshall, H. (1952). *J. Amer. Chem. Soc.* 74, 1120.
Winstein, S., and Takahashi, J. (1958). *Tetrahedron* 2, 316.
Winstein, S., Grunwald, E., and Jones, H. W. (1951). *J. Amer. Chem. Soc.* 73, 2700.
Winstein, S. Fainberg, A. H., and Grunwald, E. (1957). *J. Amer. Chem. Soc.* 79, 4146.
Winstein, S., Smith, S., and Darwish, D. (1959a). *Tetrahedron Lett.* 16, 24.
Winstein, S., Smith, S., and Darwish, D. (1959b). *J. Amer. Chem. Soc.* 81, 5511.
Yamataka, H., Kim. S.-G., Ando, T., and Yukawa, Y. (1973). *Tetrahedron Lett.* 4767.

The Reactivity–Selectivity Principle and its Mechanistic Applications

A. PROSS

Department of Chemistry, Ben Gurion University of the Negev, Beer Sheva, Israel

1. Introduction	69
2. Theoretical Considerations	71
Basis for the Reactivity–Selectivity Principle	71
Reactivity	76
Selectivity	78
3. Mechanistic Applications of the Reactivity–Selectivity Principle	82
Proton Transfer	83
Solvolysis	96
Carbene Reactions	112
Electrophilic Aromatic Substitution	116
Free Radical Reactions	121
4. Conclusion	126
References	127

1. INTRODUCTION

The reactivity-selectivity principle has long been part of the chemist's intuition. A qualitative statement of the principle is that highly reactive species are unselective in their choice of reactants compared to stable and therefore unreactive species. As a general proposition it is certainly incorrect for there are many clear exceptions. For example, highly reactive atomic oxygen reacts about one hundred times more rapidly with 2,3-dimethyl-2-butene than with ethylene, revealing high selectivity (Cvetanovic, 1959). A more limited interpretation, however, does have considerable experimental support. That is, in a particular reaction series, an increase in the reactivity of one of the reactants will result in a corresponding decrease in the selectivity of that species. In addition to the many existing examples of reactivity-selectivity relationships there are also notable failures as well. As a result, the reactivity-selectivity principle has taken on a rather unusual status. On the one hand very little has been reported on the theoretical basis of the principle and certain workers (Johnson, 1975; Kemp and Casey, 1973; Gilbert and Johnson, 1974) have raised the possibility that the principle may in fact be invalid. On the other hand in certain mechanistic areas, e.g. electrophilic substitution and carbene chemistry, workers derive considerable mechanistic information by the use—often implicit—of the principle, which serves as a key probe into the stability and therefore the structure of highly active species. Comparison of the selectivities of a species produced by a number of different methods often enables the similarities as well as the differences in these species to be gauged. Constant selectivity in such a study is particularly diagnostic of the intermediacy of one distinct species only. The observation of a reactivity-selectivity relationship for a given reaction series suggests a certain uniformity in mechanism; a sudden break or the total failure to obtain such a relationship suggests the opposite. Furthermore, in view of the basic nature of the assumptions on which the reactivity-selectivity principle is based, reactivity-selectivity relationships serve as a probe into some of the fundamental tenets of theoretical chemistry. The subject is therefore particularly significant and the uncertainty surrounding the principle, because of conflicting attitudes, unsatisfactory.

This review, therefore, attempts to survey the current status of reactivity-selectivity relationships by exploring the basis and variety of such relationships. The part they have played in a number of selected areas of mechanistic interest is also discussed. No attempt is made to cover every facet of mechanistic study. Rather, the aim has been to point out different reactivity-selectivity relationships that have been noted in such studies and to focus attention on the use of such relationships as a mechanistic tool. In addition, an attempt has been made to outline the limitations of the reactivity-selectivity principle. This serves not only to increase the utility of the principle but also strengthens the basis and justification for its use by rationalizing apparent failures that have been reported.

2. THEORETICAL CONSIDERATIONS

Basis for the Reactivity-Selectivity Principle

One of the focal points of mechanistic interest has been into the nature of the transition state. A postulate which bears heavily on this topic and which is now most commonly referred to as the Hammond postulate (Hammond, 1955) has become central in the study of transition state structure. Hammond's postulate may be stated as follows: the interconversion of two states of similar energy on a reaction pathway will involve only a small amount of structural reorganization. A precise interpretation of this postulate leads only to the limited conclusion that transition states of highly exothermic reactions are similar in structure and energy to reactants, while for strongly endothermic reactions transition states resemble products.

Leffler (1953, 1963a) has proposed a more general relationship (hence referred to as the Leffler-Hammond postulate) which represents an extension of the Hammond postulate since it treats the whole spectrum of reaction types. Thus the transition state is viewed as changing gradually from reactant-like in highly exothermic reactions, to intermediate in character for thermoneutral reactions, to product-like for endothermic reactions. In addition to this proposal, which relates the transition state structure to that of the products and reactants, a free energy relationship (1) which relates changes in

$$\delta G^{\ddagger} = \alpha \delta G_P + (1 - \alpha)\delta G_R \tag{1}$$

transition-state energy to that of products and reactants was incorporated into the scheme. Here δ is an operator which indicates the difference introduced into the function it precedes as a result of some perturbation (e.g. a substituent or medium change), $G^{\ddagger}$, G_P and G_R are the free energies of the transition state, product, and reactant respectively, and α is a mixing factor, limited to values between 0 and 1, which specifies the contribution of reactant and product free energy differences to that of the transition state. A value of α close to 1 indicates that free energy changes in the product are largely reflected in the transition state; in contrast free energy changes in the reactants will have almost no influence on $\delta G^{\ddagger}$. For values of α close to zero the reverse is true. Equation (1) may be rearranged to give the rate-equilibrium relationship (2), which states

$$\delta \Delta G^{\ddagger} = \alpha \delta \Delta G^0 \tag{2}$$

that for a given reaction, a perturbation of the free energy change, ΔG^0, will be only partially reflected in the transition state. This effect is illustrated in Fig. 1. For a reaction in which a perturbation affects only the products, it is readily seen that on moving along the reaction co-ordinate from reactant to product, the free energy difference between the two pathways gradually increases towards its maximum value. In the transition state, the magnitude of the perturbation will equal $\alpha\delta\Delta G^0$, and, therefore, α reflects the distance of the transition state along the reaction co-ordinate.

The rationale for the Leffler–Hammond postulate and the free energy relationship expressed in (1) and (2) may be found in a model

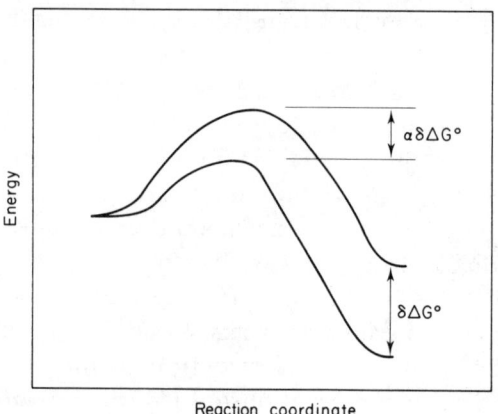

Figure 1. The effect of a perturbation on the free energy of the products and the transition state for a given reaction. $0 \leqslant \alpha \leqslant 1$.

first proposed by Evans and Polanyi (1936) and Bell (1936, 1941) and more recently discussed by Dewar (1969).

For a general substitution reaction, represented by eqn (3), the effect of perturbations on transition state structure and energy may

$$A + B - C \rightarrow A - B + C \qquad (3)$$

be qualitatively estimated by considering the intersection of potential energy curves, as shown in Fig. 2.

Curves I and II represent the energy profiles for the dissociation of $B - C$ and $A - B$ respectively. The reaction pathway (represented by curve III) may be considered as being approximated by the portions of these two curves below their point of intersection.

For a series of reactions in which BC combines with $A_1, A_2, \ldots, A_n$ to give $A_1 B, A_2 B, \ldots, A_n B$ whose dissociation energy is variable,

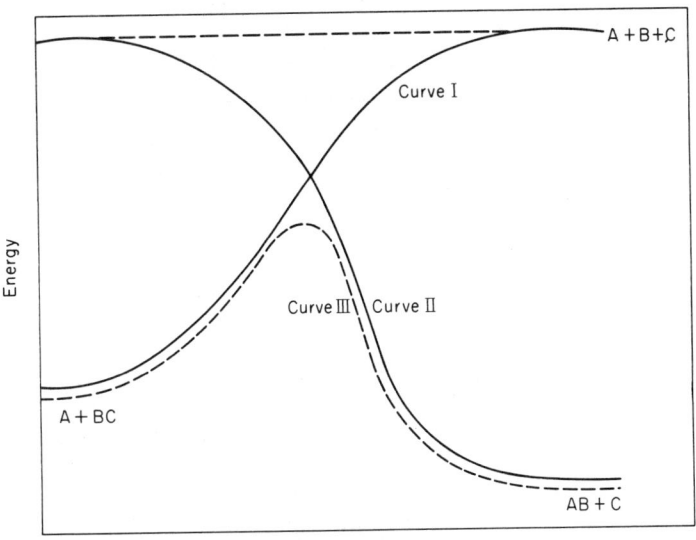

Figure 2. The Bell-Evans-Polanyi treatment for the reaction A + BC → AB + C indicating the reaction profile. (Taken from Dewar, 1969.)

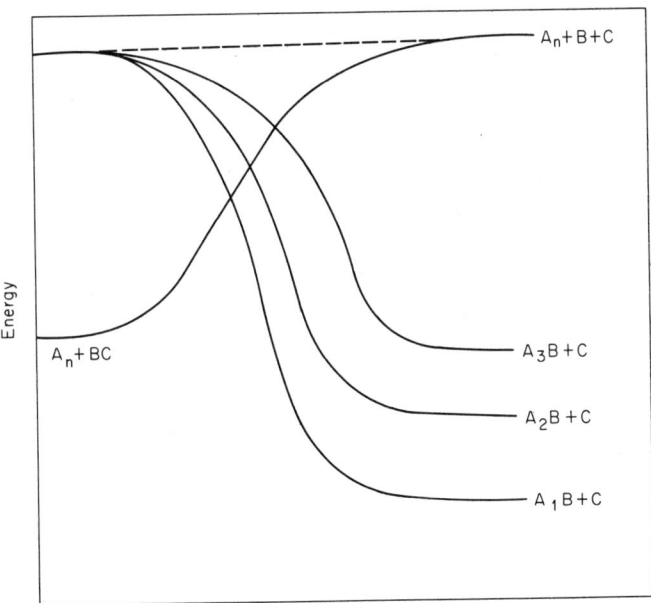

Figure 3. The effect of product stability on the transition state for the reactions A_n + BC → A_nB + C. (Taken from Dewar, 1969.)

a series of dissociation profiles may be obtained as illustrated in Fig. 3. The intersection of each dissociation profile with the dissociation curve for the species BC, indicates the approximate position of the transition state for each reaction of the series. As Dewar (1969) has noted, this treatment leads to a number of conclusions.

1. Product stabilization brings about a corresponding reduction in the activation energy.
2. As each reaction in the series becomes more exothermic, so the transition state increasingly resembles the reactants in accordance with the Leffler–Hammond postulate.
3. Due to the variation in the slopes of the dissociation curves, a continual increase in stability of the product brings about a progressively smaller stabilization of the transition state.

This final point signifies that the value of α in the rate-equilibrium relationship (2) is not constant but decreases as the reaction becomes increasingly exothermic. It should be noted however that since the Bell–Evans–Polanyi model and the Hammond postulate are couched in energy terms the assumption that free energy changes (ΔG^0) are proportional to energy changes (ΔE^0) is inherent in eqns (1) and (2).

On the basis of the Leffler–Hammond postulate the theoretical justification for the reactivity–selectivity principle may be observed. First, let us define the selectivity S, of a species A, in its reaction with two competing reagents X and Y, as indicated by (4), where k_X

$$S = \log k_X/k_Y \qquad (4)$$

and k_Y are the rate constants for reaction of A with X and Y respectively. Since $\Delta G^{\ddagger}_{AX} - \Delta G^{\ddagger}_{AY}$ (the difference in the free energies of activation for reaction of A with X and Y respectively) is linearly related to $\log k_X/k_Y$, the selectivity of A will be proportional to $\Delta G^{\ddagger}_{AX} - \Delta G^{\ddagger}_{AY}$. The schematic representation of Fig. 4 illustrates the fact that for A, a reactive species, this difference is relatively small, while for B, a more stable species, the selectivity (proportional to $\Delta G^{\ddagger}_{BX} - \Delta G^{\ddagger}_{BY}$) will be correspondingly larger. This idea may be expressed in different terms. Since the interaction between A and X or Y in the transition state is as yet weak, because it occurs early along the reaction co-ordinate, A will exhibit low preference for Y over X. The corresponding transition state for B, however, is more advanced along the reaction co-ordinate, resulting in greater interaction between B and X or Y, and as a consequence B will exhibit high selectivity.

It is important to note that the above presentation, justifying the reactivity-selectivity principle, is based on a number of fundamental assumptions. First, it is assumed that the Leffler-Hammond postulate is valid, which in turn implies that the reaction under consideration obeys a rate-equilibrium relationship [eqn (2)]. This assumption often cannot be verified since for reactions of highly active species such as carbenes, free radicals, carbonium ions, etc., equilibrium constants are generally not measurable. However it follows that for reactions which do not conform to a rate-equilibrium relationship, no reactivity-selectivity relationship is expected. Also, in Fig. 4, the difference in the free energy of the

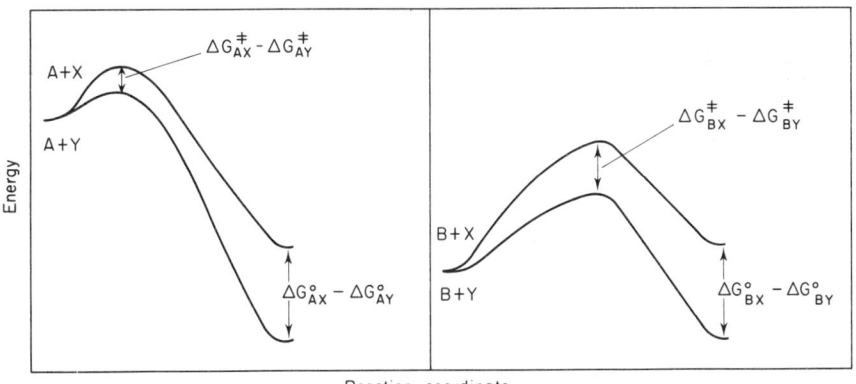

Figure 4. The effect of a change in reactivity on selectivity: (a) a highly reactive species, A, and (b) an unreactive species, B. Selectivity is proportional to $\Delta\Delta G^{\ddagger}$.

products from A, $(\Delta G^0_{AX} - \Delta G^0_{AY})$ is represented as equal to that from B, $(\Delta G^0_{BX} - \Delta G^0_{BY})$. As has been noted by Leffler and Grunwald (1963a), if $\Delta G^0_{BX} - \Delta G^0_{BY}$ is significantly smaller than $\Delta G^0_{AX} - \Delta G^0_{AY}$, the reactivity-selectivity principle will fail; the selectivity of B will be smaller than that for A. In practice, failure of the reactivity-selectivity principle may on occasion be attributed to this effect and a number of such examples are discussed on pp. 110-2.

In line with the Leffler-Hammond postulate, which is a differential analogue of the Hammond postulate, the rationale for a differential reactivity-selectivity principle has been demonstrated, i.e., the marginal stabilization of a particular species will result in a corresponding increase in its selectivity. Recently, the view has been expressed that there is no substantive evidence for such behaviour and that most quoted examples of the reactivity-selectivity principle may be explained by the fact that the principle is valid only on reaching the diffusion limit (Kemp and Casey, 1973). The argument was put that increasing the reactivity of a species to the point that its reactions are diffusion controlled will clearly reduce its selectivity. In the course of the discussion on specific reaction types, results which cast doubt on the validity of this view are presented, and key experimental data which led to its formulation are reinterpreted.

Reactivity

In order to obtain a reactivity-selectivity relationship, a means of quantifying reactivity is required. This is problematical since there is no absolute measure of reactivity. The reactivity of a species is judged to be high or low in comparison to another species and with respect to some standard reagent; a change in reaction conditions may well invert the reactivity order. For example, the partial rate factor for the detritiation reaction of toluene-p-t is greater than that for t-butylbenzene-p-t in the sulphuric acid-water system whereas the opposite order is observed in the trifluoroacetic acid-water system (Gold, 1964; Eaborn and Taylor, 1961). Also, it has been found that the nucleophilic reactivity of halide ions towards some standard substrate is quite different in protic and aprotic solvents (Weaver and Hutchison, 1964). The effect of medium changes can be eliminated by considering a reactive species in a given solvent as a single system, but the choice of reference substrate may also determine the reactivity order. The nucleophilic reactivity order for anions in methanol increases in the order $CN^- < CH_3O^- < N_3^-$ in reactions with cationic substrates such as triarylmethyl cations (Ritchie, 1972), whereas for a neutral substrate such as methyl iodide the opposite order, $N_3^- < CH_3O^- < CN^-$ is obtained (Alexander et al., 1968). Kice et al. (1974) have reported that the nucleophilic order toward sulphenyl and sulphonyl sulphur are quite different. Also, it has recently been found that the reactivity of benzenoid and heteroaromatic compounds measured by the rate of acid catalysed hydrogen exchange at a pH of 0 and the rate of nitration in 75% sulphuric acid do not correlate (Clementi et al., 1975). This occurs in spite of the linear free energy relationship exhibited by both reactions with the change of a single structural parameter. Clearly, the use of a reactivity parameter in a quantitative fashion requires certain caution; the reference compound should always be stipulated. As to the reactivity parameter itself, the most readily available parameter is the rate constant with the standard substrate (Leffler and Grunwald, 1963a). It should be noted, however, that the comparison of rate constants may yield different reactivity orders at different temperatures. It is apparent, therefore, that the determination of reactivity in this way is inherently imprecise but often, in view of the considerable rate ranges studied, the values obtained are of sufficient utility.

For very rapid reactions absolute rate constants are often unavailable. In such cases relative rate constants obtained by competitive reactions may serve instead. In the case where the process studied occurs after the rate determining step of a multi-stage reaction, the rate constant for the rate determining step may on occasion be used. For example, in solvolysis reactions the reactivity of intermediate carbocations are gauged by their rate of formation, not by their rate of reaction which is usually unavailable. This measure of reactivity is based on the assumption that the rate of formation of a carbocation is a direct function of its stability, i.e., that stable carbocations are formed more rapidly than unstable ones (Bethell and Gold, 1967). Raber et al. (1971a) have pointed out that such a use of reactivity fails to take into account differences in ground state energies. While these differences are generally not expected to be significant, in certain cases rates of carbocation formation as a measure of reactivity must be considered invalid. Thus the solvolysis of cis and trans α-bromo-β-t-butyl-p-methoxystyrenes, [1] and [2], which yield the same vinyl cation [3], results in a ratio k_{cis}/k_{trans} of 1640,

$$p\text{-CH}_3\text{OC}_6\text{H}_4\diagdown\!\!\!\!\!\diagup\!\!\text{C(CH}_3)_3 \qquad p\text{-CH}_3\text{OC}_6\text{H}_4\diagdown\!\!\!\!\!\diagup\!\!\text{H}$$
$$\text{Br}\diagup\!\!\text{C=C}\diagdown\!\!\text{H} \qquad\qquad \text{Br}\diagup\!\!\text{C=C}\diagdown\!\!\text{C(CH}_3)_3$$

[1] \qquad\qquad [2]

$$p\text{-CH}_3\text{OC}_6\text{H}_4-\overset{+}{\text{C}}\!\!=\!\!\text{C}\diagup\!\!\!\!^{\text{C(CH}_3)_3}_{\text{H}}$$

[3]

equivalent to a free energy difference of activation of 5·2 kcal mol^{-1} (Rappoport et al., 1973c). In cases where such ground state effects are significant an alternative reactivity parameter must be chosen. Thus it can be seen that reactivity may be quantified either by using absolute or relative rate constants, or in certain cases by the rate of formation of a reactive species. Whatever method is used, however, it should be recognized that, because of the inherent limitations, the measure of reactivity is of necessity approximate. In certain extreme cases a particular measure may be quite inadequate, and an alternative procedure should be sought.

Selectivity

Determination of selectivity

The definition of the selectivity of a species A in its reaction with two reagents X and Y, as has been noted earlier, may be expressed as log k_X/k_Y. Used in this way, selectivity may be determined in two possible ways. Individual rate constants may be measured directly or the ratio of rate constants obtained by product analysis. The latter method is based on the following reasoning (Swain *et al.*, 1953b; Ogston *et al.*, 1948): If a species A reacts competitively with two reagents, X and Y, to give AX and AY respectively, then the rate constant ratio k_X/k_Y may be related to the concentration ratio [X]/[Y] by eqn (5). The integral form of (5) is (6) provided that the

$$\frac{d[AX]}{d[AY]} = \frac{k_X[X]}{k_Y[Y]} \qquad (5)$$

$$\frac{[AX]}{[AY]} = \frac{k_X[X]}{k_Y[Y]} \qquad (6)$$

relative concentrations of X and Y remain constant. Since the concentrations of AX and AY may be determined by product analysis and the concentrations of X and Y are known, the selectivity may be deter-

$$S = \log \frac{k_X}{k_Y} = \log \frac{[AX][Y]}{[AY][X]} \qquad (7)$$

mined from eqn (7). This provides a method of obtaining the selectivity without the need to measure directly either relative or absolute rate constants.

Application of linear free energy relationships to selectivity

It may be recalled that the selectivity parameter serves as such because it reflects the distance of the transition state along the reaction co-ordinate. It follows that any alternative parameter which will also indicate the position of the transition state may also serve in the same way. The various proportionality factors associated with the different linear free energy relationships do just this, and therefore they may be utilized as additional selectivity factors. The common linear free energy relationships and the selectivity parameters they produce are listed in Table 1.

The use of the selectivity parameters listed in Table 1 to indicate transition state structure is only common with the parameters α and ρ. In principle, however, the other selectivity factors, m and s, may be used in the same way, although the influence of changes in mechanism on these factors in particular is sufficiently great to override the influence of selectivity. This point will be discussed in more detail on p. 109.

TABLE 1

List of Key Linear Free Energy Relationships

Name	Equation	Reactivity parameter	Selectivity parameter
Hammett[a]	$\log k/k_0 = \sigma\rho$	σ	ρ
Swain-Scott[b]	$\log k/k_0 = ns$	n	s
Brønsted[c]	$\log k = -\alpha pK_a + \log G$	pK_a	α
Winstein-Grunwald[d]	$\log k/k_0 = mY$	Y	m

[a] Hammett, 1970.
[b] Swain and Scott, 1953a.
[c] Brønsted and Guggenheim, 1927.
[d] Grunwald and Winstein, 1948.

Quantitative expressions of the Hammond postulate may be derived from these parameters (Miller, 1959; Cordes and Jencks, 1962; Jencks, 1969a). Thus, for example, combination of the Hammett and Brønsted eqns (8) and (9) respectively, for a series of i substrates reacting with j general acid catalysts leads to (10).

$$\log k_{i,j} = \log k_{0,j} + \sigma_i \rho_j \tag{8}$$

$$\log k_{i,j} = \log G_i^A - \alpha_i pK_{aj} \tag{9}$$

$$\log k_{0,j} + \sigma_i \rho_j = \log G_i^A - \alpha_i pK_{aj} \tag{10}$$

Applying eqn (10) to two general acid catalysts, 1 and 2, and subtracting leads to (11).

$$\log \frac{k_{0,1}}{k_{0,2}} + \sigma_i(\rho_1 - \rho_2) = \alpha_i(pK_{a2} - pK_{a1}) \tag{11}$$

Since

$$\log \frac{k_{0,1}}{k_{0,2}} = \alpha_0 (pK_{a2} - pK_{a1}) \quad (12)$$

substituting (12) into (11) leads to (13).

$$\frac{\sigma_i}{\alpha_0 - \alpha_i} = \frac{pK_{a2} - pK_{a1}}{\rho_2 - \rho_1} \quad (13)$$

Since the expression on the left-hand side of (13) depends only on the substrates while the expression on the right-hand side depends only on the acid catalysts, the two expressions may be separated to yield eqns (14) and (15).

$$\sigma_i = C(\alpha_0 - \alpha_i) \quad (14)$$

$$pK_{a2} - pK_{a1} = C(\rho_2 - \rho_1) \quad (15)$$

Equation (14) states that increasing the reactivity of a substrate by changing the substituent and hence the substituent constant, σ_i, will decrease the selectivity, α_i. Equation (15) shows that an increase in the acidity of an acid catalyst in a reaction which shows a Hammett relation will bring about a corresponding decrease in the magnitude of ρ. More generally, this means that an increase in any one of the reactivity parameters listed in Table 1 will bring about a decrease in all the selectivity parameters, except for the one with which it is related in the linear free energy relationship. Thus a suitable combination of equations will demonstrate that an increase in the reactivity parameter σ should bring about a reduction in the selectivity parameters α, s and m but not ρ, since ρ is constant for the reaction series. There is, of course, an obvious paradox in this result. The question arises why should not the selectivity parameter associated in a particular linear free energy relationship with its corresponding reactivity parameter equally reflect changes in reactivity. In other words the combination of linear free energy relationships and application of the reactivity-selectivity principle lead to the conclusion that linear free energy relationships should not be linear at all. A consistent model would indicate that a plot of log k for a particular reaction against any of the reactivity parameters, Y, pK_a, σ and n should show curvature such that at high reactivity the slope should be smaller (i.e. low selectivity) than for low reactivity (Johnson, 1975). For the Brønsted relationship such curvature is in fact now well documented (see p. 83). However, plots of log k against the other reactivity parameters, Y, n and σ (as well as a number of other parameters, such as N_+ (Ritchie, 1972) and the H_0 acidity function) generally give linear correlation.

For the reactivity parameters Y, n, σ^+ (but not σ) and N_+ the lack of curvature is not unexpected. This is because these parameters are defined with respect to the *rate* of some standard reaction (solvolysis of t-butyl chloride, substitution of methyl iodide, solvolysis of cumyl chlorides, combination reaction of nucleophiles with a standard electrophile). Therefore the resultant plot is of the type $\log k$ vs. $\log k$, while the curvature shown in a typical Brønsted plot (Figure 5) results from a plot of $\log k$ vs. $\log K$. This curvature is due to a gradual change from a reactant-like transition state, which is insensitive to a perturbation in the reactivity parameter, to a product-like transition state in which equilibrium perturbations are largely reflected in the transition state (and hence the rate). A $\log k - \log k$ plot is not expected to show this effect and hence is not expected to show curvature.

Unfortunately this explanation excludes the Hammett equation based on the reactivity parameter σ. Since σ is defined with respect to the dissociation constant of the substituted benzoic acids (i.e. on the basis of equilibrium constants), a plot of $\log k$ against σ is actually a $\log k - \log K$ correlation and is expected to show curvature. In fact, the evidence for linear Hammett plots is overwhelming, to the extent that curvature, when found, is attributed to a change in the mechanism or in the identity of the rate determining step (Leffler and Grunwald, 1963b). The rationale for this apparent contradiction between theory and practice remains obscure (Johnson, 1975). Possible contributing factors may be the limited reactivity range ($\sim 10^4$ for a maximal ρ value of about 4) covered in Hammett correlations based on σ (rather than σ^+). This is particularly true when one considers the fact that the nitro group which often extends the reactivity range of Hammett studies by two orders of magnitude may not correlate well with the remaining points (Kevill *et al.*, 1973). Furthermore, application of the basic principle of Marcus theory (discussed on pp. 83-5) to Hammett plots suggests that the degree of curvature is dependent not only on the reactivity range studied but to the "intrinsic" barrier (defined as the free energy of activation for the thermoneutral reaction) as well. Reaction series involving large "intrinsic" barriers are expected to show less curvature than ones involving small "intrinsic" barriers. For proton transfer reactions, where the observed curvature is often not very marked, "intrinsic" barriers of only 1-5 kcal mol^{-1} are generally obtained (Kreevoy and Oh, 1973). Larger "intrinsic" values in reactions for which Hammett studies have been conducted would be expected to reduce the degree of curvature even further, so as to give rise to plots which, within experimental error, might be best described as linear. Clearly, further work needs to be done to clarify this troublesome point.

Kinetic isotope effects as a measure of selectivity

A further measure of transition state structure (and hence of utility as a selectivity parameter) is the kinetic isotope effect, in particular for hydrogen isotopes, where many data are available. This subject has been extensively reviewed, most recently by More O'Ferrall (1975), Bell, (1973), Thornton and Thornton (1970), and Van Hook (1970). However, the use of kinetic isotope effects as selectivity parameters appears to be of limited application. Westheimer's model for proton transfer reactions (Westheimer, 1961) indicates that for a reactant-like transition state k_H/k_D will approach 1, and will increase to a maximum value for symmetrical transition states ($k_H/k_D \sim 7$) and drop to a value close to 1 (0·7-1·5) for product-like transition states. This means that, in practice, use of hydrogen isotope effects will not distinguish between reactant-like and product-like transition states, making the observation of reactivity-selectivity relationships somewhat uncertain. However, for a reaction series whose reactivity range results in transition states which fall within the reactant-like and symmetrical limits, the monotonic increase in k_H/k_D from a value of 1 for highly exothermic reactions to a value of about 7 for thermoneutral reactions may be utilized as a measure of selectivity. Certain hydrogen abstraction reactions by radicals (p. 123) appear to fall into this category.

The uses (and possible misuse) of the selectivity parameters discussed above are examined for specific reaction types in Section 3.

3. MECHANISTIC APPLICATIONS OF THE REACTIVITY-SELECTIVITY PRINCIPLE

This section shows the part reactivity-selectivity relationships have played in a number of areas of mechanistic interest. The discussion considers a number of different aspects simultaneously.

Thus the observation of reactivity-selectivity relationships is reported and the utilization of the principle pointed out. In addition, an attempt is made to indicate the validity of the various measures of selectivity as well as the possible rationalizations of the observed breakdown in the principle which have been reported for these specific areas.

Proton Transfer

Evidence for curvature in Brønsted plots

Proton transfer reactions have played an important part in the study of the reactivity-selectivity principle. This is because of the mechanistic significance attributed to the Brønsted eqn (16) in which

$$\log k_a = \alpha \log K_a + \text{constant} \qquad (16)$$

k_a is the specific rate constant of an acid or base catalysed reaction with a given acid or base, K_a is the acid or base dissociation constant and α is termed the Brønsted coefficient (Brønsted et al., 1923, 1927). Detailed discussion on many aspects of proton transfer may be found in recent reviews (Caldin and Gold, 1975; Kresge, 1975, 1973) and the monograph by Bell (1973).

Since (16) is a rate-equilibrium relationship [equivalent to the relationship (2) discussed earlier], α is considered to reflect the degree of proton transfer in the transition state and hence is a measure of selectivity. Values of α close to 0 are associated with exothermic reactions in which the degree of proton transfer in the transition state is as yet small. Similarly, values of α close to 1 are associated with endothermic reactions in which the degree of proton transfer in the transition state is almost complete.

For a reaction series studied over a wide reactivity range, i.e. those in which the transition state varies from reactant-like to product-like, a plot of $\log k_a$ against $\log K_a$ (or ΔpK_a, the dissociation constant difference between catalyst and substrate) is expected to give a curve. This is shown in Fig. 5. The tangential slope at any point on the curve yields a corresponding value of α; as the reactivity increases the value of α decreases.

Marcus theory (Marcus, 1968, 1969), originally developed to interpret the rates of electron transfer reactions, has been successfully applied to proton transfer reactions as well. The theory relates

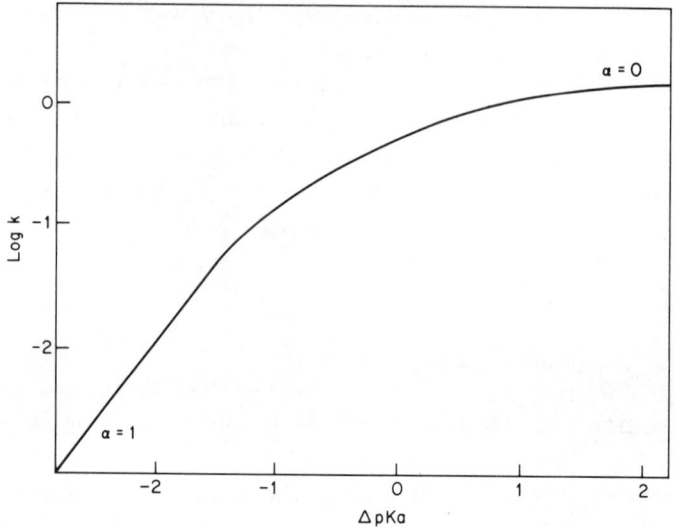

Figure 5. The effect of an increase in reactivity on the Brønsted parameter, α.

the free energy of activation, $\Delta G^{\ddagger}$, to the standard free energy of reaction, ΔG^0 as shown in eqn (17). The relation assumes that

$$\Delta G^{\ddagger} = w^r + \lambda/4 + \Delta G^0/2 + (\Delta G^0)^2/4\lambda \qquad (17)$$

proton transfer is a multiple step process where w^r is the energy of formation of the encounter complex between the substrate and catalyst as a result of solvent reorganization, ΔG^0 is the free energy change in the actual proton transfer step and $\lambda/4$ is termed the "intrinsic" barrier and equals the free energy of activation for the proton transfer step when $\Delta G^0 = 0$. This formulation represents a quantitative expression of the Hammond postulate because $\Delta G^{\ddagger}$ and ΔG^0 are found to be linked for a reaction series which shares a constant "intrinsic" barrier.

Since the Brønsted coefficient, α, is defined as $d\Delta G^{\ddagger}/d\Delta G^0$, differentiation (Albery et al., 1972c) of (17) gives (18). The degree of

$$\alpha = \tfrac{1}{2} + \Delta G^0/2\lambda \qquad (18)$$

curvature in the Brønsted plot may be found by taking the second derivative of $\Delta G^{\ddagger}$ to yield (19). Thus the curvature present in a

$$\frac{d\alpha}{d\Delta G^0} = \frac{1}{2\lambda} \qquad (19)$$

Brønsted plot is found to be dependent on the "intrinsic" barrier to proton transfer. Fast reactions (small $\lambda/4$) will show significant curvature while slow reactions (large $\lambda/4$) will show only slight curvature.

Considerable work has been invested in the experimental verification of Marcus theory. Kreevoy and Konasewich (1971) have studied the hydrolysis of the diazoacetate ion catalysed by a series of phenols and carboxylic acids (20). Proton transfer takes place in the

$$N_2CHCOO^- + HA \rightarrow {}^+N_2CH_2COO^- + A^-$$
$$^+N_2CH_2COO^- \rightarrow HOCH_2COO^-$$
(20)

first step which is rate determining. Catalysis of this reaction by alkylammonium ions has been studied as well (Kreevoy and Oh, 1973). Albery et al. (1972a, 1972b, 1972c) have investigated the reactions of ethyl diazopropionate and 3-diazo-2-butanone while Hassid, Kreevoy and Laing (1975) have recently studied the hydrolysis reaction of diphenyldiazomethane. These studies indicate that Marcus theory may be used to account for a wide range of proton transfer reactions though it should be noted that the large number of parameters present in the Marcus equation allows considerable latitude in the fitting of the experimental data to the theory. Recent accounts of the application of Marcus theory to proton transfer reactions have been published by Kresge (1973, 1975) Crooks (1975) and More O'Ferrall, (1975).

Apart from these studies of curved Brønsted plots which have been utilized to verify Marcus theory, many additional examples of curved plots are known; together they provide considerable evidence in support of an inverse relationship between reactivity and selectivity in proton transfer reactions.

Eigen (1964) found that a plot of ΔpK_a against the rate constant for proton transfer between acetylacetone and a series of bases gave a curved plot. It should be noted, however, that Eigen's explanation for curvature is quite different from the one based on Marcus theory and the reactivity-selectivity principle. The curvature discussed by Eigen is attributed to a change from a rate-determining proton transfer to a diffusion controlled reaction which is independent of the catalyst pK_a.

Bell (1959) has observed that the experimental data for the halogenation of a series of carbonyl compounds also yield a reactivity-selectivity relationship. Thus reactive compounds show a low measure of selectivity while unreactive compounds show a high measure of selectivity. The data are listed in Table 2. Bell et al.

TABLE 2

Catalytic Coefficients, Brønsted Coefficients and Acidities for the Halogenation of Various Carbonyl Compounds[a]

Substrate	$\log R^b$	β	pK_a
CH_3COCH_3	−8·56	0·88	20·0
$CH_3COCH_2CH_2COCH_3$	−7·85	0·89	18·7
CH_3COCH_2Cl	−5·29	0·82	16·5
CH_3COCH_2Br	−5·03	0·82	16·1
$CH_3COCHCl_2$	−3·78	0·82	14·9
$CH_2COCHCO_2C_2H_5$ $(CH_2)_3$	−1·76	0·64	13·1
$CH_3COCH_2CO_2C_2H_5$	−1·06	0·59	10·5
$CH_2COCHCO_2C_2H_5$ $(CH_2)_4$	−0·60	0·58	10·0
$CH_3COCH_2COC_6H_5$	−0·45	0·52	9·7
$CH_3COCH_2COCH_3$	−0·24	0·48	9·3
$CH_3COCHBrCOCH_3$	+0·26	0·42	8·3

[a] Data from Bell, 1959.
[b] R is the catalytic coefficient obtained using the anion of a hypothetical acid of $pK_a = 4$ and indicates the reactivity of the substrate.

(1940, 1949) also measured the rate of enolization for a number of ketones and ketoesters. A similar relationship to that presented in Table 2 was noted. The slope of the Brønsted plot was found to be related to the rate of enolization. High reactivity was associated with a small slope.

Kresge *et al.* (1971a) have observed a reactivity–selectivity relationship in proton transfer reactions, though of poor correlation.

TABLE 3

Brønsted Coefficients and Catalytic Coefficients for Various Vinyl Ethers[a]

Ether substrate	α	$\log R^b$
Ethyl isopropenyl	0·64	0·67
Ethyl cyclopentenyl	0·63	0·39
Ethyl cyclohexenyl	0·58	−0·58
Methyl cyclohexyl	0·66	−0·89
Phenyl isopropenyl	0·61	−1·73
Ethyl vinyl	0·70	−2·34
Phenyl vinyl	0·84	−5·30

[a] Data from Kresge *et al.*, 1971a.
[b] R is the catalytic coefficient using a hypothetical acid catalyst of $pK_a = 4$ and indicates substrate reactivity.

The hydrolytic reactivity of a series of vinyl ethers is presented in Table 3 as a function of the Brønsted slope α, for each vinyl ether. As pointed out by Kresge, the variation in selectivity is small considering the large reactivity range studied.

Cram and Kollmeyer (1968) found that the Brønsted slope for the based catalysed exchange of a series of hydrocarbons in dimethyl sulphoxide changes from 0·5 to 1·0 over some 10 pK_a units. It should be noted that this variation was based on a limited number of points.

TABLE 4

Brønsted Coefficients[a] and Basicities of Aromatic Substrates[b]

Substrate	pK_a	α_A	α_B
1,3,5-Trimethoxybenzene	−5·7	0·554	0·56
1,3,5-Triethoxybenzene	−4·8	0·539	0·55
1,3-Dimethoxybenzene (2-position)		0·604	
1,3-Dimethoxybenzene (4-position)	−9·0	0·639	0·62
1,3-Dihydroxy-2-methylbenzene	−7·5	0·539	0·59
1,3-Dimethoxy-2-methylbenzene	−9·3	0·652	0·62
Anisole	−15·3	0·706	0·72
Benzene	−23·0	0·930	0·87
Azulene	−1·7	0·545	0·47
Guaiazulene-2-sulphonate	−0·6		0·47
4,6,8-Trimethylazulene	+0·5		0·45
Guiazulene	+1·5		0·43

[a] Brønsted coefficients, α_A, obtained by a plot of kinetic acidity dependence against thermodynamic acidity dependence and α_B, obtained by a plot of rate of hydrogen exchange of the aromatic substrates against substrate pK_a.
[b] Data from Kresge et al., 1971b.

A reactivity–selectivity relationship has been also observed in aromatic proton exchange reactions (Kresge et al., 1971b). The selectivity parameter, α, was determined in two different ways. A plot of the kinetic acidity dependence of an aromatic substrate against the thermodynamic acidity dependence gave rise to a value of α, termed α_A, for each substrate. A second value of α, termed α_B was obtained by a plot of rate of hydrogen exchange of the aromatic substrate against the pK_a for the particular substrate, for a series of aromatic compounds. The latter method is a conventional Brønsted plot, although here the basicity of the substrate is varied rather than that of the catalyst. The slope of the curve at any point gives the alternative value of α, α_B. The data are presented in Table 4.

It can be seen that there is excellent agreement between α_A and α_B. Also there is a clear correlation between the value of α and substrate basicity. Weakly acidic substrates (e.g. benzene whose pK_a value is -23) show large α values, whereas more strongly acidic substrates show low α values, in accord with the reactivity–selectivity principle.

Reasons for the failure to obtain curved Brønsted plots

The examples discussed above indicate that α may be used as a selectivity parameter and that reactivity–selectivity relationships are obtained as a result of curvature in the Brønsted plot. However, in view of the many cases in which such curvature has not been observed, the general validity of α as a measure of transition state structure has been questioned. Often, linear Brønsted plots are obtained. Thus Pearson and Dillon (1953) obtained a linear rate-equilibrium relationship for the ionization of a number of pseudo acids over a pK_a range of 8 units; over a larger reactivity range, considerable scatter was observed. Streitwieser *et al.* (1971) have studied the detritiation reaction of a series of hydrocarbons by sodium methoxide and obtained two straight lines with slopes of 0·37 and 0·58 over a range of 18 pK_a units. The weak carbon bases studied by Kresge *et al.* (1971a) and listed in Table 3 each show constant Brønsted coefficients despite the large pK_a span studied. Dehydration of 1,1-diols have long been noted for their linear Brønsted plots (Brønsted *et al.*, 1923, 1927; Bell, 1959, 1966). The base catalysed elimination of a series of substituted benzisoxazoles not only results in linear Brønsted plots but the values of β obtained for the different substrates show only a slight variation despite the large reactivity range covered. Thus the reasons for the conflicting views on the significance of the Brønsted coefficient become apparent.

Jencks (1972, 1969b) has proposed a model which maintains the significance of α as a measure of transition state structure and yet provides a very elegant rationalization of the fact that linear Brønsted plots are often obtained over large reactivity ranges, both by variation in catalyst pK_a and substrate reactivity. Jencks noted that the acid catalysed nucleophilic addition to carbonyl compounds (21) yielded linear Brønsted plots (in apparent violation of the reactivity–selectivity principle); yet as the basicity of the nucleophile N was increased the Brønsted slope decreased (in agreement with the

$$\diagdown_{\diagup}\!\!N + \overset{\diagdown}{\underset{\diagup}{C}} = O + HA \rightarrow {}^+N-\overset{|}{\underset{|}{C}}-O-H + A^- \tag{21}$$

reactivity-selectivity principle). Illustrative data are presented in Table 5.

A proper use of the reactivity-selectivity principle requires that the introduction of a perturbation should not change the reaction pathway but only the relative energy levels. Expressed differently, the reactivity-selectivity principle may be applied only if no change in the reaction mechanism takes place with the introduction of a perturbation. This may be seen by reference to the contour diagram taken from Jencks' work (Fig. 6).

TABLE 5

Brønsted Coefficients for the Acid Catalysed Nucleophilic Addition of Various Nucleophiles to Acetaldehyde

Nucleophile (pK_a)	α	Ref.
RS^- (ca. 10)	0	a
$X-C_6H_4S^-$ (3-6)	0·2	b
H_2NCONH_2 (0·2)	0·45	c
H_2O (−1·7)	0·54	d
RSH (ca. −7)	~0·7	a

[a] Lienhard and Jencks, 1966.
[b] Barnett and Jencks, 1967.
[c] do Amaral et al., 1966.
[d] Bell et al., 1956.

An increase in the acidity of the catalyst (HA) will lower the right-hand side of the diagram since the acid is now present as its conjugate base. Stabilization of the product makes the transition state occur earlier along the reaction co-ordinate and therefore resemble the starting materials more closely in accordance with the Hammond postulate. However, lowering the bottom right-hand corner as well, which represents a perturbation perpendicular to the reaction profile, causes the transition state to be displaced perpendicular to the reaction profile and *towards* the bottom right-hand corner, in what has been termed an "anti-Hammond" effect (Thornton, 1967; More O'Ferrall, 1970). This occurs because the transition state lies at a saddle point (i.e. at a maximum parallel to the reaction profile but at a minimum perpendicular to it). As a consequence, stabilization of a

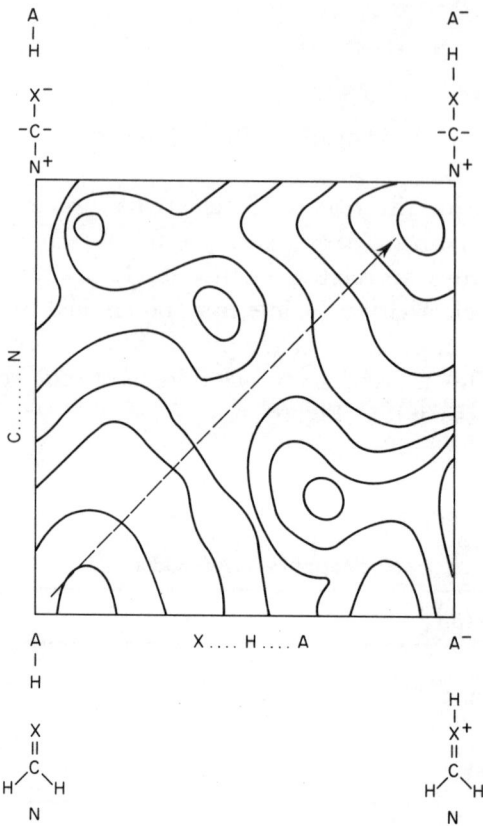

Figure 6. Contour diagram for the acid-catalysed nucleophilic addition to the C = X group. (Taken from Jencks, 1972.)

potential intermediate which does not actually lie on the reaction profile causes the transition state to move towards the intermediate in contrast to the normal Hammond behaviour.

The important result of this analysis is that these two effects, parallel and perpendicular to the reaction co-ordinate, operate in concert with respect to the carbon–nucleophile bond distance but oppose each other with respect to the degree of proton transfer. This means that while increasing the strength of the acid catalyst does advance the transition state with respect to carbon–nucleophile bond formation, the degree of proton transfer is expected to show only slight change. This is in accord with the constant value of α observed with increasing acid strength. A similar analysis for an increase in nucleophilicity shows that with strong nucleophiles the degree of proton transfer in the transition state is smaller, in line with the data in Table 5.

The data obtained by Kemp and Casey (1973) for the elimination reaction of substituted benzisoxazoles (22) which were interpreted as

$$B + X\text{-benzisoxazole} \longrightarrow BH^+ + \text{product} \qquad (22)$$

evidence for the inapplicability of the reactivity–selectivity principle are shown in Table 6,

The failure of the reactivity–selectivity principle appears to be on two scores. First, constant β values were obtained over a wide pK_a

TABLE 6

Brønsted Coefficients for Substituted Benzisoxazoles[a]

Substituents	β
5-MeO	0·66
H	0·74
6-MeO	0·72
5-Cl	0·69
6-Cl	0·72
6-NO$_2$	0·71
5-NO$_2$	0·74
5,7-Dinitro	0·65

[a] Data from Kemp and Casey, 1973.

range, and secondly, substituted benzisoxazoles, which show widely varying reactivity exhibit similar Brønsted coefficients. The diagram in Fig. 7 rationalizes these results as well as maintaining the validity of the reactivity–selectivity principle.

The two axes represent the breaking of the C—H and N—O bonds. By arguments analogous to those used for the acid catalysed nucleophilic addition to carbonyl compounds, it can be seen that strengthening the base catalyst will lower the top half of the contour diagram where the base appears as its conjugate acid. As a result, the Hammond and anti-Hammond types of behaviour are opposed with respect to C—H bond breaking, and therefore the small variation in the value of β over large pK_a ranges is expected. Similarly, the introduction of substituents into the benzisoxazole system will also result in opposing Hammond and anti-Hammond behaviour. For

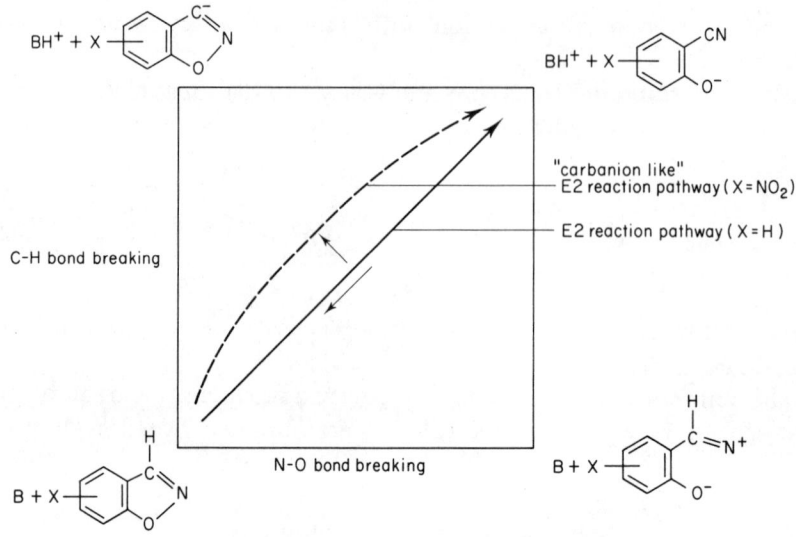

Figure 7. Energy surface for the base-catalysed elimination of substituted benzisoxazoles.

example, the introduction of a nitro group will stabilize the carbanion intermediate in the top left-hand corner as well as the cyanophenoxide ion in the top right-hand corner resulting in a cancelling effect on the degree of proton transfer in the transition state. Thus the apparent failure of the reactivity-selectivity principle may be attributed to a change in the mechanistic pathway from concerted E2 to a "carbanion like" E2 reaction for electron withdrawing substituents. This pathway is represented by the dotted line in Fig. 7.

In certain cases, only anti-Hammond effects will be obtained as a result of some perturbation, and, as a result, a simultaneous increase in both selectivity and reactivity is expected to be induced (More O'Ferrall, 1973). The elimination reactions of the compounds listed in Table 7 fall into such a category (Hudson and Klopman, 1964).

TABLE 7

Brønsted Coefficients and Rate Constants for Elimination from Alkyl bromides[a]

Substrate	$10^3 k$ (EtO$^-$)	β
$CH_3(CH_2)_2CHBr(CH_2)_2CH_3$	0·3	0·39
$C_6H_5CH_2CH_2Br$	84	0·54
$p\text{-}NO_2C_6H_4CH_2CH_2Br$	7300	0·67

[a] Data summarised by More O'Ferrall, 1973 and taken from Hudson and Klopman, 1964.

Substituent effects have a significant influence only on the carbanion intermediate. Since the reaction is basically an E2 elimination, substituent effects will result in a failure of the reactivity-selectivity principle.

The preceding discussion has suggested that α (or β) may be considered a measure of transition state structure even if the expected reactivity-selectivity relationship is not observed. There is, however, strong evidence to suggest that the Brønsted coefficient does not always reflect the degree of proton transfer in the transition state.

Failure of the Brønsted coefficient as a measure of transition state structure

The clearest example of the danger in using α as a measure of transition state structure is illustrated in the work of Bordwell *et al.* (1969, 1970, 1975). In the rate-equilibrium relationship for the deprotonation of a series of nitroalkanes the unprecedented Brønsted slopes of 1·61 for 1-aryl-2-nitropropanes and 1·37 for 1-arylnitroethanes were obtained. The simple exposition of the mechanistic significance of α disallows values greater than 1. This, coupled with the fact that the transition state for the proton transfer is not product-like (as established by alternative criteria) indicates at best that, in at least some cases, α does not reflect the selectivity of a particular reaction. Several attempts to rationalize these anomalous results have been made.

Kresge (1970) has proposed that the assumption inherent in the Brønsted relation, namely that perturbation of the transition state energy is dependent on the magnitude of the perturbation in both the product and reactant, is not always justified. It was suggested that new interactions absent from initial and final states may occur in the transition state. This may be seen by considering the reaction (23) for which an α value greater than 1 was obtained.

$$\text{HO}^- + \text{H}-\overset{\overset{\text{R}}{|}}{\text{C}}\text{Me}-\text{NO}_2 \longrightarrow$$

$$[\text{HO}\text{---}\overset{\delta-}{\text{H}}\text{---}\overset{\overset{\text{R}}{|}}{\text{C}}\text{Me}-\overset{\delta-}{\text{NO}_2}]^{\ddagger} \longrightarrow \text{HO}-\text{H} + \overset{\overset{\text{R}}{|}}{\text{C}}\text{Me}=\text{NO}_2^- \quad (23)$$

An interaction between the hydroxide group and R in the transition state is absent in the initial and final states. Kresge reasoned that, if the substituent R interacts more strongly with the hydroxide group than with the fully charged nitro group, then values of α greater than 1 are expected. It was further suggested (Kresge and Chiang, 1973) that the deviation of points from a Brønsted plot using both charged and uncharged acid catalysts may also be explained by transition state interactions absent from both initial states. Thus the acid catalysed hydrolysis of ethyl vinyl ether showed that no single Brønsted correlation occurs. Positively charged, negatively charged, and neutral acids each give a separate correlation. Kresge attributed this dispersion to intermolecular electrostatic forces between the developing positive charge on the vinyl ether and the polar groups present in the acid catalyst. The dispersion observed in the base catalysed decomposition of nitramide using dipositive, neutral, negative and dinegative catalysts (Bell, 1941) was explained by Kresge in a similar way. More O'Ferrall (1975) following on from Bell (1965) has extended this idea by pointing out that concerted elimination reactions, for which no rate–equilibrium relationship need occur, are analogous to proton transfers in nitroalkanes, as shown in eqn (24). The buildup of negative charge on the carbon

$$B + CH_3NO_2 \longrightarrow \left[\overset{\delta+}{B} \text{---} H \text{---} \overset{\delta-}{CH_2} - \overset{+}{N} \underset{O}{\overset{O^-}{\diagup}} \right]^{\ddagger} \longrightarrow BH^+ + CH_2 = \overset{+}{N} \underset{O^-}{\overset{O^-}{\diagup}} \quad (24)$$

atom in the transition state, being greater than that in the product, is of course expected to result in a greater substituent effect on the rate constants than on the corresponding equilibrium constants.

An alternative view has been presented based on Marcus theory (Marcus, 1968, 1969). If $\Delta G^{\ddagger}$ for a reaction is expressed in terms of ΔG^0 and the "intrinsic" barrier, $\lambda/4$, [eqn (17)], then it can be shown that the Brønsted slope will lie between 0 and 1 for a family of reactions which shares the same "intrinsic" barrier. In such cases the magnitude of α will reflect the position of the transition state along the reaction co-ordinate. Since $\Delta G^{\ddagger}$ for nitroalkanes is relatively high (i.e. $\lambda/4$ is large), changes in R [eqn (23)] are expected to alter the magnitude of $\lambda/4$. In such an event α ceases to be a measure of transition state structure.

There are additional factors which may invalidate the use of α as a measure of transition state structure. Murdoch (1972) has demonstrated that for multi-step reactions, even when proton transfer is rate-determining, the value of α obtained may be greatly influenced by the additional steps. An interesting feature of Murdoch's analysis is that where intermediate steps are extensively reversible (termed by Murdoch "internal return"), a linear Brønsted plot may result over a wide reactivity range. This is significant in view of the fact that a substantial proportion of Brønsted plots are linear. Bell (1959) and Gold and Waterman (1968) have observed that two different values of α may be obtained for one set of reactions (one by changing the pK_a value of the carbon base, the other through variation of the pK_a value of the acid catalyst). As pointed out by Gold, at least one of these values cannot be a measure of the degree of proton transfer in the transition state. This result contrasts with that obtained by Kresge et al. (1971b) and referred to earlier (Table 4).

Hanna et al. (1974) have noted that the rate of coupling of substituted diazonium salts with 2-naphthol-6,8-disulphonic acid leads to different β values for alicyclic amine bases ($\beta = 0.52$) than for heteroaromatic bases ($\beta = 0.30$). A simple variation in reactivity would be expected to give the inverse order because of their relative basicities. The explanation proposed for this anomaly is similar to that proposed by Kresge and Chiang (1973) for vinyl ether hydrolysis. This discussion might lead one to conclude that the use of α as a measure of transition state structure is limited to certain reaction types for any of the above mentioned reasons. However, even the study of one reaction type, aromatic hydrogen exchange, leads to conflicting results and conclusions.

As mentioned earlier, Kresge et al. (1971b) have obtained results indicating that α is a reliable measure of transition state structure in aromatic proton exchange reactions. Gilbert and Johnson (1974) have observed essentially no change in the value of α for a number of substituted acetophenones in their exchange reactions with aqueous sulphuric acid although the reactivity range extended over some seven orders of magnitude. Furthermore, although the acetophenones are of equivalent reactivity to, or less reactive than benzene, the values of α observed for the acetophenones (~0.5) are much smaller than that for benzene (1.01). Challis and Millar (1972), who also employed the method of obtaining α by both changes in substrate and catalyst for a series of indoles, obtained significantly different values by the two methods. Acetic acid and hydronium ion catalysed exchange of substituted indoles gave linear plots of slope 0.75 and 0.67 respectively, while variation in the strength of the acid catalyst for the exchange reaction of 2-methylindole gave an α value of 0.46. Exchange data for naphthalene are also inconsistent. Thus

Stevens and Strickler (1973) obtained an α value of about 1 for the dedeuteration of naphthalene in sulphuric acid while Johnson, Katritzky and O'Neill (1974) observed values of 0·83 at the 1 position and 0·64 at the 2 position for the mechanistically identical detritiation reaction.

In conclusion, it is apparent that the use of the Brønsted coefficient as a measure of selectivity and hence of transition state structure appears to be based on extensive experimental data. However, the many cases where this use of the Brønsted coefficient is invalid suggest that considerable caution be used in drawing mechanistic conclusions from such data. The limitations on the mechanistic significance of α require further clarification, but the first steps in defining them appear to have been taken. The influence of change in the "intrinsic" barrier and variable intermolecular interactions in the transition state, both of which will result in a breakdown of the rate-equilibrium relationship, as well as "internal" return appear to be some of the key parameters which determine the magnitude of the Brønsted coefficient *in addition to* the degree of proton transfer.

Solvolysis

Reactivity-selectivity relationships in solvolysis reactions

The use of reactivity-selectivity relationships in the study of solvolysis reactions has been somewhat limited, although during recent years the potential of such relationships as a mechanistic tool has become apparent.

Swain *et al.* (1953b) noted that a qualitative relationship exists between the stability of a carbocation and its selectivity. For example, the selectivity of a number of carbocations in aqueous solution and in the presence of azide ion was enhanced with increasing carbocation stability; the ratio k_N/k_W, where k_N and k_W are the specific rate constants with azide ion and water respectively, was found to increase from 3·9 for the t-butyl cation to 170 for the diphenylmethyl cation, to 240 for the 4,4'-dimethyldiphenylmethyl cation, and to 280,000 for the highly stabilized triphenylmethyl cation. Sneen *et al.* (1966a) observed that this relationship could be quantified. It was found that a plot of log (k_N/k_W) against log k (where k is the solvolytic rate constant) for a number of alkyl chlorides gave a linear correlation. Sneen made the first attempt to utilize such a relationship as a mechanistic tool. The selectivity of

2-octyl mesylate (Sneen and Larsen, 1966b) was found to be much greater than that predicted by the reactivity–selectivity correlation. The rationale for this result was that nucleophilic attack on the 2-octyl mesylate substrate occurred, not on the free carbonium ion, but on the ion pair. Raber, Schleyer *et al.* (1971a) extended Sneen's original relationship to include a number of less reactive substrates. Their extended stability–selectivity plot is shown in Fig. 8.

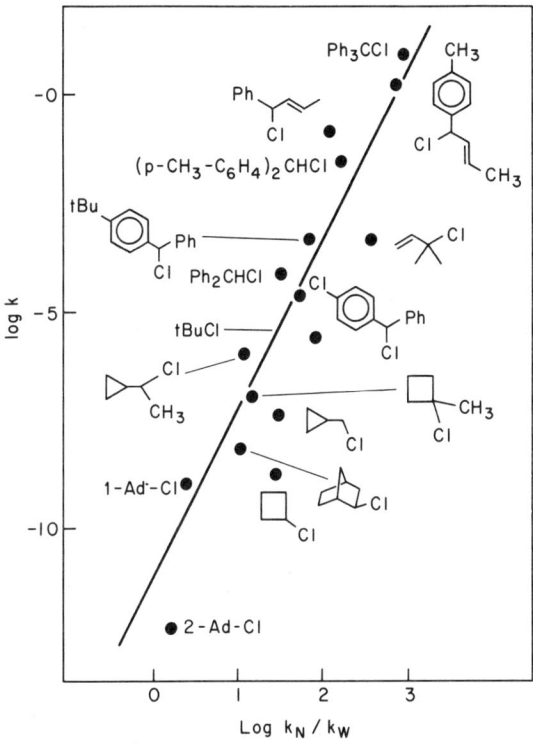

Figure 8. Plot of reactivity against selectivity (log k_N/k_W) for alkyl chlorides in aqueous dioxan containing azide ions. (Taken from Raber *et al.*, 1971a.)

Certain limitations were noted however. First, the plot did not predict the zero value of selectivity for highly unstable ions, expected from the reactivity–selectivity principle. Secondly, as discussed on p. 77, solvolytic rate constants are only a very approximate measure of substrate reactivity. Furthermore, the possibility exists of solvent sorting, in which the concentration of potential nucleophiles around the reaction intermediate differs from that in the bulk solution. Such a phenomenon would, of course, seriously diminish the significance of selectivity as a measure of reactivity, and, while such a possibility has not been entirely excluded, it has been considered and rejected (Harris et al., 1974a). The observed correlation of two different measures of selectivity (Pross and Koren, 1975) reinforces this view.

Raber et al. (1971b) noted that, in addition to 2-octyl mesylate, a number of primary and secondary substrates which also underwent solvolysis with substantial nucleophilic solvent assistance all showed considerably higher selectivities than expected from the reactivity-selectivity relationship illustrated in Fig. 8. They concluded that, while the failure of these points to correlate with the carbocations did point to a mechanistic difference between the two groups, the conclusion

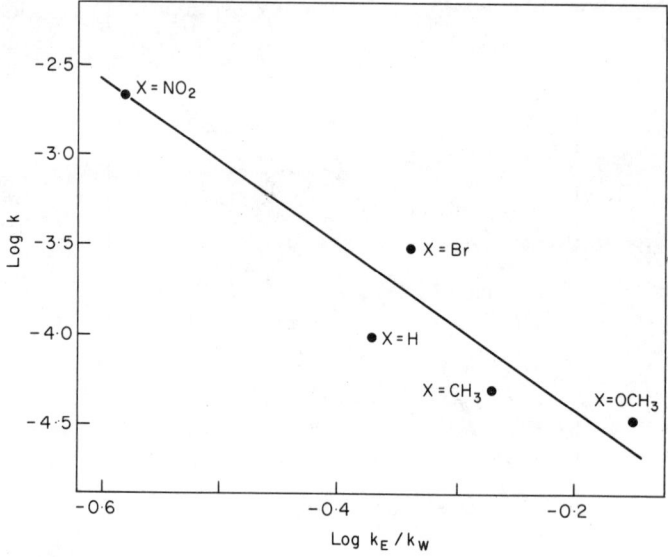

Figure 9. Plot of reactivity against selectivity (log k_E/k_W) for 2-adamantyl p-X-benzenesulphonates in 70% ethanol.

reached by Sneen, suggesting that 2-octyl mesylate solvolyses via ion-pair intermediates, was questionable. The possibility of nucleophilic attack on the neutral substrate rationalized the observed lack of correlation equally well. Harris et al. (1974b, 1972) observed a reactivity-selectivity relationship in the solvolysis of a number of 2-adamantyl arenesulphonates using ethanol and water as competitive nucleophiles. Their data are illustrated in Fig. 9 in which k_E and k_W represent the specific rate constants with ethanol and water respectively.

An interesting feature of these results is that ethanol is less nucleophilic than water toward the adamantyl derivatives, in contrast to the normal nucleophilic order (Bentley et al., 1972). This inverted order was interpreted as indicative of product formation from the solvent-separated ion pair (Harris et al., 1972, 1974b). It may be understood by considering the solvolysis scheme proposed by Winstein et al. (1956, 1958, 1965) illustrated in (25) which shows

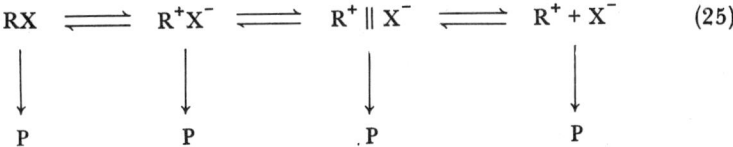

that the solvolysis of a substrate, RX, may yield the product, P, via a number of possible intermediates. Nucleophilic attack may potentially occur at any of the four possible stages. For 2-adamantyl derivatives, attack on the neutral substrate, RX, and intimate ion pair, R^+X^-, is unlikely because of the hindered nature of the adamantyl skeleton to back-side attack. Since the formation of a free secondary carbonium ion is not expected, the only effective nucleophilic pathway is by front-side attack on the solvent-separated ion pair. The greater nucleophilicity of water in the solvolysis of adamantyl derivatives may therefore be attributed to the greater stabilization of the anionic leaving group in the transition state by water than by ethanol. This is presumably due to the greater hydrogen-bonding ability of water compared to ethanol. For back-side nucleophilic attack, differences in hydrogen bonding ability are not expected to influence the relative nucleophilicities. This explanation is supported by the observation that the inversion in the relative nucleophilicity of ethanol and water occurs for those substrates in which back-side attack is either impossible or unlikely (Pross, 1975a).

The reactivity-selectivity relationship observed by Harris et al. (1973, 1974a) and illustrated in Fig. 10 provides additional support for this view. Thus 1-adamantyl, 2-adamantyl, and exo-2-norbonyl chlorides, which are hindered to back-side nucleophilic attack, show negative selectivity values. The reason for the negative selectivity value exhibited by 4,4'-dichlorodiphenylchloromethane and for its failure to conform to the reactivity-selectivity correlation is obscure. However, the result itself suggests that product formation is predominantly via the solvent-separated ion pair.

The intriguing feature of the plot in Fig. 10 is that the points for exo-2-norbonyl, 1-adamantyl and 2-adamantyl chlorides actually contradict the reactivity-selectivity principle. This is because in crossing the selectivity value of zero, the selectivity begins to increase again while the reactivity continues to decrease. The crossover from positive to negative selectivity (the reference to positive and negative selectivity is of course quite arbitrary; inverting the rate ratio changes the sign) indicates that the simplistic interpretation given to the plot in Fig. 10 is inadequate. That is, the view that the decreasing selectivity noted in Fig. 10 results from increasingly reactive carbocations is inconsistent with the observed crossover and indicates instead that some basic mechanistic difference must exist between points on each side of the zero selectivity line. A more satisfactory explanation is that nucleophilic attack takes place on several or all of

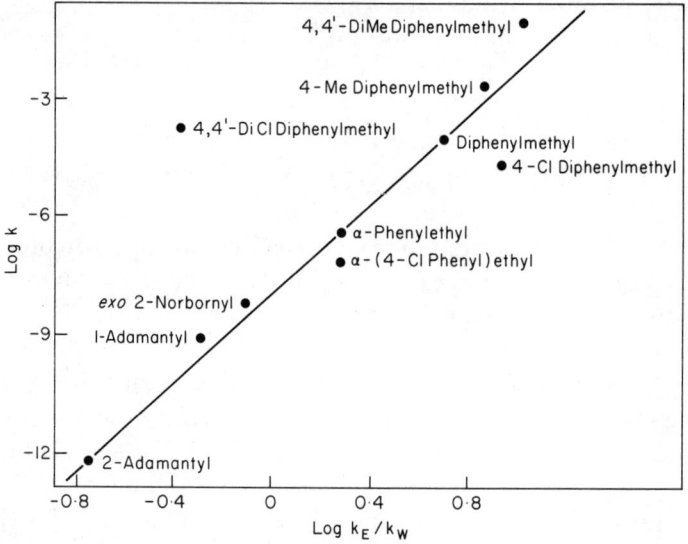

Figure 10. Plot of reactivity against selectivity (log k_E/k_W) for alkyl chlorides in 70% aqueous ethanol.

the possible solvolytic intermediates, and that, as the reactivity of the substrate is varied, the relative proportions of attack at the different stages will change (Harris *et al.*, 1974a). For points where water shows greater nucleophilicity than ethanol, the product determining step involves predominantly the solvent-separated ion pair. Further proof for the view that the change in selectivity may derive from a change in the proportion of nucleophilic attack on the various ion-pair intermediates was given by Ritchie (1971). The methanolysis of 4,4'-dimethoxydiphenylmethyl mesitoate in the presence of azide ion gave selectivity values (from competitive attack of methanol and azide) which were dependent on the azide ion concentration.

Since selectivity values should be concentration independent, this variation was attributed to the ability of the azide ion to influence the steady state concentration of the various solvolytic intermediates.

The correlation of two alternative measures of selectivity has recently been proposed as a mechanistic tool in solvolysis reactions (Pross and Koren, 1975). Thus a plot of the selectivities for a series of substrates measured both by competition between azide ion and water (k_N/k_W), and by ethanol and water (k_E/k_W) was found to be linear and of unit slope. This result strongly reinforces the view that

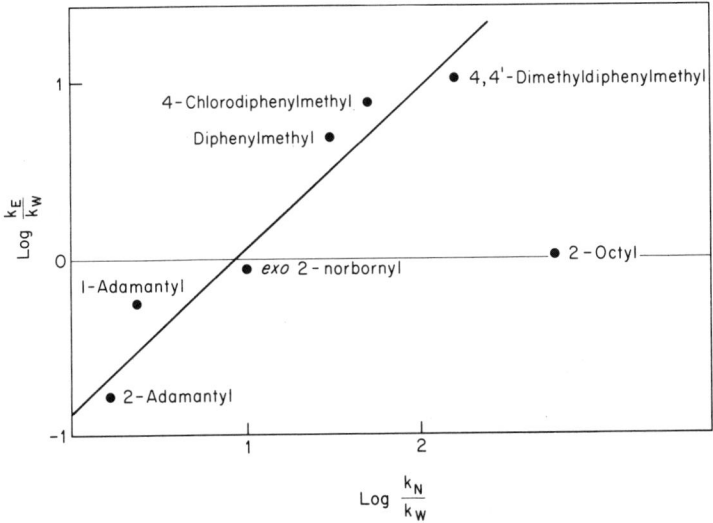

Figure 11. Plot of selectivity based on the competition between ethanol and water in 70% aqueous ethanol (log k_E/k_W) against the selectivity based on competition between azide ion and water (log k_E/k_W) for a number of alkyl chlorides.

selectivity values are not arbitrary but signify the ability of a species to differentiate between different reagents. The plot is illustrated in Fig. 11. However the point for 2-octyl chloride falls wide of the line. Thus, compared to diphenylmethyl derivatives for example, 2-octyl chloride exhibits far greater selectivity as measured by competitive attack of azide and water, yet the data for solvolysis in aqueous ethanol indicate that 2-octyl chloride shows far lower selectivity than diphenylmethyl derivatives (essentially no selectivity at all). This result suggests that the solvolysis mechanism of 2-octyl derivatives is dependent on the nucleophilic strength of the medium. In aqueous ethanol in the presence of weak nucleophiles, the low selectivity observed is attributed to attack on a highly reactive species, i.e., the intimate ion pair, in accordance with the S_N2 (intermediate) mechanism proposed by Bentley and Schleyer and discussed in the first contribution of this Volume. In the presence of azide ion, a strong nucleophile, the high selectivity observed indicates that nucleophilic attack takes place by a traditional, concerted S_N2 reaction without the formation of ion-pair intermediates. This view is supported by the data for the solvolysis of cyclobutyl mesylates in the presence of either azide or borohydride ions (Majerski et al., 1969). The data are in Table 8.

The interesting feature of these results is that the more substituted mesylate exhibits lower selectivity than the less substituted mesylate. This is contrary to what one would expect based on ion-pair stabilities. Thus it appears that a concerted S_N2 attack occurs, and

TABLE 8

Selectivity of Mesylates in Aqueous Diglyme in their Reaction with Azide Ion, Borohydride Ion and Water[a]

Mesylate	$k_{N_3^-}/k_{H_2O}$	$k_{BH_4^-}/k_{H_2O}$[b]
☐—OMs	28·2	56·7
☐⟨OMs, CH₃	15·0	12·7

[a] Data from Majerski et al., 1969.
[b] Data calculated from final concentration of BH_4^-. Values obtained using the initial BH_4^- concentration are significantly different due to the extent of the hydrolysis reaction of BH_4^-.

this contributes in particular to the overall product-determining step of the less substituted derivative.

Selectivity data may be also employed to confirm or rule out the existence of a single intermediate in a number of closely related reactions. Thus the ability of the leaving group to influence the selectivity of the competitive attack of water and borohydride in the reaction of a number of diphenylmethyl derivatives testifies to the fact that attack occurs at the ion pair stage and not on the free carbocation (Bell and Brown, 1966). The dependence of the ratio k_N/k_W on the leaving group in the solvolysis of a number of triphenylmethyl derivatives leads to the same conclusion regarding the triphenylmethyl system (Hill, 1965).

A reactivity-selectivity relationship based on the use of an ambident nucleophile has been observed by Okamoto and Kinoshita

(1974). The solvolysis of alkyl chlorides and aralkyl *p*-nitrobenzoates in phenol, an ambident nucleophile, gave C- and O-alkylated phenols. The logarithm of the ratio of product concentrations was found to be a linear function of the logarithm of the solvolytic rate constant for the particular substrate.

The influence of solvation on the reactivity–selectivity principle

The previous discussion might lead to the conclusion that solvolytic reactions invariably obey the reactivity–selectivity principle. In fact, a considerable amount of data has recently been presented in which the expected inverse relationship was not observed.

Ritchie (1972; Ritchie and Wright, 1971; Ritchie and Fleischhauer, 1972; Ritchie and Virtanen, 1972abc) has discovered a very striking correlation, termed the N_+ relationship, which represents a serious challenge to the reactivity–selectivity principle. The relationship may

$$\log k = \log k_0 + N_+ \tag{26}$$

be expressed by eqn (26), where k is the rate constant for the reaction of an electrophile with a given nucleophilic system, k_0 is a constant for a particular electrophile and N_+ is a parameter characteristic of the given nucleophilic system.

Initially, the relationship was noted for cation–anion combination reactions, but more recently Ritchie (1975; Ritchie and Virtanen, 1973; Ritchie *et al.*, 1975) has considerably extended its scope. Thus nucleophiles of large variety, including charged and neutral species as well as S-, N- and O- based nucleophiles, were found to be correlated by the relation. Furthermore, the range of electrophiles studied, which consisted only of cations in the early stages of the study, was extended to include simple esters as well. This remarkably general relation calls into question the validity of the reactivity–selectivity principle since the N_+ values are selectivity parameters ($N_+ = \log k/k_0$) and yet the values remain independent of the electrophile. In other words, electrophiles covering an extensive reactivity range exhibit the same selectivity toward each member of a large group of nucleophiles. In addition to the apparent breakdown in the reactivity–selectivity principle, no one consistent mechanism appeared able to accommodate the experimental data. Recently, a model based on solvation effects which both accommodates the experimental data and rationalises the breakdown in the reactivity–selectivity principle has been proposed (Pross, 1975b, 1976). The key element in this model is that, in the transition state of the electrophile–nucleophile reaction, both nucleophile and electrophile have undergone partial desolvation, and that this partial desolvation is the only contribution to the free energy of activation. Quantitatively, this may be expressed by equations (27) and (28). $\Delta G^{\ddagger}(E_1, N_1, S_1)$ is the free energy of

$$\Delta G^{\ddagger}(E_1, N_1, S_1) = \alpha_1 \Delta G(N_1, S_1) + \beta_1 \Delta G(E_1, S_1) \qquad (27)$$

$$\Delta G^{\ddagger}(E_1, N_2, S_2) = \alpha_2 \Delta G(N_2, S_2) + \beta_1' \Delta G(E_1, S_2) \qquad (28)$$

activation for reaction of electrophile E_1 with nucleophile N_1 in solvent S_1, $\Delta G(N_1, S_1)$ is the free energy of solvation of nucleophile N_1 in solvent S_1, $\Delta G(E_1, S_1)$ is the free energy of solvation of electrophile E_1 in solvent S_1, $\Delta G^{\ddagger}(E_1, N_2, S_2)$ is the free energy of activation for reaction of electrophile E_1 with nucleophile N_2 in solvent S_2, $\Delta G(N_2, S_2)$ is the free energy of solvation of nucleophile N_2 in solvent S_2 and $\Delta G(E_1, S_2)$ is the free energy of solvation of electrophile E_1 in solvent S_2. α_1 and α_2 represent factors whose magnitude indicates the degree of desolvation in the transition state of nucleophile N_1 in solvent S_1 and nucleophile N_2 in solvent S_2 respectively. Similarly β_1 and β_1' represent the degree of desolvation of electrophile E_1 in solvents S_1 and S_2 respectively.

Subtracting (27) from (28) leads to (29) for electrophile E_1 and, similarly to eqn (30) for a second electrophile, E_2.

$$\Delta \Delta G^{\ddagger}(E_1) = \alpha_2 \Delta G(N_2, S_2) + \beta_1' \Delta G(E_1, S_2) - \alpha_1 \Delta G(N_1, S_1)$$
$$- \beta_1 \Delta G(E_1, S_1) \qquad (29)$$

$$\Delta \Delta G^{\ddagger}(E_2) = \alpha_2 \Delta G(N_2, S_2) + \beta_2' \Delta G(E_2, S_2) - \alpha_1 \Delta G(N_1, S_1)$$
$$- \beta_2 \Delta G(E_2, S_1) \qquad (30)$$

Now the constant selectivity observed for these reactions may be expressed by equation (31).

$$\Delta \Delta G^{\ddagger}(E_2) - \Delta \Delta G^{\ddagger}(E_1) = 0 \qquad (31)$$

Substituting (29) and (30) in (31) gives (32).

$$\beta_2' \Delta G(E_2, S_2) - \beta_2 \Delta G(E_2, S_1) = \beta_1' \Delta G(E_1, S_2) - \beta_1 \Delta G(E_1, S_1) \qquad (32)$$

If we assume that $\beta_1 = \beta_1'$ and that $\beta_2 = \beta_2'$, eqn (32) simplifies to (33) for any electrophile.

$$\beta[\Delta G(E, S_2) - \Delta G(E, S_1)] = \text{constant} \qquad (33)$$

Equation (33) suggests that the failure to observe a reactivity-selectivity relationship for electrophile-nucleophile combination reactions is due to the counter influence of the solvent. The parameter β represents the "inherent reactivity" of the electrophile and is large for unreactive electrophiles, while for reactive electrophiles β is small. Now, it is the reactive electrophiles which are

TABLE 9

Selectivities ($\log k_N/k_{H_2O}$) of Methyl Benzenesulphonate, Methyl Nitrate and Methyl Perchlorate with Nucleophiles, N[a]

N	$\log (k_N/k_{H_2O})$		
	PhSO$_3$Me[b]	MeNO$_3$[c]	MeClO$_4$[d]
PO$_4^{3-}$	4·375	4·233	4·226
CO$_3^{2-}$	3·702	3·637	3·632
OH$^-$	3·630	3·570	3·563
PhO$^-$	3·616	3·476	3·469
HPO$_4^{2-}$	2·947	2·945	2·938
AcO$^-$	2·327	2·203	2·296
HCO$_3^-$	2·289		2·178
SO$_4^{2-}$	1·848		
NO$_3^-$	0·954		
CN$^-$	4·301		4·306
N$_3^-$	3·345	3·273	3·266
NO$_2^-$	3·271	3·227	3·093
S$_2$O$_3^{2-}$	5·165	5·140	5·233
SO$_3^{2-}$	5·335	4·744	4·737
SCN$^-$	3·500	3·819	3·683
I$^-$	3·383	3·690	3·683
Br$^-$	2·563	2·793	2·786
Cl$^-$	2·009	1·734	1·727
H$_2$O	0	0	0

[a] Data taken from Koskikallio, 1972.
[b] Data obtained at 25°.
[c] Data obtained at 90°.
[d] Data obtained at 0°.

expected to be strongly solvated [*i.e.*, for which $\Delta G(E, S_2) - \Delta G(E, S_1)$ will be relatively large]. Since solvation energy and "inherent reactivity" operate in opposite directions in their influence on selectivity, the constant selectivity values are effectively rationalized, and the basic integrity of the reactivity-selectivity principle is maintained.

Koskikallio (1972) has found that the selectivity of certain methyl derivatives which undergo nucleophilic substitution remains essentially constant despite large differences in reactivity. The data are shown in Table 9. The methyl halides, in contrast, show selectivity differences, but in the opposite direction to that predicted by the reactivity-selectivity principle. Methyl iodide is more selective than methyl bromide which is in turn more selective than methyl chloride. The data are presented in Table 10. Koskikallio interpreted this data

TABLE 10

Selectivities ($\log k_N/k_{H_2O}$) of Methyl Halides with Nucleophiles, N[a]

	$\log k_N/k_{H_2O}$		
N	MeCl	MeBr	MeI
$S_2O_3^{2-}$	6·35	6·64	7·44
I^-	4·61	4·91	5·63
Br^-	3·09	$(4·45)^b$	4·48
Cl^-	$(2·69)^b$	2·83	3·36
F^-	1·55	1·67	1·71
OH^-	4·14	4·27	4·69
CN^-	$(5·20)^b$	$(5·45)^b$	5·65
SCN^-	$(4·65)^b$	$(4·87)^b$	5·42
SO_3^{2-}			7·49
N_3^-		4·00	
H_2O	0	0	0

[a] Data taken from Koskikallio, 1972 and obtained at 25°.
[b] Values in brackets obtained by linear interpolation.

as being consistent with a mechanism involving ion triplets as well ion pairs [eqn (34)]. The constant selectivity was attributed to the

$$RX \underset{k_{-1}}{\overset{k_1}{\rightleftarrows}} R^+X^- \underset{k_{-2}}{\overset{k_2[N^-]}{\rightleftarrows}} N^-R^+X^- \overset{k_3}{\longrightarrow} N^-R^+ \overset{k_4}{\longrightarrow} NR$$

(34)

rates of reactions of nucleophiles with the intimate ion pair being anion-independent, although no physical justification was proposed to support this contention. The selectivity changes noted for methyl halides were attributed to a change in the relative magnitude of the individual rate constants, i.e. for methyl halides k_3 is approximately equal to k_{-2} since the nucleophilicities of the incoming nucleophile and the leaving group are now comparable.

An analysis of these results in terms of solvent effects leads to the observation of similarities with Ritchie's work on the N_+ relation. Thus the constant selectivities obtained in the solvolysis reactions of certain methyl derivatives (Table 9) may indicate the existence of a basic similarity between the rate-determining process in these reaction and in the electrophile-nucleophile combination reactions correlated by the N_+ relation. The failure of the methyl halides to conform to this pattern might suggest that their substitution reactions are fundamentally different, and that the free energy of activation is dependent on factors other than desolvation.

What is clear, is that solvent effects may play a substantial, complex, and yet often subtle role in the solvolytic reaction mechanism, and that further study is required to increase the limited understanding which has been achieved to date. The original reactivity-selectivity relationships obtained by Sneen et al. (1966a) and Raber, Schleyer et al. (1971a) are now seen to be entirely fortuitous. The N_+ correlation has demonstrated that the selectivities of cations are independent of their stability. Hence, the observed relationships are a result of the averaging of selectivities of attack at the different ion pair stages, and are not simple reactivity-selectivity relationships at all.

Common ion rate depression as a measure of selectivity

The phenomenon of common ion rate depression (Ingold, 1969) in the solvolysis of substrates yielding stable carbocations may also be utilized as a measure of selectivity. Thus for the simplified solvolysis scheme represented in eqn (35), the observed rate constant, k_{obs}, is given by eqn (36), where $\alpha = k_{-1}/k_2$. Since α measures

$$\text{RX} \underset{k_{-1}}{\overset{k_1}{\rightleftarrows}} \text{R}^+ + \text{X}^- \xrightarrow[\text{SOH}]{k_2} \text{ROS} \qquad (35)$$

$$k_{obs} = k_1/(1 + \alpha[\text{X}^-]) \qquad (36)$$

the ability of the carbocation to discriminate between the anion, X^-, and the solvent, HOS, it may function as a selectivity parameter. A study of α values for a series of substituted diphenylmethyl chlorides did in fact indicate that the more stable carbocations (i.e. those possessing electron donating groups) yielded the higher values of α as expected. (Bailey et al., 1966). The data are listed in table 11. The observation of common ion rate depression in the solvolysis of 2,2-diphenyl-1-anisyliodoethylene by iodide ion was used to challenge the widespread view held at that time that vinyl cations are formed very sluggishly because of their innate instability (Miller and Kaufman, 1968). The high selectivity of vinyl cations as indicated by common ion rate depression suggested that ground state stabilization and not carbocation instability was responsible for the observed lack of reactivity.

Rappoport et al. (1975, 1973a, 1972) have studied the factors governing the selectivity of vinyl cations and concluded that both

TABLE 11

Relative Reactivity and α Values for Substituted Diphenylmethyl Chlorides[a]

Substituents	Solvent	α	k_{rel}[b]
4-NO_2	70% acetone	0·71	0·00045
4-Ph, 4'-NO_2		1·52	0·01
H		2·08	1·00
4-Ph		8·43	8·15
4-OMe, 4'-NO_2	85% acetone	350	0·095
4-OMe, H		700	1·00
4-OMe, 4'-Ph		1100	3·04
4-OMe, 4'-Bu^t		1000	4·68
4-OMe, 4'-Me		1500	7·33
4-OMe, 4'-An		2300	10·3
4-OMe, 4'-PhO		2300	25·2
4-OMe, 4'-MeO		3700	25·2

[a] Data from Bailey et al., 1966.
[b] k_{rel} values are the relative solvolytic rate constants at zero ionic strength.

electronic and steric effects influence the magnitude of the observed selectivity. Thus vinyl carbocations stabilized by the anisyl group show greater selectivity than the corresponding phenyl stabilized carbocations. Data taken from the work by Rappoport and Houminer (1973b) which illustrate this point for a number of vinyl substrates are presented in Table 12.

These results for both aliphatic and vinyl cations suggest that the selectivity of attack of different nucleophiles on a cation depends upon the nature of the cation and that more stable cations exhibit greater selectivity. Ritchie's work, discussed earlier, indicates that cation selectivity is cation independent. The reason for this apparent conflict is not clear.

TABLE 12

Competition Factors, α, for Phenyl- and p-Methoxyphenyl-stabilised Vinyl Cations: $R^1 XC=CR^2 R^3$

Solvent	R^1	R^2	R^3	X	α	Ref.
70% acetone	An	Ph	Ph	OTs	9	a
	Ph	Ph	Ph	OTs	0·6	
70% DMF	An	Ph	Ph	I	30-50	b
	Ph	Ph	Ph	I	0	
85% acetone	An	An	H	Cl	2300	c
	Ph	An	H	Cl	700	
Trifluoroethanol	An	An	An	Br	78	d
	Ph	Ph	Ph	Br	46	

[a] Rappoport and Kaspi, 1972.
[b] Miller and Kaufman, 1968.
[c] Bailey et al., 1966.
[d] Rappoport and Houminer, 1973b.

The influence of bulky substituents in the β positions of vinyl substrates also appears to increase the selectivity of the vinyl cation (Rappoport and Gal, 1973a, 1970). This effect was attributed to shielding of the cationic intermediate from the incoming nucleophile. Thus, in the acetolysis of vinyl bromides, common ion return for two β substituents decreases in the order: anisyl, anisyl > anisyl, H > methyl, methyl > H, H (Rappoport and Gal, 1970).

Linear free energy relationship parameters as a measure of selectivity

The discussion on the possible use of the proportionality constants associated with linear free energy relationships as selectivity parameters (p. 79) might lead to the expectation that, for solvolytic reactions in particular, the Winstein-Grunwald m value (which is a measure of a substrate's sensitivity to solvent ionizing power; Grunwald and Winstein, 1948; Fainberg and Winstein, 1956) may be of utility. In actual fact this use of the m value is essentially of no practical significance since m is far more effective as a means of estimating changes in mechanism. The m value cannot simultaneously operate both as a measure of mechanistic type and as a measure of selectivity; the two are incompatible. As was indicated for the Brønsted coefficient, any change in mechanism changes the significance of the selectivity parameter and brings about a breakdown in the reactivity-selectivity principle. Since one of the key uses of the m value is to estimate the degree of nucleophilic solvent assistance in the rate determining step of solvolysis reactions (that is, as a parameter which characterizes the reaction mechanism), no correlation between reactivity and selectivity can be expected. Thus the data listed in Table 13 and which imply the failure of the reactivity-selectivity principle, result from the mechanistic variation within the S_N1-S_N2 spectrum due to changes in the magnitude of the nucleophilic solvent assistance (Pross and Koren, 1974).

For cases where no change of mechanism occurs, the use of m as a measure of transition state structure may also be unreliable. For example, the observed m values for 1-adamantyl chloride, bromide and tosylate are 1·28 (Schleyer and Nicholas, 1961), 1·08 (Raber et al., 1970), and 0·97 (Kevill et al., 1970), respectively. This appears to

TABLE 13

Rates and Grunwald-Winstein m Values for the Solvolysis of Secondary Tosylates

R =	Pinacolyl	3-Methyl-2-butyl	2-Butyl	2-Propyl
m value[a]	0·73	0·56	0·47	0·36
Rate constant[b] × 10^6 (sec^{-1})	1·27[c]	7·1[a]	3·81[e]	2·94[e]

[a] Data taken from Pross and Koren, 1974.
[b] Data for 80% aqueous ethanol at 25°.
[c] Data obtained for pinacolyl brosylate[d] and assuming a OBs/OTs ratio of 5·0 in aqueous ethanol.
[d] Shiner et al., 1969.
[e] P.v. R. Schleyer, private communication.

exemplify the reactivity-selectivity principle since the order of decreasing m values corresponds to the order of increasing reactivity. However, m values reflect not only the degree of charge separation but also the ability of the leaving group to disperse developing charge in the transition state. Since the tosylate group can disperse negative charge more effectively than halogen anions, the dependence of the transition state on solvent stabilization is reduced, resulting in a lower m value.

The use of the Hammett ρ constant as a measure of selectivity (and hence of transition state structure) in solvolysis reactions also appears to be unreliable. This is because ρ reflects not only the extent of bond making or breaking in the transition state but also the ability of a system to transmit the substituent effect to the reaction centre. This ability is variable and may in some cases even be dominant. Nishida (1967a, b), for example, noted that while a plot of the rate of solvolysis for a number of disubstituted diphenylmethyl chlorides against the sum of the substituent constants was linear for a large proportion of the compounds studied (suggesting that substituent constants are additive), substrates which were substituted by either strongly electron-attracting or electron-withdrawing groups failed to conform to the correlation. In fact, series of p-methoxy, p-methyl and unsubstituted diphenylmethyl chlorides each gave separate Hammett correlations with respect to a second substituent. The observed slopes were found to be 0·6, 0·8 and 1·0 respectively. This result appears at first sight to be an example of a series of reactions conforming to the reactivity-selective principle, in that the highly reactive p-methoxy

TABLE 14

Solvolysis of Substituted 1-Arylethyl Derivatives, ArCHXCH$_3$, in 80% Ethanol–H$_2$O at 25°[a]

Aryl Group	Sidechain Position	Substituent Position	Leaving Group (X)	ρ	Δq
Benzo[b] thiophen	2	6	OPNB	−3·96	0·1246
Benzofuran	2	6	OPNB	−4·41	0·1377
Benzene	1	4	Cl	−5·82	0·2109
Pyridine	2	5	Cl	−6·05	0·2212
Thiophen	2	5	OPNB	−7·06	0·2051
Thiazole	5	2	Cl	−6·14	0·2030
Thiazole	2	5	Cl	−6·68	0·2268
Furan	2	5	OPNB	−7·42	0·2763

[a] Data taken from Forsythe and Noyce, 1972.

substituted derivatives exhibit lower selectivity as indicated by the value of ρ. However, it appears that the weaker dependence of the p-methoxy derivatives stems not only from an earlier transition state, which requires less substituent stabilization, but also from the lower electronic demands of the free carbonium ions as a result of the stabilizing influence of the methoxy substituent. Shatenshtein (1962) has compared this behaviour to the levelling effect of strong acids or bases in acid-base reactions. In terms of the diagram in Fig. 4, the selectivity of a reaction is determined not only by $\Delta G^{\ddagger}_{AX} - \Delta G^{\ddagger}_{AY}$ but also, indirectly, by $\Delta G^0_{AX} - \Delta G^0_{AY}$ whose magnitude may change within a reaction series. A case where this variable transmission or levelling effect is dominant is discussed by Forsythe and Noyce (1972). The solvolysis of substituted 1-arylethyl derivatives, ArCHXCH$_3$, was studied for a wide variety of aromatic and heteroaromatic compounds. The data are listed in Table 14. No correlation was observed between ρ and the reactivity of the unsubstituted ring system relative to benzene. On the other hand, the value Δq, which is the difference in charge at the substituent site between the unsubstituted molecule and the cationic intermediate, is related to the corresponding ρ value for the particular system. This suggests therefore, that the magnitude of ρ is dominated by the pattern of charge distribution within each ring system, which in turn determines the energy differences between the various cationic intermediates rather than the extent of bond breaking in the transition state. M.o. studies (Noyce and Nichols, 1972) confirm this view. The investigation of the transmission of substituent effects in heterocyclic compounds, studied by both m.o. calculations and solvolysis data, also suggests that variable transmission determines the magnitude of ρ to a significant extent (Noyce et al., 1972; Noyce and Pavez, 1972).

Carbene Reactions

Reactivity-selectivity relationships in carbene chemistry are not common since the key reactivity parameter, the absolute rate constant, is generally not known. However, in spite of this, the reactivity-selectivity principle is a basic tool in mechanistic studies of carbenes. This is because a major technique for obtaining information regarding the transient carbene intermediate is by relative rate studies (Moss, 1973) in which, for example, two or more olefins compete for the carbene species. A product analysis then enables relative rate constants, which are equivalent to a selectivity parameter for the carbenes involved, to be obtained. Once these relative rate data (i.e. carbene selectivity data) are known, direct application of the reactivity-selectivity principle gives information about the relative stability of different carbenes. Highly selective carbenes are assumed to be more stable than unselective carbenes. Skell and Cholod (1969b) rationalized this inverse relationship between reactivity and selectivity as being due to a decrease in $\Delta H^{\ddagger}$ causing a corresponding decrease in $\Delta \Delta H^{\ddagger}$ in analogy to the variation in free energy shown in Fig. 4, although no justification for this assumption was presented.

In view of the limitations of the reactivity-selectivity principle, this extreme dependence on relative reactivity studies may be criticised. However, it must be admitted that many of the con-

TABLE 15

Relative Rate and Selectivity Data for the Addition of Dichlorocarbene to Substituted Styrenes[a]

Substrate	ρ value	Relative Rate[a]	Reference
(1) $Z-C_6H_4CH=CH_2$	−0·619	1·00	Seyferth et al., 1968
(2) trans-$Z-C_6H_4CH=CHCH_3$	−0·53	1·52	Sadler, 1969
(3) $Z-C_6H_4C(CH_3)=CH_2$	−0·378	5·86	Sadler, 1969

[a] Relative rates calculated from data in Moss, 1973 (Tables 39, 40 and 41).

clusions which have been reached in this way form a well-defined and self-consistent picture. This is true, in particular, of the conclusions concerning the effect of structure on carbene stability, lending indirect, but compelling support for the assumption on which they are based.

Reactivity-selectivity relationships are obtainable in carbene chemistry provided an independent selectivity parameter is found. Under such circumstances the relative rate data then serve as a measure of relative reactivity. This has been done with a number of Hammett studies. Thus the addition of dichlorocarbene to compounds [1]-[3] (listed in Table 15) indicates that a reactivity-selectivity relationship is obtained using ρ as a measure of selectivity. A more reactant-like transition state is obtained for the more reactive substrate in accordance with the Leffler-Hammond postulate.

However, such a reactivity-selectivity relationship is not always observed. For example, the selectivity data for reaction of substituted phenylcarbenes with isobutene and trans-2-butene are listed in Table 16 (Moss, 1973). Interestingly, the data appear to support the reactivity-selectivity principle. An electron donating substituent, Z, is thought to stabilize the arylcarbene. Thus the data in Table 16 were interpreted (Moss, 1973; Closs and Moss, 1964) as signifying

TABLE 16

Selectivity of Attack of Arylcarbenes ($Z-C_6H_4-CH$) on trans-2-Butene and Isobutene[a]

Substituent Z	m-Cl	p-Cl	H	p-CH_3	p-CH_3O
Ratio $\frac{k_{trans\text{-butene}}}{k_{isobutene}}$	0·8	0·9	1·1	1·2	1·3

[a] Data taken from Moss, 1973.

that the selectivity increase results from a corresponding increase in carbene stability. In fact, the data cannot be justified on the basis of the reactivity-selectivity principle, since the selectivity parameter passes through the value of 1. This value represents a limiting figure, and any value less than 1 merely indicates a growing inverse selectivity. Thus m-chlorophenylcarbene and p-methoxyphenylcarbene actually exhibit very similar selectivities. The reason for the inversion of selectivity in this system is obscure since for carbene reactions such changes are generally due to changing steric effects and these appear to be unlikely for this particular system. The very minor change in selectivity and the closeness of all the values to the limiting figure of 1 suggest that no particular significance should be attached to this variation.

TABLE 17

Selectivity Indices for Some Carbenes[a]

Carbene	Selectivity Index, m_{CYX}
CH_2	0·1
$(CH_3)_2C=C=C$:	0·77
CBr_2	0·82
CCl_2	(1·00)
$CFCl$	1·51
CF_2	1·92

[a] Data taken from Skell and Cholod, 1969b.

The use of the reactivity-selectivity principle to obtain a structure-reactivity correlation has been extensive. Thus it was suggested by Skell and Garner (1956) and by Hine and Ehrenson (1958) that the reactivity of a carbene, CXY, is moderated by the ability of the groups X and Y to provide resonance stabilization. This interpretation was placed on a quantitative basis by Skell and Cholod (1969b) and by Moss and Mamantov (1970). In order to obtain a reliable selectivity parameter for different carbenes, a carbene selectivity index, m_{CXY} was defined as the least-squares slope of a plot of log $(k_i/k_{isobutene})$ for CXY against log $(k_i/k_{isobutene})$ for CCl_2, where k_i is the rate constant for reaction with different substituted and unsubstituted butenes. The resulting selectivity factors (Table 17) signify the relative selectivity of a carbene CXY in comparison to CCl_2.

The increasing stability of the carbenes, based on m_{CXY} values, is attributed to increasing overlap of the vacant p orbital with the lone pairs associated with the substituent atoms. Thus the order of ability of halogen atoms to stabilize carbenes decreases in the order $F > Cl > Br > H$. Attempts to relate the m_{CXY} values with the substituent parameters of X and Y such as σ^+ and σ_I have met with some success. Thus a plot of m_{CXY} against a function of $-\Sigma_{X,Y}\sigma^+$ and $\Sigma_{X,Y}\sigma_I$ was found to be linear (Moss and Mallon, 1975), suggesting that the ability of an atom to stabilize a carbene is dependent on the ability of that atom to *donate* electrons to the vacant p orbital by conjugation, and to *withdraw* electronic charge inductively. This may be understood by considering the possible transition state for the reaction [4]. The carbene is stabilized by π donation as represented

$$\begin{array}{c} R_1 \diagdown \delta+ \diagup R_3 \\ C\text{---}C \\ R_2 \diagup \diagdown R_4 \\ \vdots \\ C^{\delta-} \\ \diagup \diagdown \\ X Y \end{array}$$

[4]

by the limiting structures [5] and [6]. The inductive withdrawal will presumably stabilize the transition state since negative charge accumulates on the carbenic carbon.

$$\underset{[5]}{\overset{\ddot{C}}{\underset{XY}{\diagup\diagdown}}} \longleftrightarrow \underset{[6]}{\overset{C^-}{\underset{\overset{+}{X}Y}{\diagup\diagdown}}}$$

An interesting point is that, while substitution of a halogen for a hydrogen atom in CH_2 increases the selectivity and therefore the stability of the carbene, introduction of a second halogen results in considerably greater selectivity than expected from additivity considerations (Moss and Przybyla, 1969; Moss and Mamantov, 1970). No explanation for this effect appears to have been offered; however, *ab initio* m.o. calculations by Radom and Stiles (1975) have indicated that geminal dihalogen substitution in alkanes brings about stabilization of the ground state by halogen hyperconjugation. This interaction in highly reactive carbene species could well be responsible for their unexpected stability.

The use of selectivity data as a means of identifying intermediates has been employed effectively. For example, in the dichlorocyclopropanation of olefins, the selectivity of attack using CCl_2 generated by different methods (the pyrolysis and α elimination of chloroform) was found to be almost identical (Skell and Cholod, 1969a). This was interpreted as indicating the attack of free CCl_2 and not some carbenoid species, such as occurs with the Simmons–Smith cyclopropanation reagent derived from zinc and methylene iodide (Simmons et al., 1964).

The use of selectivity data as a measure of carbene stability must be conducted with caution; limited data may lead to incorrect conclusions. For example, CCl_2 shows no selectivity with styrene and 2,4,6-trimethylstyrene (Nefedov and Shafran, 1967) whereas CBr_2 favours styrene by a factor of 3. The apparent conclusion is that CBr_2 is more stable and hence more selective than CCl_2. However, steric factors, which are more important in the reactions of CBr_2, are responsible for this effect. Thus, CBr_2 attacks cyclohexene in preference to styrene by a factor of 2·4 : 1 while CCl_2 shows the opposite preference 1 : 1·4. The true relative stability is more clearly discernible in the relative rate of carbene attack on partly substituted ethylenes. Thus $k_{isobutene}/k_{cyclohexene}$ for CCl_2 is 8·32 while for CBr_2 the corresponding figure is 3·72 (Skell and Garner, 1956; Doering and Henderson, 1958).

All the factors governing the magnitude and significance of selectivity have yet to be established. The inversion of relative selectivity as measured by two different methods has been noted (Moss and Mallon, 1975). Thus m_{CF_2}/m_{CCl_2} was found to be 1·48, based on alkylethylenes. This suggests that CF_2 is more selective and therefore more stable than CCl_2. However, the Hammett ρ parameter obtained from the reaction of carbenes with substituted styrenes may also serve as a selectivity parameter; for the styrenes ρ_{CF_2}/ρ_{CCl_2} was found to be 0·92 which suggests that CCl_2 is the more stable carbene. Since differential steric factors are not expected for these reactions, the reason for this contradiction is not clear. It has been suggested (Moss and Mallon, 1975) that carbene attack on alkylethylenes and styrenes involves significantly different transition states. Further study is clearly required to establish whether this is the true explanation or whether additional limitations on the use of selectivity as a measure of relative stability need to be delineated.

Electrophilic Aromatic Substitution

Selectivity has been used extensively as a mechanistic criterion in aromatic substitution reactions. This is a consequence of the fact that, as in carbene chemistry, absolute rate constants are often unattainable, due to the uncertainty surrounding the exact nature of the attacking electrophile and, in particular, its concentration. Therefore, much of the mechanistic analysis has been made on the basis of selectivity parameters obtained from relative rate studies.

The use of selectivity and its significance in electrophilic substitution was first recognized by Brown and Nelson (1953). A linear correlation between two different measures of selectivity was observed and expressed by eqn (37) where p_f, the partial rate factor,

$$S_f = \log p_f/m_f = b \log p_f \tag{37}$$

is a measure of the ability of an electrophile to discriminate between the *para* position of any monosubstituted benzene and any single position in a benzene molecule, m_f is the corresponding factor for the *meta* position of the monosubstituted benzene, b is a proportionality constant and S_f is termed the Selectivity Factor. The ratio p_f/m_f is an intramolecular measure of selectivity and indicates the ability of the particular reagent to distinguish between the *meta* and *para* positions of the monosubstituted benzene. An extensive review on the Selectivity Relationship [eqn (37)] has been published by Stock and Brown (1963). The relationship considerably strengthens the significance of the selectivity parameter since it indicates that the selectivity values obtained are not dominated by additional factors but are a true measure of the discriminating ability of a particular electrophilic species. Thus, if a reagent is capable of discriminating effectively between the *meta* and *para* position of a monosubstituted benzene, it will also discriminate effectively between the *para* position of the monosubstituted benzene and any position in benzene itself. The Selectivity Relationship is not only applicable to benzene derivatives. Thiophen has also been found to give an analogous correlation between intermolecular and intramolecular selectivity (Clementi, 1970).

Early examples of reactivity-selectivity relationships in aromatic substitutions are limited since, in the absence of absolute rate data, it is often difficult to assign relative reactivity to the different electrophiles. For certain cases where the relative reactivity order may be assumed, a reactivity–selectivity relationship was noted. For example, bromination with the reactive species Br^+ results in lower selectivity than with the less reactive species Br_2 (de la Mare and Harvey, 1956; Brown, 1957). However, it appears that no general reactivity-selectivity relationship exists in electrophilic aromatic substitution reactions, for there exist slow, unselective reactions such as aromatic mercuration as well as fast, selective reactions such as Friedel Crafts acetylation. Thus as was noted earlier, reactivity-selectivity relationships may be obtained only within a restricted family of reactions.

More recently, Olah et al. (1970) have varied the reactivity of the electrophile in a systematic way and observed a number of convincing examples of reactivity-selectivity relationships. The Friedel-Crafts benzylation reaction on benzene and toluene was conducted for a large number of substituted benzyl chlorides. The data are in Table 18 and indicate that the selectivity of electrophilic attack decreases as the substituent becomes more electron withdrawing.

TABLE 18

Relative Rates of $TiCl_4$-catalysed Benzylation and $AlCl_3$-catalysed Benzoylation of Toluene and Benzene[a]

$XC_6H_4CH_2Cl$	k_t/k_b	Benzoyl chloride	k_t/k_b
X = p-NO_2	2·5	C_6F_5COCl	16·1
o-F	4·8	2,4-$(NO_2)_2C_6H_3COCl$	29·0
m-F	4·6	3,5-$(NO_2)_2C_6H_3COCl$	38·9
p-F	8·7	p-$NO_2C_6H_4COCl$	52·0
o-Cl	4·6	2,5-$F_2C_6H_3COCl$	96·2
m-Cl	6·4	C_6H_5COCl	153·5
p-Cl	6·2	p-$CH_3C_6H_4COCl$	164·4
H	6·3	p-FC_6H_4COCl	170·0
o-CH_3	19·1	2,4,6-$(CH_3)_3C_6H_2COCl$	196·0
m-CH_3	7·8	p-$CH_3OC_6H_4COCl$	233·0
p-CH_3	29·0		
2,4,6-$(CH_3)_3$	37·4		
o-CH_3O	60·3		
m-CH_3O	13·2		
p-CH_3O	97·0		

[a] Data taken from Olah et al., 1970, Olah and Kabayashi, 1971.

Since such substituents are expected to decrease electrophile stability, the observed trend represents a particularly clear example of a reactivity-selectivity relationship. Isomer distribution of the benzylated toluenes, while not correlating well with the relative rate data, do show a similar trend.

A similar effect has been noted for Friedel-Crafts benzoylation of toluene and benzene (Olah, 1971; Olah and Kobayashi, 1971). The data are also in Table 18.

Sulphonylation as well as sulphonation of toluene and benzene (Table 19) also indicate that the reactivity-selectivity principle is obeyed (Olah, 1971).

TABLE 19

Relative Rates of $AlCl_3$-catalysed Methyl- and Phenyl-sulphonylation and Sulphuric Acid Sulphonation of Toluene and Benzene[a]

Sulphonylating agent	k_t/k_b	Sulphonating agent (% H_2SO_4)	k_t/k_b
CH_3SO_2Cl	3·9	77·8	106
p-$NO_2C_6H_4SO_2Cl$	2·8	81·5	57
p-$ClC_6H_4SO_2Cl$	7·5	84·3	47
$C_6H_5SO_2Cl$	9·0	89·1	25
p-$CH_3C_6H_4SO_2Cl$	17·0	99–100 in nitrobenzene	5·1
p-$CH_3OC_6H_4SO_2Cl$	83·0		

[a] Data taken from Olah, 1971.

The Hammett ρ value may also be used as a selectivity parameter in electrophilic substitution reactions because a direct relation between ρ and the Brown selectivity factor, S_f, exists. Since p_f^Y and m_f^Y are defined as $k_{p\text{-}Y}/k_H$ and $k_{m\text{-}Y}/k_H$ respectively, combining the Selectivity Relationship (37) and the Hammett equation leads to eqn (38) which shows that the proportionality constant between the selectivity factor and ρ is dependent only on the substituent constants, σ_p^+ and σ_m^+ for a particular substituent, Y.

$$S_f = \rho(\sigma_{p-Y}^+ - \sigma_{m-Y}^+) \qquad (38)$$

However, ρ as a selectivity parameter must be used with caution. An effect identical to that which was observed for solvolytic reactions (pp. 110–3) occurs in electrophilic substitution as well. The data in Table 20 illustrate this point.

TABLE 20

Hammett ρ Values for the Bromination and Hydrogen Exchange of Various Aromatic Systems

Bromination[a]	ρ	Hydrogen Exchange	ρ
Benzenes	−11·61	Benzenes[b]	−7·5
Polymethylbenzenes	−10·7	Phenols[c]	−4·9
Anisoles	−5·6	Anilines[c]	−3·3
N,N-Dimethylanilines	−2·2	N,N-Dimethylanilines[c]	−3·2

[a] Dubois et al., 1972.
[b] Clementi and Katritzky, 1973.
[c] Clementi, Johnson and Katritzky, 1974.

It is readily seen that the more activated aromatic nuclei exhibit smaller ρ values in what appears to be a simple application of the reactivity-selectivity principle. However, while ρ is expected to reflect the degree of bond formation in the transition state, it must also reflect the ability of the ring system to transmit the substituent effect. Thus highly activated rings which can stabilize the incipient positive charge effectively will exhibit a lower transmitting ability toward substituents, resulting in a lower ρ value. The additional influence of a change in transition state structure will tend to reduce the magnitude of ρ even further. The danger in drawing conclusions regarding the relative positions of the transition states of different

TABLE 21

Relative Rates and Isomer Distribution for the Nitration of Toluene and Benzene[a]

Nitrating agent	Solvent	k_t/k_b	% ortho	% meta	% para
$NO_2^+PF_6^-$	CH_3NO_2	1·6	68·2	2·0	29·8
$NO_2^+BF_4^-$	Sulfolane	1·7	65·4	2·8	31·8
$NO_2^+BF_4^-$	CH_3CN	2·3	69	2	29
HNO_3	80% H_2SO_4	4·8			
	77% H_2SO_4	5·0			
	75·3% H_2SO_4	7·2			
	68·3% H_2SO_4	17·2	60	3	37
HNO_3	CH_3NO_2	21	58·5	4·4	37·1
	$(CH_3CO)_2O$	23	58·4	4·4	37·2
CH_3COONO_2	CH_3CN	44	63	2	35
HNO_3	Sulfolane, H_2SO_4	37	61·6	2·9	35·5

[a] Data taken from Olah, 1971.

ring systems by the comparison of ρ values has been discussed by Johnson and Schofield (1973). A number of additional examples of this effect, which should not be classified as reactivity-selectivity relationships, are known. Thus the proton exchange of pyridinium and 1-hydroxypyridinium cations follows this pattern of behaviour (Clementi et al., 1974). Trifluoroacetylation of substituted furans, thiophens and pyrroles gives progressively lower ρ values for this same reason (Clementi and Marino, 1972).

In their study of aromatic substitution reactions, Olah and his coworkers have based much of their mechanistic analysis on the reactivity-selectivity principle. It was observed that electrophilic substitution reactions conducted with powerful nitrating agents showed low substrate selectivity as measured by k_t/k_b (ratios of rate constants for reaction with toluene and benzene respectively) but at the same time high positional selectivity as measured by *ortho/para* isomer ratios (Olah *et al.*, 1961, 1962, 1965, 1971). The data are in Table 21 (Olah, 1971). Thus it appears that while k_t/k_b values may range from 1-2 for unselective nitrating agents to values of 20-40 for more selective nitrating agents, positional selectivity is almost unchanged. Olah proposed that in certain cases of electrophilic substitution the rate-determining step is π-complex formation followed by σ-complex formation as an additional step. By such an account, π-complex formation determines the intermolecular selectivity parameter while subsequent σ-complex formation determines the positional selectivity. Since the steps governing intermolecular and positional selectivity are not the same, the breakdown in the Brown Selectivity Relationship occurs. It should be noted, however, that this conclusion has been criticized as a result of uncertainty surrounding the reliability of the experimental data on which it is based (Ridd, 1971; Tolgyesi, 1965).

Free Radical Reactions

Reactivity-selectivity relationships play an important part in free radical chemistry for the same reasons as in carbene chemistry and electrophilic substitution. Absolute rate constants for free radical reactions are not generally available (and when they are known they are often associated with large systematic errors), and the use of relative rate studies is an important technique in the study of free radical reactions. A comprehensive monograph dealing with various

TABLE 22

Selectivity of Radicals Towards C–H Bonds[a]

Radical (X)	Me	Prim.	Sec.	Tert.	D(X–H)	$E^{\ddagger}$(CH$_4$)
F	0·5	1	1·2	1·4	135	1·2
Cl	0·004	1	4·3	7·0	103	3·8
H		1	4·8	40	103	12
OH	0·04	1	7	45	118	3·7
CH$_3$	0·09	1	10	80	104	17·3
CF$_3$	0·06	1	8·3	90	105	11
Br	0·002	1	80	1700	87	18
I	0·006	1	18	–	71	34

[a] Data taken from Kerr, 1973.

aspects of free radical chemistry has recently been published (Kochi, 1973).

The selectivity of attack of radicals at different hydrogens has been extensively utilized as a tool to characterize the reactivity and nature of the various radicals. The data in Table 22 show that there is a qualitative relation between the selectivity of a radical, i.e., its relative ability to abstract methyl, primary, secondary and tertiary hydrogens, and its reactivity represented by the Arrhenius activation energy, $E^{\ddagger}$, for the reaction with methane (Kerr, 1973). This relationship is in accord with the reactivity-selectivity principle. The fact that the ease of abstraction of the different hydrogens runs in the same sequence as C—H bond strengths (i.e., methyl > primary > secondary > tertiary), suggests that the activation energy is related to

TABLE 23

α Values[a] Derived from the Polanyi Relationship for various radicals

Radical	·CN	·CH$_3$	·CF$_3$	H·	·OH	ROO·	F·	Cl·	Br·	I·
α	0·34	0·44	0·40	0·45	0·22	0·63	0	0·45	0·66	0·89

[a] Data from Afanas'ev, 1971.

the enthalpy change of the reaction. This was expressed by Evans and Polanyi (1938) in what is now termed the Polanyi relation, (39).

$$E^{\ddagger} = \alpha \Delta H^0 + C \qquad (39)$$

It is readily noted that this relation represents an energy analogue to the rate-equilibrium relationship (2) discussed earlier. However, it also suffers from the same limitations in that it is not general, but only correlates a limited reaction series. Thus the attack of methyl radicals on aliphatic C—H bonds is correlated by (39) where $\alpha = 0.5$ and $C = -38.5$ (Trotman-Dickenson, 1965), indicating that the transition state is about half way between reactants and products. However, the use of non-homogeneous series of compounds results in the failure of (39) to correlate enthalpy and activation energy (Benson and DeMore, 1965). This is exemplified by the fact that activation energies for the hydrogen abstraction reactions of CH$_3$—H, CH$_3$O—H, and Cl—H by CH$_3$· are very different (14·5, ca. 8 and 3·8 kcal mol^{-1} respectively) while the corresponding enthalpy changes are similar. It appears, therefore, that other factors such as polarity and steric bulk play an important role in determining radical reactivity. None the less, the Polanyi relationship yields values of α

for different radicals which do appear to be related to the reactivity of those radicals. The data are listed in Table 23 (Afanas'ev, 1971). Thus the highly reactive fluorine atom yields a transition state which is reactant-like in contrast to the more stable iodine atom which produces a product-like transition state.

Kinetic isotope effects, as has been noted, are an unsatisfactory measure of selectivity since theoretical considerations suggest that for transfer processes their magnitudes are at a maximum for symmetrical transition states rather than increasing monotonically between the extreme exothermic and endothermic states (Westheimer, 1961). However, it appears for radical reactions, that kinetic isotope effects may be utilized as selectivity parameters, and that they reflect the relative degree of hydrogen transfer in the

TABLE 24

Deuterium Isotope Effects (k_H/k_D) in the Side Chain Halogenation of Toluenes[a] in CCl$_4$ at 77°

Reagent	Substrate		
	p-Chlorotoluene	Toluene	p-Methoxytoluene
N-Bromosuccinimide	5·08	4·86	3·22
Bromine	5·22	4·59	
Chlorine	1·44	1·30	

[a] Data from Wiberg and Slaugh, 1958.

transition state (Wiberg and Slaugh, 1958). Thus the data in Table 24 show that the deuterium isotope effect for the side chain halogenation of toluene is dependent on the reactivity of the system. Both increasing the reactivity of the halogenating agent (down a column) and increasing the reactivity of the substrate (along a row) cause a decrease in the kinetic isotope effect. For reactions in which the reactivity is enhanced, the earlier bond breaking which results will bring about a reduced kinetic isotope effect if the transition states for the series are all more reactant-like than the symmetrical case in which the proton lies midway between donor and acceptor. The data suggest that this is the case in accord with the reactivity-selectivity principle.

The stability of radicals may be estimated by examining the selectivity of the radical precursor. Thus Walling and Padwa (1963) studied the photodecomposition of t-alkyl hypochlorites in the presence of cyclohexane. The alkoxy radicals formed undergo competitive reactions (40) and (41). The ratios k_1/k_2 for different R substituents are shown in Table 25. The results show that the

$$R(CH_3)_2CO\cdot \xrightarrow{k_1} CH_3COCH_3 + R\cdot \longrightarrow RCl \qquad (40)$$

$$R(CH_3)_2CO\cdot \xrightarrow[C_6H_{12}]{k_2} R(CH_3)_2COH + \cdot C_6H_{11} \longrightarrow C_6H_{11}Cl \qquad (41)$$

preference of the alkoxy radical to undergo β-scission increases with the stability of the alkyl radical expelled. The anomalous position of the benzyl radical which would be expected to exhibit the largest k_1/k_2 ratio was attributed to the incursion of a chlorine atom chain which resulted in a higher yield of cyclohexyl chloride. Thus the cyclohexyl radical is formed not only by hydrogen abstraction by the alkoxy radical but by chlorine atoms as well. This chlorine atom

TABLE 25

Ratios of Rate Constants for β-Scission of (k_1) and Proton Abstraction from Cyclohexane (k_2) by Alkoxy Radicals[a] in Carbon Tetrachloride at 40°

Radical precursor	Radical product	k_1/k_2
$(CH_3)_3CO\cdot$	$CH_3\cdot$	0·02
$(CH_2Cl)_3CO\cdot$	$CH_2Cl\cdot$	0·12
$C_2H_5(CH_3)_2CO\cdot$	$C_2H_5\cdot$	2·09
$i\text{-}C_3H_5(CH_3)_2CO\cdot$	$i\text{-}C_3H_5\cdot$	76·4
$C_6H_5CH_2(CH_3)_2CO\cdot$	$C_6H_5CH_2\cdot$	1·98
$t\text{-}C_4H_9(CH_3)_2CO\cdot$	$t\text{-}C_4H_9\cdot$	>300

[a] Data from Walling and Padwa, 1963.

chain may well operate to a small extent for the other substrates, but Walling and McGuinness (1969) have concluded that under the particular conditions used its contribution is unimportant.

The variation in selectivity of radicals in different solvents has been interpreted as being due to radical-solvent interaction which changes the reactivity of the radical. Thus the selectivity of chlorination of 2,3-dimethylbutane which may react at either a tertiary or a primary position increases in aromatic solvents (Table 26; Russell, 1958, 1960). Since the effect appears to be proportional to the basicity of the aromatic substrate, it was concluded that aromatic solvents yield a complexed chlorine atom which is consequently less reactive and therefore more selective in its reactions. Confirmation of this came from the finding that the increased selectivity of the photochlorination of 2,5-dimethylhexane in aromatic solvents was due to an increase in the activation energy of the reaction (Russell,

TABLE 26

Effect of Different Diluents on the Selectivity
of Chlorine Atoms in their Reaction with
Primary and Tertiary Hydrogen in
2,3-Dimethylbutane[a]

Diluent (Conc. = 4M)	$k_{primary}/k_{tertiary}$
None	3·7
Nitrobenzene	4·9
Benzoyl chloride	6·4
Methyl benzoate	10·2
Chlorobenzene	10·2
Benzene	14·6
Anisole	18·4
1-chloronaphthalene	33

[a] Data taken from Russell, 1958, 1973.

1973). The use of the kinetic isotope effect as a selectivity parameter also indicates that the chlorination reaction is solvent dependent (Brown and Russell, 1952). Thus the chlorination of toluene, in toluene instead of carbon tetrachloride, increases the isotope effect k_H/k_D, indicating that the reaction in toluene is more selective and hence that the attacking species is less reactive in this solvent.

Finally, as has been noted earlier, selectivity parameters serve as an effective means of identification of reactive intermediates. The many

TABLE 27

Selectivity of Free Radical Initiated Chlorinating Agents with 1-Butyl Chloride

| Chlorinating agent | % attack | | | | Attacking radical | Ref. |
	C-1	C-2	C-3	C-4		
Cl_2, 34°	7	24	51	18	Cl·	a
$C_6H_5ICl_2$, 40°	5	21	61	12	Cl·	a
HOCl, CCl_4, H_2O, 40°	7	23	44	26	Cl·	a
Cl_2, 80°	5	21	52	23	Cl·	a
Cl_2, H_2SO_4, HOAc	5	23	46	26	Cl·	b
SO_2Cl_2, 80°	6	21	43	30	Cl·, SO_2Cl·	a
t-BuOCl, 40°	21	20	44	15	t-BuO·	a
$(CH_3)_2NCl$, H_2SO_4, HOAc	4	10	77	8	$(CH_3)_2NH^+$·	b
Cl_2O, 40°	22	20	51	8	ClO·	a

[a] Tanner and Nychka, 1967.
[b] Minisci et al., 1970, Spanswick and Ingold, 1970.

possible reaction mechanisms for free radical chlorination, for example, may be differentiated in this way. The relevant data are presented in Table 27. It is readily seen that different radicals exhibit different selectivities, and thus those reactions which involve chlorine atoms as the active intermediate may be identified (Russell, 1973; Tanner and Nychka, 1967; Spanswick and Ingold, 1970; Minisci *et al.*, 1970).

4. CONCLUSION

This work has attempted to illustrate that in spite of many apparent failures the reactivity-selectivity principle is fundamentally valid. The discussion has indicated that the failures in the principle may stem either from unsatisfactory use of reactivity and selectivity parameters or simply because, like every other principle, it is bound by certain limitations. Some of these limitations have been pointed out though others undoubtedly remain. One of the significant limitations of the reactivity-selectivity principle concerns the type of chemical process involved. It appears that the reactivity-selectivity principle only applies to processes which obey a rate-equilibrium relationship. Processes which fail to obey such a relationship (e.g. the proton transfer reaction of nitroalkanes) will fail to yield meaningful measures of selectivity, and a subsequent failure of the principle will result. Secondly, the principle will only operate for relatively simple processes, such as those in which a single bond is formed and a single bond is broken in the transition state (e.g. proton transfer, carbene addition, aromatic electrophilic substitution). Processes in which more than one bond is involved in bond making and bond breaking (e.g. E2 eliminations, in which two bonds are formed and two are broken in the transition state) will not necessarily conform to the reactivity-selectivity principle. Reactions in which solvation factors contribute substantially to the overall energy change (e.g. solvolytic reactions) may also bring about a breakdown in the principle. This is because of a compensation effect which occurs in the interplay between solvation energy and the "inherent" reactivity of the system.

While considerable evidence has been presented in support of the reactivity-selectivity principle and a number of apparent failures have been rationalized, other systems exist where this failure remains unexplained (Johnson, 1975). This suggests that further limitations to the practical application of the principle may exist in addition to those discussed. Clearly, further work in this area is required, particularly with regard to the theoretical basis of the Hammett equation whose generality is puzzling in view of the apparent contradiction involved in its application (Johnson, 1975, 1973).

In spite of these uncertainties, however, the utility of the reactivity-selectivity principle has been illustrated for a number of diverse areas of mechanistic interest. Such applications are being extended to other areas as well. For example, Olah has recently studied the mechanism of electrophilic addition to multiple bonds using selectivity data and concluded that the transition states of the bromine addition to alkenes are of a π-complex nature (Olah and Hockswender, 1974). Finally the large number of reactivity-selectivity relationships which have been discovered offer considerable experimental support for the various expressions and formulations of the Hammond postulate whose profound effect on modern mechanistic chemistry is now beyond question.

ACKNOWLEDGEMENTS

I wish to thank Dr. R. A. More O'Ferrall and Professor Z. Rappoport for comments and helpful discussions.

REFERENCES

Albery, W. J., Campbell-Crawford, A. N., and Hobbs, K. S. (1972a). *J. C. S. Perkin II*, 2180.
Albery, W. J., Curran, J. S., and Campbell-Crawford, A. N. (1972b). *J. C. S. Perkin II*, 2185.
Albery, W. J., Campbell-Crawford, A. N., and Curran, J. S. (1972c). *J. C. S. Perkin II*, 2206.
Alexander, R., Ko, E. C. F., Parker, A. J., and Broxton, T. J. (1968). *J. Amer. Chem. Soc.* **90**, 5049.
Afanas'ev, I. B. (1971). *Russ. Chem. Rev.* **40**, 216.
Bailey, T. H., Fox, J. R., Jackson, J., Kohnstam, G., and Queen, A. (1966). *Chem. Comm.* 122.
Barnett, R., and Jencks, W. P. (1967). *J. Amer. Chem. Soc.* **89**, 5963.
Bell, H. M., and Brown, H. C. (1966). *J. Amer. Chem. Soc.* **88**, 1473.
Bell, R. P. (1936). *Proc. Roy. Soc., Ser. A.* **154**, 414.

Bell, R. P. (1941). "Acid-Base Catalysis", Oxford University Press, London, p. 85.
Bell, R. P. (1959). "The Proton in Chemistry", Cornell University Press, Ithaca, N.Y. Chapt. 10.
Bell, R. P. (1965). *Disc. Faraday Soc.* **39**, 16.
Bell, R. P. (1966). *Adv. Phys. Org. Chem.* **4**, 1.
Bell, R. P. (1973). "The Proton in Chemistry", 2nd Ed. Chapman and Hall, London.
Bell, R. P., and Lidwell, O. (1940). *Proc. Roy. Soc., Ser. A* **176**, 88.
Bell, R. P., Gelles, E., and Moller, E. (1949). *Proc. Roy. Soc., Ser. A* **198**, 308.
Bell, R. P., Rand, M. H., and Wynne-Jones, K. M. A. (1956). *Trans. Faraday Soc.* **52**, 1093.
Benson, S. W., and DeMore, W. B. (1965). *Ann. Rev. Phys. Chem.* **16**, 397.
Bentley, T. W., Schadt, F. L., and Schleyer, P. von R. (1972). *J. Amer. Chem. Soc.* **94**, 992.
Bethell, D., and Gold, V. (1967). "Carbonium Ions: An Introduction", Academic Press, London, p. 88.
Bordwell, F. G., Boyle, W. J., Jr., Hautala, J. A., and Yee, K. C. (1969). *J. Amer. Chem. Soc.* **91**, 4002.
Bordwell, F. G., Boyle, W. J., Jr., and Yee, K. C. (1970). *J. Amer. Chem. Soc.* **92**, 5926.
Bordwell, F. G., and Boyle, W. J., Jr., (1971). *J. Amer. Chem. Soc.* **93**, 512.
Bordwell, F. G., and Boyle, W. J., Jr., (1975). *J. Amer. Chem. Soc.* **97**, 3447.
Brønsted, J. N., and Pedersen, K. (1923). *Z. Phys. Chem.* **108**, 185.
Brønsted, J. N., and Guggenheim, E. A. (1927). *J. Amer. Chem. Soc.* **49**, 2554.
Brown, H. C., and Nelson, K. L. (1953). *J. Amer. Chem. Soc.* **75**, 6292.
Brown, H. C., and Russell, G. A. (1952). *J. Amer. Chem. Soc.* **74**, 3995.
Brown, H. C., and Stock, L. M. (1957). *J. Amer. Chem. Soc.* **79**, 1421.
Caldin, E. F., and Gold, V. (1975). "Proton Transfer Reactions", Chapman and Hall, London.
Casey, M. L., Kemp, D. S., Paul, K., and Cox, D. (1973). *J. Org. Chem.* **38**, 2294.
Challis, B. C., and Millar, E. M. (1972). *J. C. S. Perkin II*, 1618.
Clementi, S., Linda, P., and Marino, G. (1970). *J. Chem. Soc. (B)* 1153.
Clementi, S., and Marino, G. (1972). *J. C. S. Perkin II*, 71.
Clementi, S., and Katritzky, A. R. (1973). *J. C. S. Perkin II*, 1077.
Clementi, S., Johnson, C. D., and Katritizky, A. R. (1974). *J. C. S. Perkin II*, 1294.
Clementi, S., Katritzky, A. R., and Tarhan, H. O. (1975). *Tetrahedron Lett.* 1395.
Closs, G. L., and Moss, R. A. (1964). *J. Amer. Chem. Soc.* **86**, 4042.
Cordes, E. H., and Jencks, W. P. (1962). *J. Amer. Chem. Soc.* **84**, 4319.
Cram, D. J., and Kollmeyer, W. D. (1968). *J. Amer. Chem. Soc.* **90**, 1791.
Crooks, J. E. (1975). In "Proton Transfer Reactions" (Caldin, E. F. and Gold, V., ed.), Chapman and Hall, London. P. 153.
Cvetanovic, R. J. (1959). *J. Chem. Phys.* **30**, 19.
de la Mare, P. B. D., and Harvey, J. T. (1956). *J. Chem. Soc.* 36.
do Amaral, L., Sandstrom, W. A., and Cordes, E. H. (1966). *J. Amer. Chem. Soc.* **88**, 2225.
Dewar, M. J. S. (1969). "The Molecular Orbital Theory of Organic Chemistry", McGraw-Hill Inc., New York, N.Y., p. 284-288.
Doering, W. v. E., and Henderson, W. A. Jr. (1958). *J. Amer. Chem. Soc.* **80**, 5274.
Dubois, J. E., Aaron, J. J., Alcais, P., Doucet, J. P., Rothenberg, F., and Uzan, R. (1972). *J. Amer. Chem. Soc.* **94**, 6823.

Eaborn, C., and Taylor, R. (1961). *J. Chem. Soc.* 247
Eigen, M. (1964). *Angew. Chem., Int. Ed. Engl.* 3, 1.
Evans, M. G., and Polanyi, M. (1936). *Trans. Faraday Soc.* 32, 1340.
Evans, M. G., and Polanyi, M. (1938). *Trans. Faraday Soc.* 34, 11.
Fainberg, A. H., and Winstein, S. (1956). *J. Amer. Chem. Soc.* 78, 2770.
Forsythe, D. A., and Noyce, D. S. (1972). *Tetrahedron Lett.* 3893.
Gilbert, T. J., and Johnson, C. D. (1974). *J. Amer. Chem. Soc.* 96, 5846.
Gold, V. (1964). In "Friedel-Crafts and Related Reactions" (G. A. Olah, ed.), Wiley-Interscience, New York, N.Y., p. 1268.
Gold, V., and Waterman, D. C. A. (1968). *J. Chem. Soc. (B)* 849.
Grunwald, E., and Winstein, S. (1948). *J. Amer. Chem. Soc.* 70, 846.
Hammett, L. P. (1970). "Physical Organic Chemistry", McGraw-Hill, New York, N.Y., 2nd Ed. p. 355.
Hammond, G. S. (1955). *J. Amer. Chem. Soc.* 77, 334.
Hanna, S. B., Jermini, C., Loewenschuss, H., and Zollinger, H. (1974). *J. Amer. Chem. Soc.* 96, 7222.
Harris, J. M., Fagan, J. F., Walden, F. A., and Clark, D. C. (1972). *Tetrahedron Lett.* 3023.
Harris, J. M., Becker, A., Clark, D. C., Fagan, J. F., and Kennan, S. L. (1973). *Tetrahedron Lett.* 3813.
Harris, J. M., Clark, D. C., Becker, A., and Fagan, J. F. (1974a). *J. Amer. Chem. Soc.* 96, 4478.
Harris, J. M., Becker, A., Fagan, J. F., and Walden, F. A. (1974b). *J. Amer. Chem. Soc.* 96, 4484.
Hassid, A. I., Kreevoy, M. M., and Liang, T. M. (1975). *Faraday Symposia* 10, 69.
Hill, E. (1965). *Chem. and Ind.* 1696.
Hine, J., and Ehrenson, S. J., (1958). *J. Amer. Chem. Soc.* 80, 824.
Hudson, R. F., and Klopman, G. (1964). *J. Chem. Soc.* 5.
Ingold, C. K. (1969a, b). "Structure and Mechanism in Organic Chemistry", 2nd Ed. Cornell University Press, Ithaca, N.Y., (a) p. 290; (b) p. 483.
Jencks, W. P. (1969a, b). "Catalysis in Chemistry and Enzymology", McGraw-Hill, New York, N.Y., (a) p. 195; (b) Ch. 3.
Jencks, W. P. (1972). *Chem. Rev.* 72, 705.
Johnson, C. D. (1973). "The Hammett Equation", Cambridge University Press, London and New York, p. 156.
Johnson, C. D. (1975). *Chem. Rev.* 75, 755.
Johnson, C. D., Katritzky, A. R., and O'Neill, B. (1974). Quoted by Gilbert and Johnson, 1974.
Johnson, C. D., and Schofield, K. (1973). *J. Amer. Chem. Soc.* 95, 270.
Kemp, D. S., and Casey, M. L. (1973). *J. Amer. Chem. Soc.* 95, 6670.
Kerr, J. A. (1973). In "Free Radicals" (J. K. Kochi, ed.), Wiley, New York, N.Y., Vol. 1., Chapter 1.
Kevill, D. N., Kolwyck, K. C., and Weitl, F. L. (1970). *J. Amer. Chem. Soc.* 92, 7300.
Kevill, D. N., Kolwyck, K. C., Shold, D. M., and Kim, C.-B. (1973). *J. Amer. Chem. Soc.* 95, 6022.
Kice, J. L., Rogers, T. E., and Warheit, A. C. (1974). *J. Amer. Chem. Soc.* 96, 8020.
Kochi, J. K. (1973). "Free Radicals", Wiley, New York, N.Y.
Koskikallio, J. (1972). *Acta. Chem. Scand.* 26, 1201.
Kosower, E. M. (1968). "Introduction to Physical Organic Chemistry", Wiley, New York, N.Y., Chapter 1.1.
Kreevoy, M. M., and Konasewich, D. E. (1972). *Adv. Chem. Phys.* 21, 241.
Kreevoy, M. M., and Oh, S. (1973). *J. Amer. Chem. Soc.* 95, 4805.

Kresge, A. J. (1970). *J. Amer. Chem. Soc.* **92**, 3210.
Kresge, A. J. (1973). *Chem. Soc. Rev.* **2**, 475.
Kresge, A. J. (1975). *Accounts Chem. Res.* **8**, 354.
Kresge, A. J., and Chiang, Y. (1973). *J. Amer. Chem. Soc.* **95**, 803.
Kresge, A. J., Mylonakis, S. G., Sato, Y., and Vitullo, V. P. (1971b). *J. Amer. Chem. Soc.* **93**, 6181.
Kresge, A. J., Chen, H. J., Chiang, Y., Murrill, E., Payne, M. A., and Sagatys, D. S. (1971a). *J. Amer. Chem. Soc.* **93**, 413.
Leffler, J. E. (1953). *Science* **117**, 340.
Leffler, J. E., and Grunwald, E. (1963a, b). "Rates and Equilibria of Organic Reactions", Wiley, New York, N.Y., (a) pp. 163-64, 157; (b) p. 187.
Lienhard, G. E., and Jencks, W. P. (1966). *J. Amer. Chem. Soc.* **88**, 3982.
Majerski, Z., Borcic, S., and Sunko, D. E. (1969). *Tetrahedron* **25**, 301.
Marcus, R. A. (1968). *J. Phys. Chem.* **72**, 891.
Marcus, R. A. (1969). *J. Amer. Chem. Soc.* **91**, 7224.
Miller, L. L., and Kaufman, D. A. (1968). *J. Amer. Chem. Soc.* **90**, 7282.
Miller, S. I. (1959). *J. Amer. Chem. Soc.* **81**, 101.
Minisci, F., Gardini, G. P., and Bertini, F. (1970). *Can. J. Chem.* **50**, 1544.
More O'Ferrall, R. A. (1970). *J. Chem. Soc. (B)* 274.
More O'Ferrall, R. A. (1973). "Notes on Organic Reactivity", Lectures given in the Spring of 1973 at the Institute of Chemistry, Hebrew University, Jerusalem.
More O'Ferrall, R. A. (1975). "Proton Transfer Reactions" (E. F. Caldin, and V. Gold, eds.), Chapman and Hall, London. p. 201.
Moss, R. A. (1973). "Carbenes", Vol. 1., (M. Jones and R. A. Moss, eds.), Wiley, New York, Ch. 2.
Moss, R. A., and Mallon, C. B. (1975). *J. Amer. Chem. Soc.* **97**, 344.
Moss, R. A., and Mamantov, A. (1970). *J. Amer. Chem. Soc.* **92**, 6951.
Moss, R. A., and Przybyla, J. R. (1969). *Tetrahedron* **25**, 647.
Murdoch, J. R. (1972). *J. Amer. Chem. Soc.* **94**, 4410.
Nefedov, O. M., and Shafran, R. N. (1967). *Zh. Obsch. Khim.* **37**, 1561.
Nishida, S. (1967a). *J. Org. Chem.* **32**, 2695.
Nishida, S. (1967b). *J. Org. Chem.* **32**, 2697.
Noyce, D. S., Lipinski, C. A., and Nichols, R. W. (1972). *J. Org. Chem.* **37**, 2615.
Noyce, D. S., and Nichols, R. W. (1972). *Tetrahedron Lett.* 3889.
Noyce, D. S., and Pavez, H. J. (1972). *J. Org. Chem.* **37**, 2620
Ogston, A. G., Holiday, E. R., Philpot, J. S. L., and Stocken, L. A. (1948). *Trans. Faraday Soc.* **44**, 45.
Okamoto, K., and Kinoshita, T. (1974). *Chem. Lett.* 1037.
Olah, G. A. (1971). *Accounts Chem. Res.* **4**, 240.
Olah, G. A., and Hockswender, T. R. Jr. (1974). *J. Amer. Chem. Soc.* **96**, 3574.
Olah, G. A., and Kobayashi, S. (1971). *J. Amer. Chem. Soc.* **93**, 6964.
Olah, G. A., Kuhn, S. J., and Flood, S. (1961). *J. Amer. Chem. Soc.* **83**, 4571, 4581.
Olah, G. A., Kuhn, S. J., Flood, S., and Evans, J. C. (1962). *J. Amer. Chem. Soc.* **84**, 3687.
Olah, G. A., and Overchuk, N. (1965). *J. Amer. Chem. Soc.* **87**, 5786.
Olah, G. A., Tashiro, M., and Kobayashi, S. (1970). *J. Amer. Chem. Soc.* **92**, 6369.
Pearson, R. G., and Dillon, R. L. (1953). *J. Amer. Chem. Soc.* **75**, 2439.
Pross, A. (1975a). *Tetrahedron Lett.* 637.
Pross, A. (1975b). *Tetrahedron Lett.* 1289.
Pross, A. (1976). *J. Amer. Chem. Soc.* **98**, 776.

Pross, A., and Koren, R. (1974). *Tetrahedron Lett.* 1949.
Pross, A., and Koren, R. (1975). *Tetrahedron Lett.* 3613.
Raber, D. J., Bingham, R. C., Harris, J. M., Fry, J. L., and Schleyer, P. von R. (1970). *J. Amer. Chem. Soc.* 92, 5977.
Raber, D. J., Harris, J. M., Hall, R. E., and Schleyer, P. von R. (1971a). *J. Amer. Chem. Soc.* 93, 4821.
Raber, D. J., Harris, J. M., and Schleyer, P. von R. (1971b). *J. Amer. Chem. Soc.* 93, 4829.
Radom, L., and Stiles, P. J., (1975). *Tetrahedron Lett.* 789.
Rappoport, Z., and Apeloig, Y. (1975). *J. Amer. Chem. Soc.* 97, 821.
Rappoport, Z., and Gal, A. (1970). *Tetrahedron Lett.* 3233.
Rappoport, Z., and Gal, A. (1973a). *J. C. S. Perkin II*, 301.
Rappoport, Z., and Houminer, Y. (1973b). *J. C. S. Perkin II*, 1506.
Rappoport, Z., Pross, A., and Apeloig, Y. (1973c). *Tetrahedron Lett.* 2015.
Rappoport, Z., and Kaspi, J. (1972). *J. C. S. Perkin II*, 1102.
Ridd, J. H. (1971). *Accounts Chem. Res.* 4, 248.
Ritchie, C. D. (1971). *J. Amer. Chem. Soc.* 93, 7324.
Ritchie, C. D. (1972). *Accounts Chem. Res.* 5, 348.
Ritchie, C. D., (1975). *J. Amer. Chem. Soc.* 97, 1170.
Ritchie, C. D., and Fleischhauer, H. (1972). *J. Amer. Chem. Soc.* 94, 3481.
Ritchie, C. D., and Virtanen, P. O. I. (1972a). *J. Amer. Chem. Soc.* 94, 1589.
Ritchie, C. D., and Virtanen, P. O. I. (1972b). *J. Amer. Chem. Soc.* 94, 4963.
Ritchie, C. D., and Virtanen, P. O. I. (1972c). *J. Amer. Chem. Soc.* 94, 4966.
Ritchie, C. D., and Virtanen, P. O. I. (1973). *J. Amer. Chem. Soc.* 95, 1882.
Ritchie, C. D., and Wright, D. J. (1971). *J. Amer. Chem. Soc.* 93, 6574.
Ritchie, C. D., Wright, D. J., Huang, D., and Kamego, A. A. (1975). *J. Amer. Chem. Soc.* 97, 1163.
Russell, G. A. (1958). *J. Amer. Chem. Soc.* 80, 4987.
Russell, G. A. (1960). *Tetrahedron.* 8, 101.
Russell, G. A. (1973). In "Free Radicals" (J. K. Kochi, ed.), Wiley, New York, N.Y., Vol. 1, Chapter 7.
Sadler, I. H. (1969). *J. Chem. Soc. (B)* 1024.
Schleyer, P. von R., and Nicholas, R. D. (1961). *J. Amer. Chem. Soc.* 83, 2700.
Seyferth, D., Mui, J. Y.-P., and Damrauer, R. (1968). *J. Amer. Chem. Soc.* 90, 6182.
Shatenshtein, A. I. (1962). "Isotopic Exchange and the Replacement of Hydrogen in Organic Compounds", Consultants Bureau, New York, Sect. III.
Shiner, V. J., Jr., Fisher, R. D., and Dowd, W. (1969). *J. Amer. Chem. Soc.* 91, 7748.
Simmons, H. E., Blanchard, E. P., and Smith, R. D. (1964). *J. Amer. Chem. Soc.* 86, 1347.
Skell, P. S., and Cholod, M. S. (1969a). *J. Amer. Chem. Soc.* 91, 6035.
Skell, P. S., and Cholod, M. S. (1969b). *J. Amer. Chem. Soc.* 91, 7131.
Skell, P. S., and Garner, A. Y. (1956). *J. Amer. Chem. Soc.* 78, 5430.
Sneen, R. A. (1973). *Accounts Chem. Res.* 6, 46.
Sneen, R. A., Carter, J. V., and Kay, P. S. (1966a). *J. Amer. Chem. Soc.* 88, 2594.
Sneen, R. A., and Larsen, J. W. (1966b). *J. Amer. Chem. Soc.* 88, 2593.
Spanswick, J., and Ingold, K. U. (1970). *Can. J. Chem.* 50, 546.
Stevens, C. G., and Strickler, S. J. (1973). *J. Amer. Chem. Soc.* 95, 3918.
Stock, L. M., and Brown, H. C. (1963). *Adv. Phys. Org. Chem.* 1, 35.
Streitwieser, A., Jr., Hollyhead, W. B., Pudjaatmaka, A. H., Owens, P. H., Kruger, T. L., Rubenstein, P. A., MacQuarrie, R. A., Brokaw, M. L., Chu, W. K. C., and Niemeyer, H. M. (1971). *J. Amer. Chem. Soc.* 93, 5088.

Swain, C. G., and Scott, C. B. (1953a). *J. Amer. Chem. Soc.* 75, 141.
Swain, C. G., Scott, C. B., and Lohman, K. H. (1953b). *J. Amer. Chem. Soc.* 75, 136.
Tanner, D. D., and Nychka, N. (1967). *J. Amer. Chem. Soc.* 89, 121.
Thornton. E. R. (1967). *J. Amer. Chem. Soc.* 89, 2915.
Thornton, E. K., and Thornton, E. R. (1970). "Isotope Effects in Chemical Reactions" (C. J. Collins, and N. S. Bowman., eds.), Van Nostrand Reinhold, New York, N. Y. Chapter 4.
Tolgyesi, W. S. (1965). *Can. J. Chem.* 43, 343.
Trotman-Dickenson, A. F. (1965). *Chem. Ind.* 379.
Van Hook, W. A. (1970). "Isotope Effects in Chemical Reactions" (C. J. Collins, and N. S. Bowman, eds.), Van Nostrand Reinhold, New York, N.Y. Chapt. 1.
Walling, C., and McGuinness, J. A., (1969). *J. Amer. Chem. Soc.* 91, 2053.
Walling, C., and Padwa, A. (1963). *J. Amer. Chem. Soc.* 85, 1593.
Weaver, W. M., and Hutchison, J. D. (1964). *J. Amer. Chem. Soc.* 86, 261.
Westheimer, F. H. (1961). *Chem. Rev.* 61, 265.
Wiberg, K. B., and Slaugh, L. H. (1958). *J. Amer. Chem. Soc.* 80, 3033.
Winstein, S., Appel, B., Baker, R., and Diaz, A. (1965). *Chem. Soc., Spec. Publ.,* No. 19, 109.
Winstein, S., Clippinger, E., Fainberg, A. H., Heck, R., and Robinson, G. C. (1956). *J. Amer. Chem. Soc.* 78, 328.
Winstein, S., and Robinson, G. C. (1958). *J. Amer. Chem. Soc.* 80, 169.

Physical Organic Chemistry of Reactions in Dimethyl Sulphoxide

E. BUNCEL and H. WILSON

Department of Chemistry, Queen's University, Kingston, Ontario, Canada K7L 3N6

1. Scope 133
2. Solute-Solvent Interactions 135
   Medium Effects. Initial State and Transition State Interactions . 135
   pK_a: Medium Dependence 142
3. Proton Transfer Processes 150
   General Considerations 150
   Symmetry of Transition States from Isotope Effects . . 151
   Brønsted Coefficients and Solvent Effects 154
   Curvature of Brønsted Plots and Medium Effects . . . 156
4. Rate Variations in DMSO as a Guide to Mechanism . . . 159
   Some Bimolecular Nucleophilic Substitution Reactions . . 160
   Ester Hydrolysis 163
   Reactions Involving Proton Transfers 167
5. Role of DMSO in Selected Mechanism Studies 174
   Stabilization and Catalytic Decomposition of Intermediates in
   S_NAr Reactions 174
   The Steric Course of Bimolecular Olefin-forming Eliminations . 182
6. Future Developments 188
7. Conclusion 191
   Appendix: A compilation of selected properties of DMSO and
   Aqueous DMSO systems 192
   References 196

1. SCOPE

Dimethyl sulphoxide (DMSO) is a dipolar aprotic compound with unusual solvent properties, synthetic versatility, and possible clinical applications. It is therefore not surprising that it has been the subject of extensive investigation resulting in the accumulation of an immense literature. Earlier works have emphasized the physical properties (Ranky and Nelson, 1961), chemical properties (Szmant, 1971), synthetic uses (Durst, 1969) and solvent characteristics (MacGregor, 1968) of DMSO. It has also been the topic for some general reviews and monographs (Kharasch and Thyagarajan, 1966; Jacob et al., 1971).

This article will avoid, as far as possible, duplication of these earlier reports and will emphasize the use of DMSO as an "inert" solvent in mechanistic investigations of heterolytic reactions. Studies where such use has led to interesting results will of course be highlighted, but others have been selected which illustrate the methods developed for mechanistic study in dipolar aprotic media. Discussion has not been restricted to reactions which give the characteristic rate variations on transfer to dipolar aprotic solvents; the effect of DMSO on product distributions is discussed, and we have also speculated on possible future uses of DMSO in mechanistic investigations.

The first section, under the heading solute–solvent interactions, considers the origin of the medium effect which is exhibited for reactions on changing from a hydroxylic solvent to a dipolar aprotic medium such as DMSO. This section is subdivided into two parts, the first concentrating on medium effects on rate processes, the second on equilibria of the acid–base variety. The section includes discussion of the methods used in obtaining and analysing kinetic and thermodynamic transfer functions. There follows a discussion of proton transfers. The methods and principles used in such studies have a rather unique character within the context of this work and have been deemed worthy of elaboration. The balance of the article is devoted to consideration of a variety of mechanistic studies featuring DMSO; many of the principles developed in earlier sections will be utilized here.

No attempt has been made to be comprehensive. The reactions discussed reflect the bias of the authors, and the choice is in no way meant to slight those investigators whose work lies outside the chosen areas of interest.

Finally, mention should be made of the earlier reviews of Parker (1962, 1969) and the monograph by Coetzee and Ritchie (1969) which, either directly or by inspiration, have been largely responsible for the accumulation of data which has made this article possible.

2. SOLUTE–SOLVENT INTERACTIONS

Medium Effects. Initial State and Transition State Interactions

The use of dimethyl sulphoxide in organic chemistry has revealed the profound influence of solvent on the course and the rates of organic reactions. The striking discovery, in the early 1960's, that certain dipolar aprotic solvents such as dimethylformamide (DMF), tetrahydrothiophene-1,1-dioxide (sulfolane), hexamethylphosphortriamide (HMPT), as well as DMSO, could result in certain instances in rate enhancements of the order of 10^{10} relative to hydroxylic solvents led to a re-appraisal of the then established theories of medium effects.

The prevailing theory of solvent effects on reaction rates at that time was first proposed by Hughes and Ingold (1935). This qualitative theory, based on considerations of solvent polarity, was able to explain a large number of rate phenomena (Ingold, 1953). The primary measure of solvent polarity was the macroscopic dielectric constant. Other workers (Laidler and Eyring, 1940; Scatchard, 1940; Amis, 1966) placed the theory on a more quantitative basis by deriving mathematical relationships between dielectric constant and various kinetic parameters using transition state theory.

It was realized that the use of the macroscopic dielectric constant as the prime measure of solute–solvent interactions at the microscopic level was bound to be an over-simplification. Consequently other empirical measures of solvent polarity were introduced: Y (Grunwald and Winstein, 1948), Z (Kosower, 1958), S (Brownstein, 1960), k_{ion} (Smith *et al.*, 1961), Ω (Berson *et al.*, 1962), E_T (Dimroth *et al.*, 1963), G (Allerhand and Schleyer, 1963), F (Dubois *et al.*, 1964), δ (Herbrandson and Neufelt, 1966) and Q (Krygowski and Fawcett, 1975). Use of these parameters has been reviewed (Reichardt, 1965; Kosower, 1968; Koppel and Palm, 1972). To the extent that these parameters are generally based upon microscopic processes, they act as probes of solute–solvent interactions such as hydrogen-bonding, charge-dipole, dipole–dipole, and dipole-induced dipole interactions. It should be emphasized that, since each parameter is experimentally based on a specific model process, great care must be taken to select the parameter appropriate to the reaction being studied.

Analysis of medium effects in terms of transition state theory requires a knowledge of the interaction of both the initial state and the transition state with the solvent. Traditionally, greater emphasis was placed on transition state interactions. However, studies in the domain of nucleophilic aliphatic substitution (Ingold, 1953), and particularly investigation of the solvolysis of t-butyl chloride (Winstein and Fainberg, 1956; Arnett et al., 1965), revealed that initial state interactions with solvent could be as important (Robertson et al., 1959; Hyne et al., 1962).

An experimental measure of the interaction of a solute species with a given solvent medium is afforded by the partial molal heat of solution, ΔH_S. For a stable species this is obtained directly by calorimetry or indirectly from the temperature coefficient of the Henry's law constant. The difference in the value of ΔH_S between the two solvents is termed the enthalpy of transfer, denoted by $\delta \Delta H_{tr}$. Combining enthalpies of transfer for the reactants, $\delta \Delta H_{tr}^R$, with transfer enthalpies of activation, $\delta \Delta H_{tr}^{\ddagger}$, using equation (1)

$$\delta \Delta H_{tr}^T = \delta \Delta H_{tr}^R + \delta \Delta H_{tr}^{\ddagger} \qquad (1)$$

enables enthalpies of transfer for the transition state, $\delta \Delta H_{tr}^T$ to be derived. The standard reference state with respect to which the enthalpy of transfer terms are evaluated may be chosen as the gas phase or as a particular pure solvent (commonly methanol or water). It should be pointed out that whereas the $\delta \Delta H_{tr}$ term is readily available for neutral solute species, corresponding values for single ions can only be evaluated using extra-thermodynamic assumptions, e.g., $\delta \Delta H_{tr}(Ph_4N^+) = \delta \Delta H_{tr}(Ph_4B^-)$ (Grunwald et al., 1960).

Entropies of transfer, $\delta \Delta S_{tr}$, are available from relationship (2).

$$\delta \Delta G_{tr} = \delta \Delta H_{tr} - T\delta \Delta S_{tr} \qquad (2)$$

The free energy term, $\delta \Delta G_{tr}$, is obtained with relative ease from Henry's law constants. Thus, complete dissection of the effect of solvent on the various thermodynamic properties is possible in favourable cases. This was in fact achieved by Arnett et al. (1965) for the solvolysis of t-butyl chloride in aqueous ethanol mixtures and revealed that the peculiar rate variation with changing solvent composition was largely caused by changes in the initial state interactions.

An equivalent approach uses activity coefficients of transfer and has been effectively applied to bimolecular reactions by Parker and co-workers (Parker, 1969).[1] The medium effect for a bimolecular

---

[1] The activity coefficient approach in the solvolysis of t-butyl chloride was considered extensively by Clarke and Taft (1962).

reaction between a nucleophile Y and an alkyl halide RX is interpreted according to (3) which is derived from absolute rate

$$\log \frac{k^S}{k_0} = \log {}^o\gamma_Y^S + \log {}^o\gamma_{RX}^S - \log {}^o\gamma_{YRX^{\ddagger}}^S \qquad (3)$$

theory. The parameters ${}^o\gamma^S$, which are measures of the medium effect, represent the solvent transfer activity coefficients of reactants or transition state; ${}^o\gamma_i^S$ is defined as in (4) and is proportional to the

$$\bar{\mu}_i^S = \bar{\mu}_i^o + RT \ln {}^o\gamma_i^S \qquad (4)$$

change in the standard chemical potential μ_i of a solute, i, on transfer from an arbitrarily chosen reference solvent, O, to a solvent S. Activity coefficients for non-electrolytes are measured by standard techniques (e.g., from solubilities), while single ion activity coefficients are obtained using extrathermodynamic assumptions, as previously indicated. This approach enables one to evaluate the ${}^o\gamma_{YRX^{\ddagger}}^S$ term, which gives an indication of whether the transition state is better solvated by the reference solvent or by the solvent under examination.

Application of these methods has revealed the limitations of certain qualitative generalizations. Thus it had been common practice to attribute the rate enhancement resulting from a change of solvent from a protic one to DMSO primarily to desolvation of the anionic nucleophiles. However, it has been shown that in a number of cases increased solvation of the transition state is the predominant factor. For example, in the reaction of p-nitrofluorobenzene with azide ion in DMSO it is found that $\delta \Delta H_{tr}(N_3^-) = -1 \cdot 0$ kcal mol^{-1} whilst $\delta \Delta H_{tr}^T = -6 \cdot 5$ kcal mol^{-1} (Cox and Parker, 1973). Similar studies, using DMSO and other dipolar aprotic media, have been carried out by Haberfield (1971), Fuchs (1974), Abraham (1974), Jones et al. (1976), and their co-workers, and will be considered in later sections.

In order for such dissection of thermodynamic parameters to be possible in general, it is clearly essential that a sufficient body of data ($\delta \Delta G_{tr}$, $\delta \Delta H_{tr}$, $\delta \Delta S_{tr}$) on individual species be available. Measurements of such quantities are increasingly being made, and a recent excellent report contains an up to date summary (Cox, 1973; see also Abraham, 1973; Cox et al., 1974). Some of the currently available data referring to transfer of ionic species from water to various dipolar aprotic solvents are presented in Tables 1 and 2.

Interpretation of this accumulated thermodynamic data for ionic solute species has been attempted in terms of various solute-solvent

TABLE 1[a]

Free Energies of Transfer of Ions ($\delta \Delta G_{tr}$) from Water to Non-aqueous Solvents at 25 °C (molar scale in kcal mol^{-1})[b,c]

	$\delta \Delta G_{tr}$							
Ion	MeOH	HCONH$_2$	NMeF	DMF	DMSO	Me$_2$CO	PC	MeCN
H$^+$	2·6	—	—	−3·4	−4·5	−0·7	—	11·1
Li$^+$	1·0	−2·3	−3·5	−5·3	−3·5	—	5·3	7·1
Na$^+$	2·0	−1·9	−1·9	−2·5	−3·3	—	2·6	3·3
K$^+$	2·4	−1·5	−2·0	−2·3	−2·9	0·7	0·8	1·9
Rb$^+$	2·4	−1·3	−1·8	−2·4	−2·6	0·5	−1·3	1·6
Cs$^+$	2·3	−1·8	−1·7	−2·2	−3·0	0·4	−3·5	1·2
Ag$^+$	1·8	−3·7	—	−4·1	−8·0	1·5	3·3	−5·2
Tl$^+$	1·0	—	—	−2·8	−6·0	—	2·0	2·2
NEt$_4^+$	0·2	—	—	−2·0	−1·2	—	—	−2·1
Ph$_4$As$^+$	−5·6	−5·7	—	−9·1	−8·8	−7·1	−8·5	−7·8
Cl$^-$	3·0	3·3	4·9	11·0	9·2	14·0	10·1	10·1
Br$^-$	2·7	2·7	3·6	7·2	6·1	10·5	7·8	7·6
I$^-$	1·6	1·8	—	4·5	3·2	6·7	4·6	4·5
N$_3^-$	2·5	2·9	—	8·2	5·7	10·9	7·1	7·0
ClO$_4^-$	1·4	—	0·4	—	—	3·6	—	1·1
OAc$^-$	3·7	—	—	14·8	11·1	—	—	13·4
BPh$_4^-$	−5·6	−5·7	—	−9·1	−8·8	−7·1	−8·5	−7·8

[a] Cox, 1973.
[b] Data calculated using the assumption $\delta \Delta G_{tr}(Ph_4As^+) = \delta \Delta G_{tr}(BPh_4^-)$.
[c] Abbreviations: NMeF, N-methylformamide; PC, propylenecarbonate.

interactions such as electrostatic (ion–dipole or dipole–dipole) interactions, hydrogen bonding, dispersion forces and the making or breaking of solvent structure. The observed variations are in accord with Ritchie's (1969) qualitative analysis of the interactions between ionic species and dipolar aprotic solvents and also agree with the more quantitative treatment of Parker (1969), using solvent activity coefficients (Table 3).

The most noticeable feature of these data is the striking increase in the free energy of small anions with high charge density on transfer to the dipolar aprotic solvents. This is consistent with the large rate increase observed in reactions involving these anions. If the anion is small, it should be a strong hydrogen bond acceptor in protic solvents. Such interactions are absent in DMSO, which accounts for these anions being much less solvated in DMSO than in water (see also footnote on p. 167). They could also explain the observation (Wooley and Hepler, 1972) that the ionization product of water decreases rapidly with increasing DMSO content in mixtures of

TABLE 2[a]

Enthalpies and Entropies of Transfer of Ions[b] from Water to Non-aqueous Solvents at 25 °C

	(a) $\delta \Delta H_{tr}$/kcal mol^{-1}						
Ion	H$_2$O	MeOH	HCONH$_2$	DMF	DMSO	PC	MeCN
Li$^+$	0	−5·2	−1·3	−7·7	−6·3	0·4	—
Na$^+$	0	−4·9	−3·9	−7·9	−6·6	−2·1	−3·1
K$^+$	0	−4·4	−4·0	−9·4	−8·3	−5·0	−5·4
Rb$^+$	0	−3·6	−4·1	−9·0	−8·0	−5·6	−6·2
Cs$^+$	0	−3·2	−4·1	−8·8	−7·7	−6·2	—
Ag$^+$	0	−5·0	−5·4	−9·2	−13·1	−3·0	−12·6
NMe$_4^+$	0	4·0	0·2	−3·9	−3·9	−4·6	−3·7
EtN$_4^+$	0	2·2	1·8	−0·2	1·0	−0·2	−0·4
BuN$_4^+$	0	5·2	6·8	4·3	6·1	3·8	4·4
Ph$_4$As$^+$	0	−0·4	−0·1	−4·7	−2·8	−3·6	−2·5
Cl$^-$	0	2·0	0·8	5·1	4·5	6·7	4·7
Br$^-$	0	0·9	−0·4	0·8	0·8	4·2	2·0
I$^-$	0	−0·5	−1·8	−3·3	−3·2	−0·2	−1·7
N$_3^-$	0	0·1	—	0·3	−0·6	3·9	2·1
ClO$_4^-$	0	−0·6	−4·8	−6·0	−4·6	−3·7	—
BPh$_4^-$	0	−0·4	−0·1	−4·7	−2·8	−3·6	−2·5

		(b) $298 \delta \Delta S_{tr}$/kcal mol^{-1}					
Ion	H$_2$O ($298 \delta \Delta S_s$)	MeOH	HCONH$_2$	DMF	DMSO	PC	MeCN
Li$^+$	−1·4	−6·7	1·0	−2·4	−2·8	−4·9	—
Na$^+$	1·0	−6·9	−1·6	−5·4	−3·1	−4·7	−6·2
K$^+$	3·2	−6·8	−2·5	−6·9	−5·4	−5·8	−7·3
Rb$^+$	4·1	−6·0	−2·8	−6·6	−5·4	−4·3	−7·8
Cs$^+$	4·3	−5·5	−2·3	−6·6	−4·7	−2·7	—
Ag$^+$	0·5	−6·8	−1·7	−5·1	−5·1	−6·3	−7·4
NEt$_4^+$	1·3	2·0	—	1·8	2·2	—	1·7
Ph$_4$As$^+$	−4·0	5·2	5·6	4·4	6·0	4·9	5·3
Cl$^-$	2·0	−1·0	−2·5	−5·9	−4·7	−3·4	−5·4
Br$^-$	3·0	−1·8	−3·1	−6·4	−5·3	−3·6	−4·6
I$^-$	4·3	−2·1	−3·6	−7·8	−6·4	−4·8	−6·2
N$_3^-$	4·5	−2·3	—	−7·9	−6·3	−3·2	−4·9
BPh$_4^-$	−4·0	5·2	5·6	4·4	6·0	4·9	5·3

[a] Cox, 1973.
[b] Values obtained by application of assumptions $\delta \Delta H_{tr}(Ph_4As^+) = \delta \Delta H_{tr}(BPh_4^-)$ and $\delta \Delta S_{tr}(Ph_4As^+) = \delta \Delta S_{tr}(BPh_4^-)$.

DMSO and water. In contrast, the free energies for large polarizable anions usually decrease on transfer to dipolar aprotic solvents, as strikingly illustrated for the case of BPh$_4^-$.

The results for the free energies of transfer of cations do not present as clear a picture. They suggest that solvents such as DMSO which contain a relatively basic oxygen atom solvate cations better than does water. To account for these solvation differences, cation-dipole interactions and solvent structure-making or -breaking effects could be invoked. Some cations (e.g., R_3NH^+) could also be stabilized in DMSO because of their ability to act as hydrogen bond donors.

DMSO has also been found to interact strongly with polarizable neutral species (Table 4). These interactions will be electrostatic in nature and can be attributed to dipole-dipole interactions or dis-

TABLE 3[a]

Solvent activity coefficients of some cations and anions at 25°C [Reference solvent: methanol (M)][b]

Cation	$\log{}^M\gamma_+^{DMSO}$	Anion	$\log{}^M\gamma_-^{DMSO}$
		Cl^-	+5·5
Ag^+	−8·2	Br^-	+3·6
		I^-	+1·3
Na^+	−3·6	N_3^-	+3·5
		SCN^-	+1·4
K^+	−4·5	BPh_4^-	−2·6
		OAc^-	+6·5
Cs^+	−4·3	ClO_4^-	−0·3
		$AgCl_2^-$	−1·3
Ph_4As^+	−2·6	$AgBr_2^-$	−2·3
		AgI_2^-	−2·7
		I_3^-	−3·6

[a] Parker, 1969.
[b] A positive value of $\log{}^M\gamma^{DMSO}$ means that the ionic species is more solvated by methanol than by DMSO and vice versa.

persion effects (Grunwald and Price, 1964). Because of its high basicity DMSO can also be expected to interact with neutral species which are strong hydrogen bond donors. Its apparent difficulty in solvating non-polar solutes may be a reflection of solvent-structure breaking; experimental evidence indicates DMSO to be a highly ordered solvent, as shown for example by the abnormally large Trouton constant (29·6 cal. deg^{-1} mol^{-1}, compared to 26·0 cal. deg^{-1} mol^{-1} for water).

It is distressing that the above discussion of solute–solvent interactions should be so qualitative when they are based on reasonably precise measurement. More detailed information on solvation phenomena can in principle be obtained from a breakdown of ΔG_{tr} into its constituent ΔH_{tr} and ΔS_{tr} (Abraham, 1973; Cox et al., 1974), but even this does not replace most of the qualitative discussion. This frustrating aspect of solution chemistry can only be overcome if independent confirmation of these interpretations of thermodynamic data can be obtained. Some small progress in this regard is being made in the fields of electrochemistry and spectroscopy.

Measurement of limiting ionic conductances and their variation with solvent can in principle offer information about the solvation

TABLE 4[a]

Solvent activity coefficients of some nonelectrolytes at 25°C
[Reference solvent: methanol (M)][b]

Solute RX	$\log{}^M\gamma_{RX}^{DMSO}$
Neon	+0·6
Krypton	+0·6
Xenon	+1·1
Ethane	+0·6
Ethylene	−0·1
Ethylene oxide	−0·3
CH_3I	−0·5
n-BuBr	+0·1
t-BuCl	+0·1
$4\text{-}NO_2C_6H_4I$	−1·1
$[2,4\text{-}(NO_2)_2C_6H_3S]_2$	−2·1
I_2	−4·1

[a] Parker, 1969.
[b] See footnote (b) in Table 3.

shell around a particular ion. From Stokes' law the limiting conductance $\lambda°$ of a singly charged ion of radius r_s moving through a continuous medium of viscosity η is given by (5). Thus $\lambda°\eta$, the

$$\lambda° = \frac{0·82}{\eta r_s} \tag{5}$$

Walden product, is inversely proportional to r_s (the size of the moving ion, including its solvation sphere) which in turn should be a measure of ion–solvent interactions. A comparison of Walden products (Cox, 1973) suggests that such an interpretation may be at least qualitatively reasonable. In water, K^+ and Cl^- have almost equal

Walden products, whereas in all dipolar aprotic solvents $\eta\lambda°(K^+) \ll \eta\lambda° (Cl^-)$. This suggests a large decrease in size of solvation shell of Cl^-, relative to K^+, on transfer from water. The quantitative significance of these electrochemical results with regard to calculated hydration numbers is in some doubt but they do appear to substantiate, at least in this particular case, conclusions drawn from thermodynamic measurements.

Spectroscopic investigations, such as infrared and nuclear magnetic resonance, should provide more information at the molecular level of the effect of solute species on the properties of a solvent. Cox (1973) has outlined the results of several such investigations which provide qualitative confirmation of the interpretations of solute-solvent interactions given above.

In the absence of any precise information on solute-solvent interactions at the molecular level we shall discuss many of the medium effects observed in chemical reactions in terms of the interactions postulated in this section. Normally they suffice to rationalize the observed results.

pK_a: Medium Dependence

The variation in pK_a of a substrate on changing the solvent medium has often been used for qualitative consideration of solute-solvent interactions. Such discussions have usually emphasized the role of anion stabilization in determining the relative magnitudes of pK_a values.

As an illustrative case, comparison of the pK_a values of acetylacetone and malononitrile in DMSO and water shows that pK_a for the former increases by 4·4 units on going from water to DMSO whilst pK_a for the latter remains virtually unchanged. One interpretation of this result has been that, whereas the respective anions would be equally desolvated in DMSO, in water the conjugate base of acetylacetone would be stabilized by hydrogen bonding to a greater degree. This interpretation attributes the pK_a difference exclusively to anion-solvent interactions. This type of argument, although quite common, can be misleading as it ignores solvent interactions with the uncharged acid molecule.

These latter interactions between the solvent and the neutral species have been shown to be important in a variety of solvents, including water, sulfolane and DMSO. This point is illustrated in the following paragraphs.

The difference in the pK_a values of *m*- and *p*-nitrophenol in aqueous medium has aroused interest. In earlier work (Hepler *et al.*, 1965), the enthalpies of ionization were found to be almost identical, and the difference in pK_a was attributed to an entropy effect. Since *m*-nitrophenol had the more negative entropy of ionization, this appeared to be consistent with a greater degree of solvent orientation around its conjugate base as expected since the negative

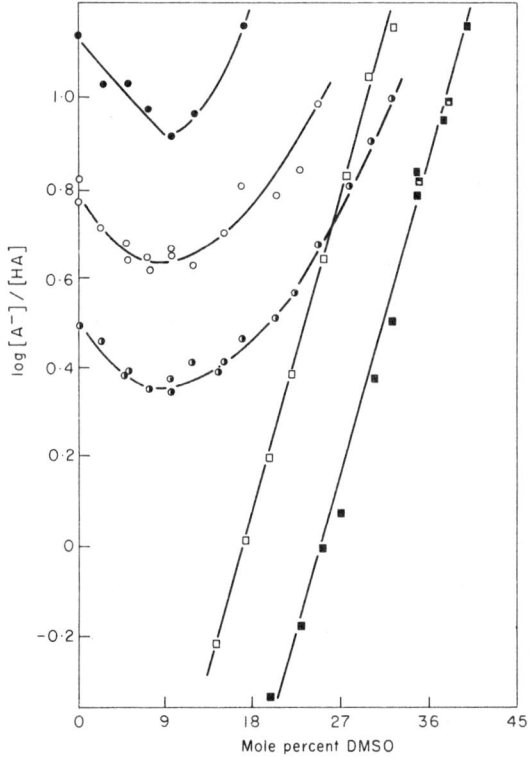

Figure 1. Degree of ionization of various phenols in aqueous DMSO as a function of solvent composition: (●) 2,6-di-isopropylphenol, (○) 2-*t*-butylphenol, (half-shaded circle) 2,4-di-*t*-butylphenol, (□) 2,6-di-*t*-butylphenol, (■) 2,6-di-*t*-butyl-4-methylphenol, (half-shaded square) 2,4,6-tri-*t*-butylphenol. (Albagli *et al.*, 1973.)

charge would be more localized in the *m*-nitrophenoxide ion. It has now been shown that the pK_a difference is actually a consequence of solvation differences in the unionized molecules. This is evident from the measured partial molar volumes of the unionized and ionized forms of the respective nitrophenols (Liotta *et al.*, 1972).

It has been found that the addition of up to 10 mole % DMSO to aqueous hydroxide *reduces* the basicity of the system toward hindered phenols, as seen in Fig. 1 (Albagli *et al.*, 1973). A similar result has been found in sulfolane mixtures (R. A. Cox et al., 1974). In the latter case it has been established that this effect is due to a marked decrease in the activity of the neutral compound (see also Halle et al., 1973).

Clearly it would be desirable to correlate the pK_a differences between solvents with measurable properties corresponding to both the uncharged and charged species. This can be approached by use of the thermodynamic relationship (6) between the pK_a difference and

$$pK_a^S - pK_a^0 = [\delta\Delta G_{tr}(H^+) + \delta\Delta G_{tr}(A^-) - \delta\Delta G_{tr}(HA)]/2\cdot 303RT \quad (6)$$

the respective transfer functions. The applicability of (6) can be demonstrated in certain systems where the required thermodynamic data are available. Thus for benzoic acid in acetonitrile and water, $pK_a^0 = 4\cdot 2$, $\delta\Delta G_{tr}(H^+) = +11$, $\delta\Delta G_{tr}(PhCO_2^-) = +9\cdot 7$, and $\delta\Delta G_{tr}(PhCO_2H) = -2\cdot 0$ kcal mole^{-1}. This leads to a predicted value for pK_a^S of 20·9, which compares with an experimental value of 20·7 found in acetonitrile.

It must be stressed that the above agreement between the predicted and experimental values for pK_a as a function of medium changes is an exceptional case. The reasons for this are two-fold. First, there is a paucity of thermodynamic measurements ($\delta\Delta G_{tr}$) for organic acids and their conjugate bases. Secondly, there is at present considerable doubt as to the validity of pK_a values measured in different solvents. These doubts arise because of variations in apparent orders of acidity in different solvents and the observations of anomalous types of ionization behaviour in a number of systems.

The pK_a values of a number of organic acids in DMSO are listed in Table 5, which inlcudes for comparison corresponding data for a hydroxylic solvent, methanol or water. The particular data in the Table have been selected mainly to illustrate the reversal which can occur on comparing pK_a for certain compounds in hydroxylic solvents and in DMSO. A more complete listing has been compiled by Ritchie (1969) and more recently by Matthews, Bordwell et al. (1975).

Additionally, examination of pK_a values in DMSO and mixtures of DMSO with hydroxylic solvents, as obtained by different methods, has revealed considerable variation. This is illustrated in Table 6 for two compounds frequently used as anchors for acidity function scales. If one is to correlate pK_a values with thermodynamic transfer functions [eqn (6)], rationalize varying orders of acidity in different solvents (Table 6), or use them in Brønsted relationships (Section 3),

TABLE 5

Acidities in DMSO and in Hydroxylic Media[a]

Acid	pK_a(DMSO)[b]	pK_a(solvent)
9-Carbomethoxyfluorene	10·3	12·9 (H_2O)[c]
p-Nitrophenol	10·4	11·2 (MeOH)
		7·15 (H_2O)
Malononitrile	11·0	11·1 (H_2O)
Benzoic acid	11·0	9·4 (MeOH)
Benzoylacetone	12·1	9·6 (H_2O)
Tris(p-nitrophenyl)methane	12·2	14·3 (H_2O)[c]
2,4-Dinitroaniline	14·8	14·7 (H_2O)[c]
Nitromethane	15·9	10·2 (H_2O)
9-Phenylfluorene	16·4	18·2 (H_2O)[c]
Indene	18·5	18·2 (H_2O)[c]
9-Cyanofluorene	8·4	14·2 (MeOH)
1,3-Difluorenylpropane	8·7	—
p-Nitrobenzoic acid	8·9	8·4 (MeOH)
p-Chlorobenzoic acid	10·1	9·1 (MeOH)
Fluoradene	10·5	>17 (MeOH)
m-Toluic acid	11·0	9·5 (MeOH)

[a] Ritchie, 1969.
[b] Measurements made by the potentiometric method (cf. Tables 6, 7).
[c] Measurements made by acidity function techniques.

TABLE 6

pK_a Values for Fluorene and 9-Phenylfluorene as Determined by Different Procedures[a]

	pK_a	
Procedure	Fluorene	9-Phenyl fluorene
Potentiometric[b]	20·5	16·4
Indicator (DMSO)[c]	22·6	17·9
Kinetic[d]	21·0	—
AF[e]	22·10,[f] 21·0[g]	18·38,[g] 18·59[g]
BO[e]	22·15	19·67
MCP[e]	22·35	20·24

[a] Note that different procedures can give pK_a values referring to different standard states.
[b] Ritchie and Uschold, 1967.
[c] Matthews et al., 1975.
[d] Earls et al., 1974.
[e] Cox and Stewart, 1975.
[f] Bowden and Cockerill, 1970.
[g] Bowden and Stewart, 1965.

it is essential to have certain criteria of their reliability. Some comment on the methods of measurement therefore seems appropriate.

The feasibility of measuring pK_a values in pure DMSO potentiometrically by means of the glass electrode was first demonstrated by Kolthoff and Reddy (1962). This method was used by Ritchie and Uschold (1967) in evaluating the pK_a values of organic acids up to *ca.* 28. As shown by these workers, solvent purity is of paramount importance in attaining success by this method.

The use of the dimsyl anion[2] (Corey and Chaykovsky, 1962) to titrate weak acids (using triphenylmethane as indicator) was first

TABLE 7[a]

Equilibrium Acidities of Carbon Acids in Dimethyl Sulphoxide[b]

Carbon Acid	$pK_a{}^b$	Carbon Acid	$pK_a{}^b$
9-Cyanofluorene	8·3	i-Propyl phenyl ketone	26·3
9-Carboxymethylfluorene	10·3₅	Acetone	26·5
Malononitrile	11·1	Diethyl ketone	27·1
Nitroethane	16·7	9-Phenylxanthene	27·9
Nitromethane	17·2	Phenylacetylene	28·8
9-Phenylfluorene	17·9[e]	Benzyl methyl sulfoxide	29·0
9-Methylfluorene	22·3	Methyl phenyl sulfone	29·0
Fluorene	22·6	Diphenyldiphenylmethane	29·4
Dibenzyl sulfone	23·9₅	Triphenylmethane	30·6
Ethyl phenyl ketone	24·4	Dimethyl sulfone	31·1
Acetophenone	24·7	Acetonitrile	31·3
1,3,3-Triphenylpropene	25·6		

[a] Matthews *et al.*, 1975.
[b] pK_a values determined by the indicator method (cf. Tables 5, 6).

introduced by Price and Whiting (1963). The method was subsequently modified as the indicator overlap technique by Steiner and co-workers (1965; 1967) and recently refined by Bordwell and Matthews (1974). At the low solute concentrations used in the Steiner-Bordwell procedure, the complicating effects of ion pair association in pK_a measurement (Szwarc *et al.*, 1974) are avoided. Values of pK_a for some carbon acids in DMSO, as determined by the indicator method, are given in Table 7 (Matthews, Bordwell, *et al.*, 1975).

[2] The ionization of CH_3SOCH_3 in basic systems has been studied by deuterium and tritium exchange (Buncel *et al.*, 1965; Albagli *et al.*, 1970) and by ^1H and ^{13}C nmr (Brauman *et al.*, 1970; Chrisment *et al.*, 1974).

A thermochemical method for comparing pK_a values of weak acids in DMSO has been described by Arnett et al. (1973). A linear relationship is observed between ΔG_i° and ΔH_i° for 30 weak acids in DMSO over a range of 20 pK_a units. Thus ΔH_i°, the enthalpy of ionization in DMSO, is a useful guide to relative acid/base strength (Arnett and Scorrano, 1976). The relationship implies, of course, that ΔS_i° is linearly related to ΔH_i°. Correlations have been drawn between acidities in DMSO and intrinsic gas-phase acidities (Arnett et al., 1975; Bordwell et al., 1975).

The use of pure DMSO for pK_a measurement has obvious advantages because of its unique properties.[3] Thus its high pK_a enables determination of acidities of very weak acids, and its large dielectric constant facilitates electrometric measurement. Nevertheless, a practical disadvantage of this system lies in the rigours associated with techniques using pure DMSO. These practical difficulties are considerably ameliorated in methods employing aqueous or alcoholic DMSO. Two such methods are available, as described below.[4]

A kinetic method for pK_a determination in aqueous DMSO has recently been developed (Earls et al., 1974 Cockerill et al., 1974). This is based on rates of isotopic exchange and is applicable to the pK_a range ca. 15–25. The method has the advantage that it does not depend on spectral changes occurring on ionization. However, the majority of pK_a values of weak acids have been obtained by the equilibrium technique, using acidity functions (AF) (Hammett, 1940; Bowden, 1966; Rochester, 1970).

Although the AF method has enjoyed wide usage for aqueous DMSO (Stewart and O'Donnell, 1964), aqueous DMF (Buncel et al., 1970) and other dipolar aprotic–protic solvent mixtures, recent results point to severe limitations inherent in the underlying assumptions of the procedure. These limitations become evident when one examines the relevant thermodynamic relationship (7), used in generating

$$pK_{HA}^S - pK_{HA_0}^S = pK_{HA}^0 - pK_{HA_0}^0 + \log \frac{{}^o\gamma_{A^-}^S \cdot {}^o\gamma_{HA_0}^S}{{}^o\gamma_{A_0^-}^S \cdot {}^o\gamma_{HA}^S} \qquad (7)$$

[3] The aspect of considering the unique properties of DMSO as a medium in performing measurements for pK_a evaluation needs to be considered separately from the advantages and disadvantages of using DMSO medium as the standard state (relative to water which has traditionally been used for this purpose; see also Table 6, footnote a).

[4] The reader is reminded that other methods for measurement of pK_a values of weak acids exist, notably Streitwieser's method using cesium (lithium) cyclohexylamide in cyclohexylamine, which has been used to measure both kinetic and thermodynamic acidities (Streitwieser and Hammons, 1965; Szwarc et al., 1974).

acidity functions by the overlapping indicator technique. In (7), HA_0 is the reference indicator acid, HA is the acid of interest, S refers to the solvent system under discussion and O refers to the standard solvent (usually water). Parker's (1969) notation is used for activity coefficients. Hammett (1940) originally assumed that the activity coefficient ratio $^o\gamma_{A^-}^S \; ^o\gamma_{HA_0}^S / ^o\gamma_{A_0^-}^S \; ^o\gamma_{HA}^S$ would be equal to unity, but this assumption breaks down if the acid indicators involved differ structurally (Hammett, 1970; Bunnett and Olsen, 1966; Yates, 1971; Arnett and Scorrano, 1975). If this ratio were indeed unity, pK_{HA}^S would be a linear function of pK_{HA}^O with unit slope. Recent measurements with indicators in DMSO are particularly pertinent in revealing the limitations of the above asumption and its implications.

Plots of pK_{HA} (DMSO) versus pK_{HA} (H_2O) are often linear, but their slopes may vary widely from unity. The slope takes a value of 1·4 for phenols in 80% DMSO–20% water (Kreevoy and Baughman, 1973), while in 98% DMSO–2% water the slope is 2·2 (Goitein and Bruice, 1972). For carboxylic acids in anhydrous DMSO the slope is 2·7 (Kolthoff et al., 1968). Straightforward application of the naive acidity function theory to data in these systems leads to highly inaccurate and misleading pK_a values. For example, benzoic and 2,6-dihydroxybenzoic acids have pK_a(DMSO) values of 11·1 and 3·1 respectively; $pK_a(H_2O)$ of benzoic acid is 4·2, so that if the activity coefficient term were unity, then $pK_a(H_2O)$ of 2,6-dihydroxybenzoic acid would be −3·8, rather than the known value of 1·2. Kreevoy and Baughman (1973) suggested that $^o\gamma_{A^-}^S / ^o\gamma_{HA}^S$ is a function of the acid strength as well as of the solvent composition and the structural type of the indicator.

Complementary results have been obtained by Albagli and coworkers (1973) in their investigation of the possibility of constructing an acidity function scale for sterically hindered phenols in basic DMSO–water mixtures. The results, shown in Fig. 1, indicate that a unique scale cannot be constructed over the whole range of solvent composition because the various ionization curves are not parallel. This is further evidence that the activity coefficient ratio can vary in complex manner with changing solvent composition.

In view of the difficulties mentioned above it is not surprising that statements such as "acidity function failure" have appeared in the literature in recent years. However, Cox and Stewart (1976) have delineated the conditions which are necessary for use of acidity function techniques for pK_a determination. Also the recent measurement of water activities in aqueous DMSO mixtures at 25° (Lam and

Benoit, 1974) has enabled them to use the linear free energy relationship (BO) approach of Bunnett and Olsen (1966), and the (MCP) modification by Marziano, Cimino and Passerini (1973), to evaluate pK_a (see also Levi *et al.*, 1974). Not surprisingly, it is found that those compounds whose ionization is closely governed by an acidity function give similar pK_a values by the BO, MCP and AF procedures. Appreciable differences result, however, for other compounds such as carbon acids, and the differences have been attributed to an incorrect anchoring of the AF scale for these acids. The chosen anchor, 9-phenylfluorene, had been assigned a pK_a value of 18·59 (Bowden and Stewart, 1965). However the BO and MCP methods give a value close to 20 (Table 6).

It should be noted at this point that there is a consistency in the pK_a values measured in pure DMSO by the various techniques. However, there is some uncertainty at present relating to pK_a values obtained by acidity function procedures in alcoholic and aqueous DMSO media. Thus one has the unexpected situation that theoretical analysis of the medium effect on pK_a is hampered because the values for a number of weak acids referring to the standard state in water are in doubt. Under these circumstances less than critical application of (6) to many weak acids, including some carboxylic acids, phenols as well as carbon acids, is inadvisable. Of course, (6) is strictly applicable to those cases in which the pK_a values can be measured in the pure hydroxylic solvents and also for those weak acids which obey the criteria outlined by Cox and Stewart (1976). Despite this difficulty there is now a large body of reliable pK_a data in both DMSO and water. Thus in principle it should be possible to account for variations, or reversals, in acidity order in terms of the thermodynamic transfer functions in (6).

It may be pointed out that the value of $\delta \Delta G_{tr}(H^+)$ from water to several dipolar aprotic solvents including DMSO has already been measured (see Table 1 and Table 24). It should thus be possible to ascribe pK_a differences to solvent interactions with the acid or its conjugate base if appropriate $\delta \Delta G_{tr}$ data are obtained; further measurement of these quantities is clearly desirable. The nature of these interactions, and their interpretation in terms of structural and electronic effects, will then have a firmer basis.

3. PROTON TRANSFER PROCESSES

General Considerations

Proton transfers are an integral part of acid and base catalysed reactions and as such they are important in chemical and biological processes (Jencks, 1969; Bender, 1971; Bell, 1973). DMSO has been found to influence many enzyme catalysed reactions, including hydrolytic processes, and its possible utility in enzymological studies has been noted (Rammler, 1971).

Studies of simpler organic reactions in DMSO or its protic mixtures have revealed that this solvent can affect the outcome of a proton transfer process in a variety of ways. Thus the stereochemical outcome of proton transfer can be retention, inversion, racemization, or a combination of these, depending on the presence or absence of DMSO (Cram, 1965; Buncel, 1975). Many such proton transfer processes are characterized by large rate enhancements due to DMSO. The rate of the methoxide-catalysed racemization of 2-methyl-3-phenylpropionitrile is increased by a factor of 10^8 on going from methanol to DMSO (Cram et al., 1959).

It is in order to enquire whether rate enhancements such as these are merely the consequence of changes in pK_a or reflect the involvement of the solvent in the transition state for proton transfer. Central to this problem of detection of such solvent involvement in the transition state is an understanding of probes such as the Brønsted equation, kinetic isotope effects, and certain related concepts.

The rates of proton transfer reactions cover a wide spectrum, from exasperatingly slow to diffusion controlled. Any theory which can rationalize this range has obvious merit. Such a rationalization is in fact accomplished, to a large degree, by Brønsted and Pedersen's (1923) relationship between rate (kinetic acidity) and pK_a (thermodynamic acidity). The relationship, known as the Brønsted equation, has the form (8) where k_B is the catalytic rate constant. The

$$k_B = G_B \left(\frac{1}{K_a}\right)^\beta \qquad (8)$$

corresponding relation for acid catalysis is (9). G_A and G_B are constants dependent on the particular system. The exponents α and β

$$k_A = G_A K_a^\alpha \qquad (9)$$

are known as the Brønsted coefficients, and have generally been assumed to have values between 0 and 1.

It has been common practice to equate the value of β with the degree of proton transfer in the transition state; β values close to 0 are taken to be indicative of reactant-like transition states and those close to 1 of product-like transition states. Any value outside these limits is inconsistent with this practice. Early investigators were only able to follow reactions within a limited rate constant range. With the development of fast reaction techniques (Eigen, 1964; Caldin, 1964) the predicted (Brønsted and Pedersen, 1923) curvature of the plots was fully established (cf. Bell and Lidwell, 1940). Pronounced curvature is in fact seen for fast proton transfers in DMSO (see p. 156).

The use of kinetic isotope effects (KIE) is recognized as a powerful tool in mechanistic investigations. Following Wilberg's (1955) review on deuterium isotope effects, an increasing number of cases came to light in which k_H/k_D values for apparently similar proton/deuteron transfer processes were found to vary over a wide range. It appeared logical to explore relationships between k_H/k_D and ΔpK, in view of the then current belief that both β and k_H/k_D provided a direct measure of the degree of proton transfer in the transition state. (ΔpK is the difference in the pK_a values of the proton-donating substrate and the conjugate acid of the attacking base.)

Westheimer (1961) presented theoretical arguments to show that for a series of related proton transfers the isotope effect will be a maximum when the transition state is symmetrical. The symmetry of the transition state has been related to ΔpK, with a maximum being expected for $\Delta pK = 0$. This variation in ΔpK can advantageously be obtained by changing the solvent rather than the reactants (commonly by substituent changes); DMSO is an ideal solvent for this purpose.

Symmetry of Transition States from Isotope Effects

In agreement with the Westheimer (1961) hypothesis, the results of early investigations showed an apparent maximum in k_H/k_D near $\Delta pK = 0$. However, Bell and Cox (1970) noted the considerable scatter in the plot of k_H/k_D versus ΔpK. They realized that for a wide variety of substrates and catalysts the transition state symmetry would not be uniquely related to ΔpK, i.e., for a given value of ΔpK there may be a varying degree of proton transfer in the transition state. An attempt was made to overcome this problem by keeping the reactants constant and effecting variation in ΔpK by varying the solvent (Jones and Stewart, 1967). This was done for the hydroxide ion catalysed inversion of (−)-menthone in aqueous DMSO over the range 0–75 mole % DMSO. From the known H_- values (H_- changes from 12 to 20; Dolman and Stewart, 1967), an estimate of ΔpK, the difference between the pK values for menthone and water in the various solvent mixtures, could be made. A maximum in k_H/k_D was observed at ca. 30–40 mole % DMSO. However the k_H/k_D values were not highly sensitive to these extreme changes in ΔpK_a. The detection of a maximum by this method is in qualitative agreement with a previous (Bell and Goodall, 1966) evaluation of the dependence of k_H/k_D on ΔpK. In the earlier study, various nitroalkanes and a series of nitrogen and oxygen bases were used to effect the ΔpK variation.

Much of the data supporting Westheimer's hypothesis was critically examined by Bordwell and Boyle (1975) and in their estimation found wanting. They redefined the maximum of the log k_H/k_D versus ΔpK curve of Bell and Goodall by taking into account secondary KIE's and data from other systems. The consequence of this treatment was that the slopes on either side of the maximum were much shallower. Bordwell concluded that either the ratio k_H/k_D is relatively insensitive to the symmetry of the transition state or that symmetry does not change over wide ranges of ΔpK.

In a later study, Bell and Cox (1971) used the same method of medium variation to effect changes in ΔpK. The rates of ionization of nitroethane and [1-^{2}H$_2$]nitroethane induced by hydroxide ion were measured in six aqueous solvents containing 0–58 mole % DMSO. The results are reproduced in Table 8 and shown graphically in Fig. 2. It should be noted that all the data of this later investigation fall on the left hand side of the curve. However, when combined with the earlier results of Bell and Goodall (1966), the predicted maximum appears. These results are not corrected for the secondary KIE, and also refer to more than one substrate and more than one base.

As in the case of Brønsted correlations, one can expect different quantitative behaviour in plots of this kind when the type of substrate and the base used is varied. It can readily be shown that superposition of all currently available data leads to large scatter in

TABLE 8[a]

Rates of Ionization of Nitroethane at 25°C

Mole Fraction DMSO	k_H/k_D	ΔpK
0	9·3	−7·1
0·13	8·4	−8·3
0·26	7·6	−9·5
0·37	6·9	−10·6
0·48	6·3	−11·4
0·58	5·9	−12·1

[a] Bell and Cox, 1971.

the plot of k_H/k_D versus ΔpK in which a sharply defined maximum is not apparent (Leffek, 1976; More O'Ferrall, 1975; see, however, Hanna et al., 1974).

A further point of interest regarding this problem has been raised by Bell et al. (1971). Calculations based on an electrostatic charge cloud model indicate that the variation in k_H/k_D is primarily determined by the tunnel correction. Different reactions will have different barrier widths, hence different tunneling probabilities, and, in the context of this hypothesis, different variations of isotope effects. The hypothesis still predicts, however, that for a given system k_H/k_D will have maximum value for the symmetrical transition state where the probability of tunneling is highest.

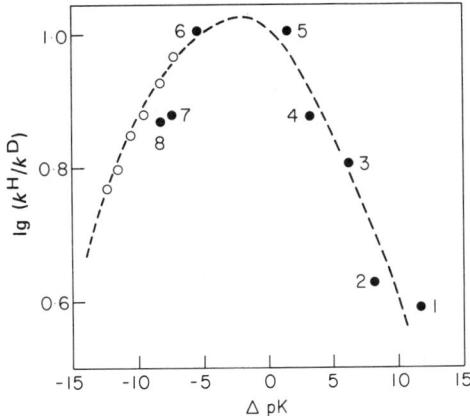

Figure 2. Relation between kinetic isotope effect and pK difference of the reacting systems: 1, MeNO$_2$ + H$_2$O; 2, MeNO$_2$ + CH$_2$ClCO$_2^-$; 3, MeNO$_2$ + MeCO$_2^-$; 4, Me$_2$CHNO$_2$ + MeCO$_2^-$; 5, Me$_2$CHNO$_2$ + pyridine; 6, MeNO$_2$ + OH$^-$; 7, nitrocyclohexane + OH$^-$; and 8, Me$_2$CHNO$_2$ + OH$^-$ (● Bell and Goodall, 1966; ○ Bell and Cox, 1971).

One of the criteria for tunneling in proton transfer reactions is the observation of abnormally large isotope effects. Among the largest is the k_H/k_D value of 45 for the reaction of 4-nitrophenylnitromethane with tetramethylguanidine in toluene (Caldin and Mateo, 1973). However, this isotope effect was found to depend on solvent polarity, decreasing to a value of 12 in the more polar acetonitrile. These results have been rationalized in terms of solvent involvement in the transition state in the case of the polar solvent, increasing the effective mass for passage across the energy barrier. This explanation is supported by the results for the racemization of 2-triphenylmethylpropionitrile in which the k_H/k_D value of 15 with t-BuOK/t-BuOH at 25° is reduced to 5 on addition of crown ether (Wong et al., 1971). In this context, the large KIE observed for the reaction of 2-nitropropane with 2,6-dimethylpyridine in the reasonably polar solvent aqueous t-butyl alcohol (Lewis and Funderburk, 1965) might seem anomalous. However, in this case the solvent may be excluded from the transition state as a consequence of steric hindrance.

The above and other recent results indicate that solvent effects and steric effects are both of importance in tunneling (Caldin and Mateo 1973; Lewis, 1975; Bell, 1974). Thus, if tunneling is primarily responsible for the variation of k_H/k_D, then the juxtaposition of these various factors, the width of the barrier and its symmetry, steric effects, and solvent involvement, would make it difficult to correlate isotope effects with a single parameter such as ΔpK.

Brønsted Coefficients and Solvents Effects

The Brønsted coefficient is generally obtained from the slope of a plot of log k_B versus pK_a. Variations in pK_a are usually brought about by changes in substituents in the base (acid) moiety. An alternative manner of effecting variation in pK_a, would be by changes in the reaction medium, as described in the previous section. The question arises, however, as to the equivalence of the two methods.

Cox and Gibson (1974) have provided data bearing on this question. They have measured rates for the acetate ion catalysed proton transfer reactions of $MeCH_2NO_2$, Me_2CHNO_2, and $MeCOCH(Me)CO_2Et$ in a series of DMSO-water mixtures, and of $MeCH_2NO_2$ in $CF_3CH_2OH\ H_2O$ mixtures. The pK_a values of the carbon acids, as well as of acetic acid, in the various solvent mixtures are known. The results are shown in Table 9.

TABLE 9[a]

Rates and Equilibria for Acetate-Catalysed Proton Transfer Reactions from Carbon Acids in Solvent Mixtures at 25°C

1. Me_2SO–H_2O mixtures.
  (i) $Me_2CH_2NO_2$; $k_0(OAc^-) = 2.70 \times 10^{-6}$ l mol^{-1} s^{-1}

x_{org}[b]	pK_{SH}	pK_{HOAc}	ΔpK[c]	log k_e/k_0[d]
0·00	7·70	4·75	3·0	0·00
0·20	–	–	–	0·98
0·39	11·0	7·1	3·9	1·82
0·50	12·1	8·0	4·1	2·30
0·69	–	–	–	3·36
0·83	–	–	–	3·37
1·00	16·2	12·0	4·2	–

  (ii) $MeCO \cdot CH(Me)CO_2Et$; $k_0(OAc^-) = 4.30 \times 10^{-3}$ l mol^{-1} s^{-1}

0·00	12·4	4·75	7·7	0·00
0·20	12·3	6·0	6·3	0·99
0·28	–	–	–	1·28
0·39	12·7	7·1	5·6	1·75

  (iii) $CH_3CH_2NO_2$; $k_0(OAc^-) = 3.26 \times 10^{-5}$ l mol^{-1} s^{-1}

0·00	8·8	4·75	4·1	0·00
0·20	10·2	6·0	4·2	1·14
0·39	11·7	7·1	4·5	1·96
1·00	16·4	12·0	4·4	–

2. CF_3CH_2OH–H_2O mixtures.
  $MeCH_2NO_2$; $k_0(OAc^-) = 3.26 \times 10^{-5}$ l mol^{-1} s^{-1}

0·00	8·8	4·75	4·1	0·00
0·08	–	–	–	–0·05
0·20	10·2	5·8	4·4	–0·30
0·32	–	–	–	–0·41
0·50	11·5	7·1	4·4	–0·60

[a] Cox and Gibson, 1974.
[b] Mole fraction of organic component.
[c] pK_{SH}–pK_{HOAc}
[d] k_e are catalytic constants in solvent mixtures, measured as rates of iodination.

Adherence to the Brønsted law would require a monotonic relationship between the rate constant ratio, log k_e/k_0, and ΔpK_a. This is clearly not the case. Moreover, the results show that β can be greater than unity or less than zero.

In these systems the anions have the negative charge largely concentrated on oxygen, so that they are suceptible to a much larger medium effect than the transition state, in which charge is dispersed. When the anions involved in a given acid-base equilibrium are comparably affected by the medium change, there will be little resulting effect on ΔpK_a, but there could well be large rate increases (see Section 2). This in turn implies a breakdown of the Brønsted relationship, and abnormal Brønsted coefficients are to be expected (Kresge 1973, 1975). Such results $(0 > \beta > 1)$ are not an artifact of this approach, but have been detected previously in studies using traditional methods (Bordwell and Boyle, 1971).

Curvature of Brønsted Plots and Medium Effects

As stated on p. 151, curvature in Brønsted plots has become apparent from studies of fast proton transfer processes. Fundamental differences in Brønsted plots have been observed for reactions of

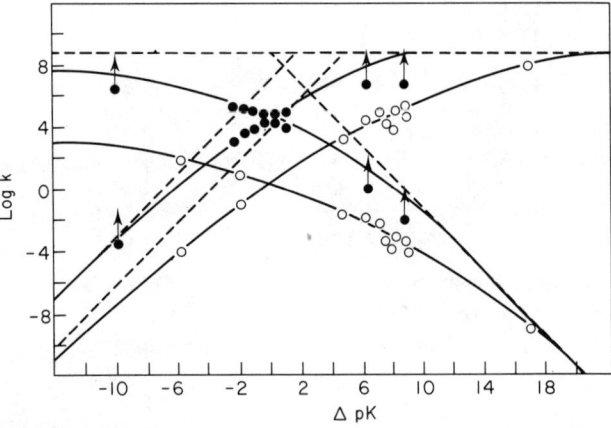

Figure 3. Extended plot of forward and reverse rates of reactions of hydrocarbons with anionic oxygen bases in methanol (open circles) and in DMSO (closed circles) (Ritchie and Uschold, 1968); The horizontal dashed line at the top of the figure indicates the diffusion-controlled limit.

carbon acids with anionic oxygen bases in methanol and in DMSO (Ritchie and Uschold, 1968). The results, illustrated in Fig. 3, show that in DMSO the proton transfer reactions approach diffusion control limits in both directions at much smaller ΔpK_a values than is the case in methanol. It is noteworthy that at the negative end of the scale (hydrocarbons more acidic than methanol), the rates in methanol appear to be levelling out considerably below the expected value for diffusion control.

Ritchie (1969) discussed this behaviour in terms of Eigen's (1964) three-step mechanism for proton transfer reactions[5] and interpreted

---

[5] For a theoretical (*ab initio*) study of the potential energy profile for proton transfer from NH_4^+ to NH_3, see Delpuech et al., 1972.

the differing results in the two media as evidence for solvent involvement in the proton transfer stage (10b).

$$AH + B \underset{k_{-a}}{\overset{k_a}{\rightleftarrows}} [AH\text{---}B] \qquad (10a)$$

$$[AH\text{---}B] \underset{k_{-b}}{\overset{k_b}{\rightleftarrows}} [A\text{---}HB] \qquad (10b)$$

$$[A\text{---}HB] \underset{k_{-c}}{\overset{k_c}{\rightleftarrows}} A + HB \qquad (10c)$$

Support for this contention of solvent involvement in the proton transfer comes from Marcus's theory (1968). The relationship (11)

$$\Delta G^{\ddagger} = \left(1 + \frac{\Delta G^0}{4\Delta G_0^{\ddagger}}\right)^2 \Delta G_0^{\ddagger} \qquad (11)$$

involves the free energy of activation, $\Delta G^{\ddagger}$, the standard free energy of reaction, ΔG^0, and the (hypothetical) intrinsic barrier, $\Delta G_0^{\ddagger}$. The intrinsic barrier is the free energy of activation for a reaction with $K = 1$. From the theory, the Brønsted coefficient is given by eqn (12)

$$\alpha = \frac{d\Delta G^{\ddagger}}{d\Delta G^0} = \left(1 + \frac{\Delta G^0}{4\Delta G_0^{\ddagger}}\right)\bigg/2 \qquad (12)$$

and the curvature, which is of interest in the present context, by (13).

$$\frac{d\alpha}{d\Delta G^0} = \frac{1}{8\Delta G_0^{\ddagger}} \qquad (13)$$

The curvature is large when the intrinsic barrier is low, which is in accord with the observation of curvature in fast reactions. The theory thus supports Kresge's (1973) statement that rapid proton transfers will give curved Brønsted plots and slow proton transfers will give linear plots, irrespective of the identity of the atoms involved.

The above treatment takes into account only the proton transfer step (eqn 10b) within the reaction complex. If the encounter and separation steps are also considered, some interesting results arise. Using the symbolism of (14) it can be shown that $\Delta G_{obs}^{\ddagger}$ and ΔG_{obs}^0 are given by (15) and (16) respectively.

$$AH + B \underset{\text{Encounter}}{\overset{W_r}{\rightleftarrows}} AH \cdot B \overset{\Delta G^0}{\rightleftarrows} A \cdot HB \underset{\text{Separation}}{\overset{W_p}{\rightleftarrows}} A + HB \qquad (14)$$

$$\Delta G_{obs}^{\ddagger} = \Delta G^{\ddagger} + W_r \qquad (15)$$

$$\Delta G_{obs}^0 = \Delta G^0 + W_r - W_p \qquad (16)$$

Equation (11) now becomes (17).

$$\Delta G^{\ddagger}_{obs} = W_r + \left\{1 + \frac{[\Delta G^0_{obs} - W_r + W_p]}{4\Delta G^{\ddagger}_0}\right\}^2 \Delta G^{\ddagger}_0 \qquad (17)$$

Analysis (Kresge, 1973) of several reactions in terms of this relationship has revealed that intrinsic barriers ($\Delta G^{\ddagger}_0$) are on the whole very small and the work terms (W_r, W_p) are consequently large. The work terms average 12 kcal mol^{-1}, whereas the diffusion-controlled formation of the encounter complex in aqueous solution requires only about 2·5 kcal mol^{-1}. It is reasonable to assign a large proportion of this work term to proper orientation of the reactants for proton transfer; reorganization of the solvent could be an important part of this orientation. The question arises as to whether it is correct to separate reagent positioning and solvent reorganization from the proton transfer. Do orientation and proton transfer occur in separate steps?

The Brønsted plots (Fig. 3) give information on this point. The higher curvature of the plot for DMSO compared to methanol is indicative of a lower intrinsic barrier to proton transfer for the dipolar aprotic solvent. Since in the extended Marcus theory the solvent effect has already been taken into account, one would expect the intrinsic barrier for proton transfer to be identical in the two systems. This is not the case. Therefore it appears that separation of the mechanism into reagent positioning with concomitant solvent reorganization is not warranted.

The reactants and products of the proton transfer stage (10b) can reasonably have different solvation requirements. The above conclusion would indicate that the change in solvation required would occur concurrently with transfer of the proton. The reacting species would thus be in equilibrium with the solvent molecules all along the reaction co-ordinate. On the other hand, the intriguing suggestion has been made by Kurz and Kurz (1972), on the basis of a mathematical model, that in the case of large bases a stepwise mechanism should be favoured for the proton transfer. First the solvent reorganizes to a configuration appropriate for the transition state, then the proton transfers, and finally the solvent relaxes to the product configuration.

Another approach used to interpret curvature of Brønsted plots has been given by Murdoch (1972). This model, which incorporates Marcus theory, shows that the diffusive steps (10a, c) of the three-stage Eigen mechanism can also influence curvature. It is shown mathematically that increased difficulty of diffusion has the same effect on curvature as lowering $\Delta G_0^{\ddagger}$. Thus the observed solvent effect on the proton transfer may merely reflect the higher viscosity of DMSO.

Using the Smoluchowski theory of diffusion controlled bimolecular reactions (Noyes, 1961), and making several simplifying assumptions, it can be shown that a diffusion controlled rate constant will be inversely proportional to the viscosity of the reaction medium. This viscosity dependence has been observed by the groups of Ivin *et al.* (1971) and Caldin (1974) in their experiments on the reactions of tertiary amines with phenols. In these papers a detailed analysis of the involvement of the solvent in the proton transfer was attempted. The results suggest diffusion-controlled formation of an encounter complex which then undergoes reorientation to form the reaction complex in the rate determining step. Viscosity effects, though important, are not the sole determining factor.

It may be noted that the acids (e.g. tetrabromophenolphthalein ethyl ester) and bases (e.g. tri-n-butylamine) used in this study are in general quite large. It would therefore be of some interest to attempt a correlation of the data with the Kurz model referred to above.

In conclusion, there is now strong evidence for the involvement of the solvent in proton transfer. However, current theories do not allow one to assess the importance of this factor in causing rate enhancements. Complete dissection of thermodynamic parameters such as would enable one to account for these effects has not been achieved so far for proton transfer processes. On the other hand, this has been done in some other reactions which are considered in the following section.

4. RATE VARIATIONS IN DMSO AS A GUIDE TO MECHANISM

DMSO has been used in a multitude of mechanistic investigations. In the present section we have chosen to highlight certain studies which illustrate the principles and methods discussed in earlier sections of this article. This will be done by reference to several contrasting situations. The systems chosen illustrate rate phenomena, both *retardation* and acceleration, resulting from use of DMSO. Various techniques for analysing these effects are presented, including the use of acidity functions and thermodynamic transfer functions, and their value as a guide to mechanisms demonstrated.

Some Bimolecular Nucleophilic Substitution Reactions

The reaction of 9-cyanofluorenyl anion (9-CFA, prepared from 9-cyanofluorene and sodium ethoxide) with alkyl and benzyl halides in ethanol is shown to be an $S_N 2$ reaction from a kinetic study (Bowden and Cook, 1968). Qualitative application of the Hughes-Ingold theory of solvent effects predicts that, for an $S_N 2$ reaction of this charge type, an increase in the dielectric constant of the medium will cause a small rate decrease. That should be the case if the reaction is studied in ethanol (dielectric constant, D, 24·2) and DMSO (D, 48·9) is added. The data for the 9-CFA alkylation are in accord with this prediction, as seen in Table 10.

Commonly, nucleophilic substitutions involving anions experience large rate increases when DMSO replaces a protic solvent (Delpuech,

TABLE 10[a]

Effect of Added Dimethyl Sulphoxide on the Rate Coefficients, k_2, for the Reaction of Benzyl Chloride with the 9-Cyanofluorenyl Anion at 35·7°

DMSO (mol %)	$10^2 k_2$ (l mol^{-1} s^{-1})	DMSO (mol %)	$10^2 k_2$ (l mol^{-1} s^{-1})
0·0	18·7	35·2	2·21
6·55	6·80	54·9	1·66
17·1	3·65	74·6	1·41

[a] Bowden and Cook, 1968.

1966; Parker, 1969). However, on p. 138 we commented on the contrasting behaviour of small and large anions on transfer from protic to dipolar aprotic solvent. In that context it was noted that $\delta \Delta G_{tr}(H_2O \rightarrow DMSO)$ for BPh_4^- is negative ($-8\cdot 8$ kcal mol^{-1}) whereas for Cl^- it is positive (9·2 kcal mol^{-1}). By analogy, one would expect 9-CFA to be similarly stabilized.

Before the effect of DMSO on the reaction rate in this alkylation can be rationalized, it is necessary to have some knowledge of charge distribution in the transition state. It was found that the relative rate coefficients for reaction of 9-CFA with the series of alkyl halides, R = Me, Et, Pr^i, cover a much smaller range than observed with other nucleophiles. This decreased selectivity is attributed to the high polarisability of the anion and its tendency to undergo significant orbital overlap with the electrophilic centre at a much greater distance than does a hard nucleophile. A symmetrical transition state in which a large degree of concerted bond making and breaking exists is postulated. This conclusion is supported by the observed curvature of the Hammett plot (log k/k_0 versus σ); both electron-withdrawing and releasing substituents increase the reaction rate of benzyl halides with 9-CFA. Qualitatively, one would expect the free energies of transfer for such a polarizable transition state and the 9-CFA to be comparable. Thus no rate acceleration is expected; the moderate observed retardation can be readily accounted for (see below).

Quantitative data are available for other systems which bear out the above discussion. The reactions of highly polarizable anions with methyl iodide in methanol and in DMF are characterized by large enthalpies of transfer of the transition state, as seen in Table 11. A

TABLE 11[a]

Enthalpies of Transfer ($\delta \Delta H^T$) of the Transition States of $S_N 2$ and $S_N Ar$ Reactions from Methanol to Dimethylformamide

Reaction	$\delta \Delta H_s$ kcal mol^{-1}	$\delta \Delta H^{\ddagger}$ kcal mol^{-1}	$\delta \Delta H^T$ kcal mol^{-1}
SCN$^-$ + CH$_3$I	−4·00	−4·3	−8·3
2,4-Dinitrophenoxide + CH$_3$I	−0·69	−6·4	−7·1
4-Nitrophenoxide + CH$_3$I	3·59	−7·7	−4·1
SCN$^-$ + 2,4-dinitroiodobenzene	−6·80	−0·45	−7·25
SCN$^-$ + 2,4-dinitrochlorobenzene	−5·26	−3·08	−8·34
N$_3^-$ + 4-nitroiodobenzene	−1·36	−5·6	−7·0
N$_3^-$ + 4-nitrofluorobenzene	−0·95	−4·8	−5·8

[a] Haberfield et al., 1969.

similar result is found for aromatic substitution reactions (Haberfield et al., 1969; Cook et al., 1966).

The data in Table 11 can be represented schematically for individual reactions as in Fig. 4. Such diagrams illustrate the factors responsible for rate enhancement. The reactions chosen illustrate cases where the effects either balance (Fig. 4a) or reinforce one another (Fig. 4b). The 9-CFA system discussed above appears to belong to the type shown in Fig. 4a, but with $\delta \Delta H$(reactants) slightly greater than $\delta \Delta H$(transition state). Of some interest is the case of the 4-nitrophenoxide ion reaction (Fig. 4b) in which the anion is destabilized, even though it is relatively large, while the transition state is stabilized. Apparently any charge localization on oxygen in the initial state is largely dissipated in the transition state.

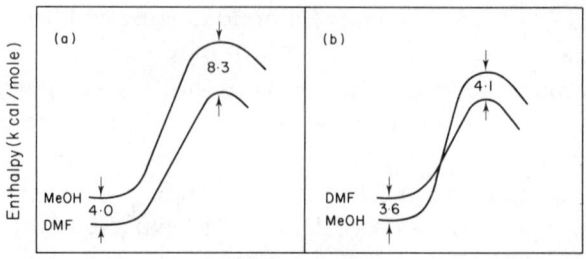

Figure 4. Schematic representation of the medium effect on the reaction of methyl iodide with (a) thiocyanate ion and (b) p-nitrophenoxide ion.

The contrast in the $\delta \Delta H_s$ value between 4-nitrophenoxide and 2,4-dinitrophenoxide ion (Table 11) is also noteworthy.

There is no need for the transition states to be charged for these effects to be observed. As noted in Section 2, dipolar aprotic solvents also interact strongly with polarizable neutral species. The data for Menschutkin reactions on transfer from methanol to DMF (Table 12;

TABLE 12[a]

Enthalpies of Transfer ($\delta \Delta H^T$) of the Transition States in the Menshutkin Reaction from Methanol to Dimethylformamide

Reaction	$\delta \Delta H_s$ kcal mol^{-1}	$\delta \Delta H^{\ddagger}$ kcal mol^{-1}	$\delta \Delta H^T$ kcal mol^{-1}
Pyridine + p-nitrobenzyl chloride	−0·95	−6·8	−7·8
Pyridine + benzyl chloride	0·03	−5·3	−5·3
Pyridine + p-methylbenzyl chloride	0·00	−3·7	−3·7
Pyridine + p-nitrobenzyl bromide	−1·45	−3·2	−4·7
Pyridine + benzyl bromide	−0·34	−1·7	−2·0
Pyridine + p-methylbenzyl bromide	−0·09	−1·4	−1·5

[a] Haberfield et al., 1971.

Haberfield et al., 1971) and a number of other media (Abraham, 1969) bear this out.

Use of the enthalpy data to account for rate differences implies that the solvent effects are more a consequence of enthalpy than entropy effects. This is supported by the results of Alexander et al. (1968). At present free energy data are available in only a limited number of cases (Abraham, 1969).[6]

[6] It should be pointed out, with respect to the kinetic studies discussed in this and in the remaining portions of Section 4, that the studies have generally not made allowance for the possible role of ion association effects (cf. pp. 182, 188).

Ester Hydrolysis

The reactions so far discussed in this section have been characterized by stabilization of the transition state on transfer from protic to dipolar aprotic media. In principle the opposite situation can also obtain, when the transition state is relatively stabilized by the protic medium. A good illustration of this effect is provided by alkaline ester hydrolysis, a reaction which has been the focus of a great deal of attention over the years (Euranto, 1969; Ingold, 1969).

Esters which hydrolyse by the $B_{Ac}2$ mechanism are subject to rate acceleration on solvent change from aqueous (alcoholic) to DMSO-containing media. The acceleration is more pronounced for aromatic than aliphatic esters (Tommila and Murto, 1963; Roberts, 1966; Balakrishnan et al., 1974). The rate increases have variously been rationalized by postulates such as desolvation of the lyate ion, increased solvation of the anionic transition state, or a catalytic effect of the DMSO. The relative importance of these various factors has been resolved by measurement of the pertinent thermodynamic parameters (heats of solution, etc.) by Haberfield (1972), Fuchs (1974) and their co-workers.

Enthalpies of transfer of reactants and transition states for the hydrolyses of ethyl acetate and ethyl benzoate are given in Table 13. A diagramatic illustration for one of these is shown in Fig. 5. It is clearly seen that a large portion of the desolvation energy for reactants is retained in the transition state. This indicates that there is considerable localization of charge on an electronegative atom in the transition state. This is in accord with rate-determining formation of a tetrahedral intermediate with negatively charged oxygen.

Knowledge of thermodynamic transfer parameters is not a prerequisite for the use of DMSO as a probe of transition state structure

TABLE 13[a]

Enthalpies of Solvent Transfer of Reactants, $\delta \Delta H_s^R$, and Transition States, $\delta \Delta H^T$ (kcal mol^{-1}) in Ester Hydrolysis

Reaction	Solvents	$\delta \Delta H_s^R$	$\delta \Delta H^T$
Ethyl acetate + OH$^-$	0·60 aq. DMSO → 0·60 aq. ethanol	−13·95	−10·0
Ethyl benzoate + OH$^-$	0·60 aq. DMSO → 0·85 aq. ethanol	−17·14	−11·0

[a] Haberfield, et al., 1972.

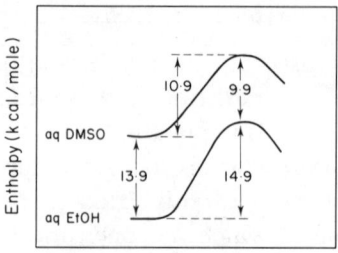

Figure 5. Schematic representation of the medium effect on the basic hydrolysis of ethyl acetate.

in those cases where modification of the substrate is examined. It is possible to correlate rate variation in different solvents with transition state variation if a constancy of reaction conditions is maintained. A case in point is the study by Balakrishnan et al. (1974a) on the hydrolysis of a series of esters (Tables 14–16).

For the alkaline hydrolysis of an ester in the two solvent systems, k_0 for aqueous ethanol and k_S for aqueous DMSO can be correlated by equation (18). For a series of esters under constant conditions of

$$\log \frac{k_S}{k_0} = \log {}^o\gamma_{Y^-}^S + \log {}^o\gamma_{RX}^S - \log {}^o\gamma_{YRX\ddagger}^S \tag{18}$$

solvent composition and temperature, the first term, $\log {}^o\gamma_{Y^-}^S$ (referring to the lyate ion), will be constant. Any change in the value of

TABLE 14[a]

Rate Data for the Alkaline Hydrolysis of Acetate Esters at 30°

	10^2 $k/1$ mol^{-1} s^{-1}		
Ester	70% EtOH (v/v)	70% DMSO (v/v)	k_S/k_0
Methyl acetate	32·5	714	22·0
Ethyl acetate	21·1	325	15·4
Isopropyl acetate	6·08	48·0	7·89
t-Butyl acetate	0·313	2·84	9·07
n-Butyl acetate	21·6	251	11·6
Cyclohexyl acetate	6·55	62·3	9·51
Cyclopentyl acetate	8·59	101	11·8
Isoamyl acetate	16·7	180	10·8
Allyl acetate	33·7	609	18·1
2-Methoxyethyl acetate	37·3	907	24·3
2-Ethoxyethyl acetate	32·9	776	23·6
Benzyl acetate	35·5	1090	30·7

[a] Balakrishnan et al., 1974a.

TABLE 15[a]

Rates of Alkaline Hydrolysis of p-Substituted Ethyl Benzoates at $25°$

Substituent	$10^2\ k/1\ mol^{-1}\ s^{-1}$		k_S/k_0
	0.69 M-DMSO	0.69 M-EtOH	
p-NO$_2$	4467	7.7	580
H	28	0.07	400
p-Cl	131	0.301	439
p-OMe	4.37	0.015	290

[a] Hojo et al., 1966.

log k_S/k_0 can thus be traced to changes in the $^o\gamma^S_{RX}$ value for the neutral ester and the $^o\gamma^S_{YRX\ddagger}$ value for the anionic transition state. Earlier studies on ester hydrolysis have indicated that substrate solvation changes are not of major importance in causing rate enhancements in dipolar aprotic solvents. In most cases the value of log $^o\gamma^S_{RX}$ tends to be ca. 0. Thus the observed changes in the values of k_S/k_0 are due largely to differential changes in transition state interactions in the two solvents.

The effect noted earlier that aromatic esters are subject to larger rate accelerations in DMSO, as evident from Tables 14 and 15, can now be explained. The transition state anion in the case of an aromatic ester will be more polarizable than that for an aliphatic ester. Moreover, the phenyl ring attached to the carbonyl group acts

TABLE 16[a]

Rate Data for the Alkaline Hydrolysis of Glycol Monobenzoates at $30°$

Ester	$10^3\ k/1\ mol^{-1}\ s^{-1}$		k_S/k_0
	80% DMSO (v/v)	80% EtOH (v/v)	
Ethylene glycol monobenzoate	626	0.625	1000
Propane-1,3-diol monobenzoate	192	1.08	177
2,2-Dimethylpropane-1,3-diol monobenzoate	219	0.986	220
Butane-1,4-diol monobenzoate	124	1.12	112
3,6-Dioxaoctane-1,8-diol monobenzoate	401	0.800	500
Ethyl benzoate	173	1.47	118

[a] Balakrishnan et al., 1974a.

as an electron sink, thereby decreasing the localization of negative charge on the carbonyl oxygen, with consequent decrease in the hydrogen bond acceptor capacity of the transition state. This will lead to a less positive value for $°\gamma_{RXY\ddagger}^{S}$, resulting in an increase of the k_S/k_0 value.

The transition state for the alkaline hydrolysis of a glycol mono-ester can be represented by [1] (Bruice and Fife, 1973). In

such a transition state the negative charge on the carbonyl oxygen atom is decreased through intramolecular hydrogen bonding. The high k_S/k_0 values for these substrates (Table 16) bear this out. Dispersal of the negative charge could also be achieved by electrostatic interaction with a positive centre as shown in [2] (Fuchs and Caputo, 1966). Thus enhanced k_S/k_0 values would be expected for such a system (Balakrishnan et al. (1974b).

It should be noted, before concluding this section, that the ester hydrolysis studies described above all refer to certain selected mixed media. If one investigates trends throughout the complete range of

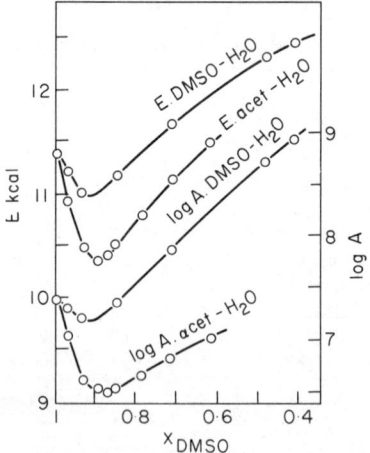

Figure 6. Variations of E and log A with solvent composition for the alkaline hydrolysis of ethyl acetate in DMSO–water mixtures (Tommila and Murto, 1963).

mixed solvent compositions, one often observes extrema in the various properties, as seen for example in Fig. 6 (Tommila and Murto, 1963). The cause of such behaviour is at present still speculative (Macdonald and Hyne, 1971; Symons, 1971). In general, comparative studies avoid the solvent compositions at which the measured property undergoes abrupt changes.

Reactions Involving Proton Transfers

Use of acidity functions

Many base catalysed isotope exchange or racemization processes involve either rate-determining proton abstraction or rapid pre-equilibrium formation of the anion (19) followed by rate-determining reaction with solvent (20) (L is an isotope of hydrogen).

$$RH + B^- \underset{k_{-1}}{\overset{k_1}{\rightleftharpoons}} R^- + HB \qquad (19)$$

$$R^- + SL \underset{k_{-2}}{\overset{k_2}{\rightleftharpoons}} RL + S^- \qquad (20)$$

Several such studies demonstrated a linear correlation between log k_{obs} and an appropriate acidity function. The most commonly used acidity function in strongly basic systems is H_-, defined by (21),

$$H_- = -\log a_{H^+} \frac{\gamma_{A^-}}{\gamma_{HA}} \qquad (21)$$

which measures the proton abstracting ability of the medium towards a neutral, weakly acidic, substrate AH. In (21), a_{H^+} is the hydrogen ion activity, while γ_{A^-} and γ_{HA} are activity coefficients. H_- has been evaluated in aqueous and alcoholic media using various alkali-metal hydroxides and alkoxides (Bowden, 1966; Terrier et al., 1969; Rochester, 1970). However a much larger range of H_- values is available by introducing a dipolar aprotic solvent such as DMF or DMSO.[7] In Fig. 7 are reproduced the H_- values for aqueous and

---

[7] The increased basicity of these systems with increasing proportion of the dipolar aprotic component is due only in part to the progressive desolvation of hydroxide ion. An additional factor is the reduced water activity in such systems, and recent data indicate that this latter factor is the more important one for the DMSO-water system (Cox and Stewart, 1975).

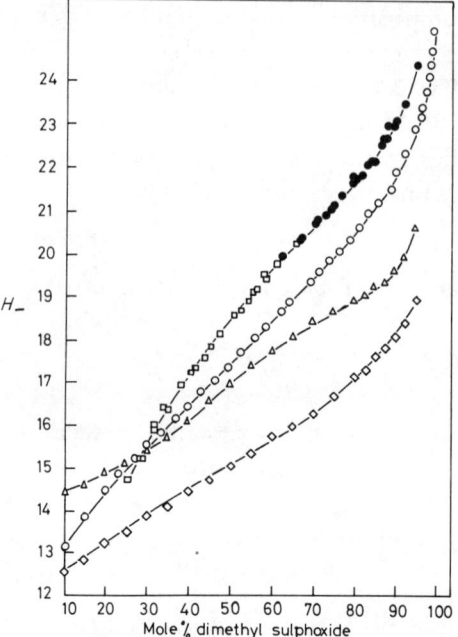

Figure 7. Variation of H_- for hydroxide, methoxide and ethoxide solutions as a function of dimethyl sulphoxide concentration. ●, Data of Bowden and Cockerill (1970): OH^-—H_2O—Me_2SO; □, data of Cockerill and Lamper (1971): OH^-—H_2O—Me_2SO; ○, data of Dolman and Stewart (1967): OH^-—H_2O—Me_2SO; △, data of Bowden and Stewart (1965): OEt^-—$EtOH$—Me_2SO; ◇, data of Stewart et al. (1962): OM^-—$MeOH$—Me_2SO. (Jones, 1973.)

alcoholic DMSO mixtures containing a constant concentration (0·01 M) of hydroxide or alkoxide ion (Jones, 1973).

The possible use of log (rate) $- H_-$ correlations as criteria of mechanism has been explored by a number of authors. For reactions such as (19) and (20) it can be shown (Kollmeyer and Cram, 1968) that relationship (22) will hold. A linear correlation between log

$$\log k_{obs} = H_- + \log \frac{\gamma_{RH}\gamma_{B^-}}{\gamma_{HB}\gamma_{\mp}} + \text{const.} \qquad (22)$$

k_{obs} and H_- would be expected if the activity coefficient term in (22) is zero, a constant, or a linear function of H_-. Under these conditions (22) reduces to (23).

$$\log k_{obs} = lH_- + C \qquad (23)$$

Interpretations of the slopes, l, in terms of the extent of proton transfer in the transition state have been attempted (Kollmeyer and

Cram, 1968; Albagli *et al.*, 1970; Kroeger and Stewart, 1970; More O'Ferrall, 1972; Streitwieser *et al.*, 1972; Jones, 1972). An analogy between this procedure and the Brønsted correlation (Section 3) is apparent and the same limitations will apply (Bowden and Cook, 1972). Nevertheless, some useful relationships are revealed from comparison of such slope values; Table 17 lists data for some carbon acids in the DMSO—H$_2$O—OH$^-$ system (Earls *et al.*, 1974).

It is seen from the data of Table 17 that, in general, there is an increase in slope with decreasing acidity of the carbon acids. However, there are some anomalies, as for example the case of nitroethane which is clearly out of line. Carbon acids of similar structures must be grouped together, as in the case of Brønsted

TABLE 17[a]

Slopes of log k vs. H_- Correlations for Various Carbon Acids[b]

Carbon Acid	pK_a	Slope
9-Phenylfluorene	18·5	0·40
Fluorene	22·1	0·56
9-t-Butylfluorene	23·4	0·74
Acetophenone	21·5	0·47
2-Phenylpropiophenone		0·49
(−)-Menthone	~21	0·48
Chloroform		0·98
1,4-Dicyanobut-2-ene	~21	0·71
Nitroethane	8·6	0·72
DMSO	~32	0·93

[a] Earls *et al.*, 1974.
[b] All results refer to DMSO—HO$^-$—H$_2$O.

correlations. In addition it should be noted that the H_- scale used in this Table is based on aromatic amine indicators. It would of course be preferable to use scales based on the respective carbon acids but such scales are at present not as firmly anchored (see p. 144).

In principle it should be possible to use the acidity function technique to distinguish between a rate-determining step involving hydroxide ion addition and one involving proton abstraction. In the former case log k_{obs} should correlate linearly with $H_- + \log a_{H_2O}$ (i.e. J_-) and in the latter case with H_-, within the limits of the approximations involved (Rochester, 1966; Terrier and Schaal, 1967; Kroeger and Stewart, 1970; Rochester, 1970). However, such a distinction is dependent upon the type of relationship existing between H_- and log a_{H_2O} for the given solvent system. This relation

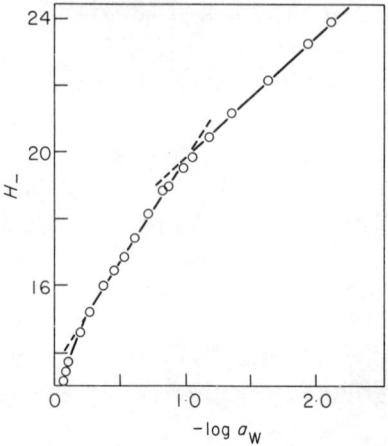

Figure 8. Relation between log a_W and H_- for aqueous DMSO (Bowden and Cook, 1971a).

is shown in Fig. 8 (Bowden and Cook, 1971a) for aqueous DMSO and reveals that, unless measurements are made over a wide range of basicities straddling $H_- = 19 \cdot 6$, linear correlations will be observed between log k_{obs} and *both* these functions. An illustrative example is given by Fig. 9 for the reaction of 1-chloro-4-nitrobenzene with base (Bowden and Cook, 1971a). The small deviations from linearity at high basicities are recognized as insufficient for this to be used as a criterion of mechanism.

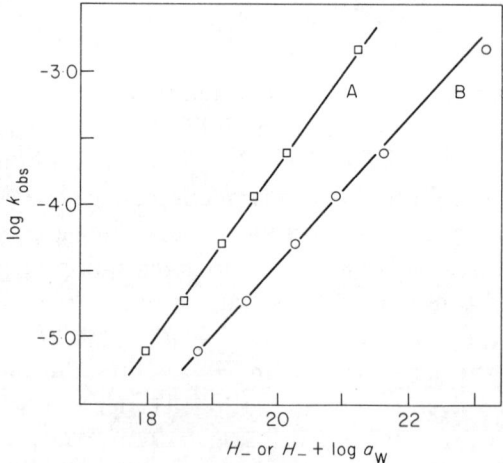

Figure 9. Relations between log k_{obs} for the hydrolysis of 1-chloro-4-nitrobenzene with (A) $H_- + \log a_W$, and (B) H_- (Bowden and Cook, 1971a).

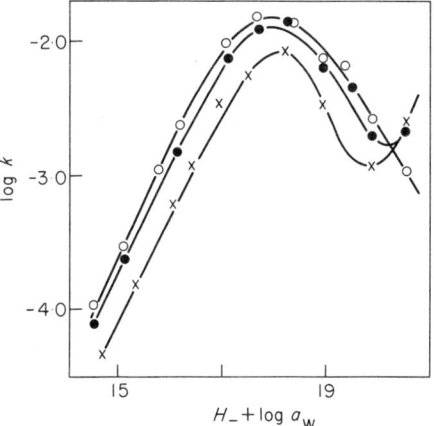

Figure 10. Relation between log k_{obs} for the hydrolysis of 1-halogeno-2,4-dinitrobenzenes with $H_- + \log a_W$. ○, Cl; ●, Br; ×, I (Bowden and Cook, 1971b).

A further complication in correlating rate coefficients with acidity functions is that in certain cases non-linear plots are observed. A startling example of this was found in the reactions of a series of 1-substituted 2,4-dinitrobenzenes with hydroxide ion in aqueous DMSO, as shown in Fig. 10 (Bowden and Cook, 1971b). This

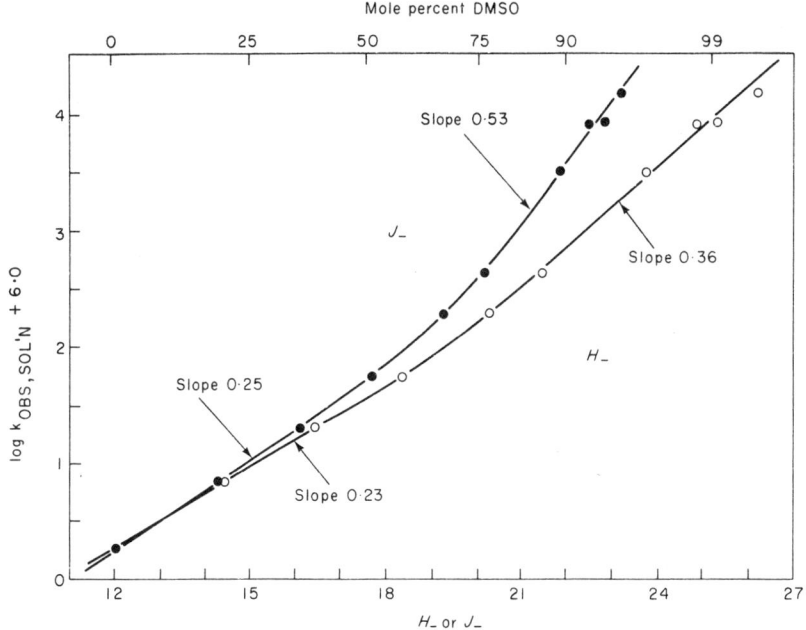

Figure 11. The medium dependence of the first order, in-solution rate constants for D_2 exchange at 65° (s^{-1}) on the $H-$ and $J-$ acidity functions (Symons and Buncel, 1973).

behaviour was a clear indication of mechanistic complexity, and was found to be caused by Meisenheimer adduct formation. In other cases deviations are less marked and can occur because of the breakdown of the approximations involved in deriving (23) or by use of an inappropriate acidity function.

The correlation of log k_{obs} with H_- or J_- for the isotopic exchange of molecular hydrogen (Symons and Buncel, 1973) illustrates the difficulties that can arise in such procedures. The relevant plots, shown in Fig. 11, do not readily distinguish between the mechanism involving proton transfer (Wilmarth, 1953) and that in which an addition complex is formed (Ritchie and King, 1968).

In summary, rate-acidity function correlations in basic systems appear to be quite informative when the acidity function used is based on an indicator system closely resembling the reactant under study. At present this requirement is only rarely met, and consequently the acidity function technique is best used in conjunction with other mechanistic criteria.

Use of thermodynamic transfer functions: D_2-OH^- exchange

The isotopic exchange between molecular hydrogen and a hydroxylic solvent under base catalysis poses a rather intriguing mechanistic problem. The two main mechanisms that had been proposed (Symons and Buncel, 1972) involve rate determining

$$HO^- + D-D \xrightarrow{slow} HOD + D^- \qquad (24a)$$

$$D^- + HOD \xrightarrow{fast} DH + OH^- \qquad (24b)$$

proton transfer (24a, b) or formation of an addition complex (25a, b). These two mechanisms could in principle be differentiated by

$$\underset{\underset{H}{|}}{HOH\text{---}O^-} + D-D \xrightarrow{slow} HOH + [HO-D-D]^- \qquad (25a)$$

$$[HO-D-D]^- + HOH \xrightarrow{fast} HOD + DH + OH^- \qquad (25b)$$

correlation with H_- or J_- respectively, as outlined in the preceding section. However, the plots of log k_{obs} versus H_- and J_- (Fig. 11) exhibit slight curvature and the approximate slopes have small magnitudes. Evidently no distinction between the mechanisms can be made as a result of these plots. Some of the limitations inherent in this procedure were indicated above. Especially noteworthy in this case is the extreme difference in structure of the substrate and the indicators used in setting up the H_- acidity function scale.

The rate constant for exchange increases by ca. 10^4 on changing the medium composition from purely aqueous to 99·5 mole % DMSO at 65°. It is significant that this increase in rate is considerably less than observed in many other reactions for which the medium effect has been evaluated. Analysis in terms of the thermodynamic and kinetic transfer functions gives information about the origin of the observed medium effect.

In Table 18 are presented the relevant data, derived from measured values of enthalpies of activation and available literature

TABLE 18[a]

Enthalpies of Transfer (kcal mol^{-1}) for Reactants and for the Transition State of the D_2–OH^- Exchange Process in the DMSO–H_2O System

Mol % DMSO	$\delta\Delta H_{tr}^{D_2}$	$\delta\Delta H_{tr}^{OH^-}$	$\delta\Delta H_{tr}^{R}$	$\Delta H^{\ddagger}$	$\delta\Delta H^{\ddagger}$	$\delta\Delta H_{tr}^{T}$
0	0	0	0	24·0	0	0
10	1·2	−0·8	0·4	—	—	—
20·2	2·1	+3·7	5·8	21·2	−2·8	3·0
40·1	2·6	10·1	12·7	17·7	−6·3	6·4
59·0	2·2	13·8	16·0	16·4	−7·6	8·4
77·9	2·1	16·0	18·1	16·9	−7·1	11·0
87·5	2·1	16·9	19·0	16·8	−7·2	11·8
96·9	2·1	17·5	19·6	18·1	−5·9	13·7

[a] Buncel and Symons, 1976.

data for the enthalpies of transfer of hydrogen (Symons, 1971) and hydroxide ion (Fuchs et al., 1974). It is apparent that there is a close parallel between the transfer enthalpies for the reactants, $\delta\Delta H_{tr}^R$, and those for the transition state, $\delta\Delta H_{tr}^T$. There is some similarity with the alkaline ester hydrolysis considered earlier, in which case the transition state enthalpy transfer is also endothermic, though not nearly to the same degree as in the present system. Evidently the destabilization of OH^- (noting that $\delta\Delta H_{tr}(OH^-)$ is the major component of $\delta\Delta H_{tr}^R$ in DMSO rich media) is largely retained in the transition state.

The above results point to a rate-determining transition state with considerable charge localization on an electronegative atom. Any proposed mechanism must be in accord with this conclusion. The relative merits of various mechanisms in the light of this finding are considered in detail elsewhere (Buncel and Symons, 1976).

5. ROLE OF DMSO IN SELECTED MECHANISM STUDIES

The use of DMSO as a mechanistic tool is not restricted to rate variation effects (Section 4). Advantage can also be taken of its unique molecular properties which enable it to stabilize certain types of structures, such as the anionic intermediates in $S_N Ar$ reactions. Moreover, as a consequence of its influence on ion association constants, it is found to affect the product distribution and the stereochemical course of bimolecular olefin-forming eliminations. These two illustrative systems which have been chosen for discussion are intended to demonstrate the versatility of this solvent in mechanistic studies and may suggest other avenues of investigation.

Stabilization and Catalytic Decomposition of Intermediates in $S_N Ar$ Reactions

It is now generally accepted that nucleophilic substitutions in aromatic substrates containing suitably placed (*ortho, para*) electron withdrawing groups (EWG) proceed by the addition-elimination mechanism (26). This mechanism was originally inferred (Bunnett

$$\underset{\substack{\text{EWG} \\ (o, p)}}{\text{Ar-X}} + Y^- \rightleftharpoons \underset{\substack{\text{EWG} \\ (o, p) \\ [I]}}{\text{Ar(X)(Y)}^-} \rightleftharpoons \underset{\substack{\text{EWG} \\ (o, p)}}{\text{Ar-Y}} + X^- \qquad (26)$$

and Zahler, 1951; Miller, 1951) from traditional reactivity studies which emphasized, for example, variations in the nature of X and Y (see Bunnett, 1958; Miller, 1968). The then novel approach of investigating such reactions in dipolar aprotic solvents (Miller and Parker, 1961) gave results which suggested that the intermediate [I] may be stabilized to a sufficient degree in such a media as to be observable.

The covalent adducts such as [3] and [4] which were known to be formed from 2,4,6-trinitroanisole and 1,3,5-trinitrobenzene, respectively, and methoxide ion in methanol (Meisenheimer, 1902) served as models for [I]. It can be noted that conclusive proof of the structure of [3] and [4] was obtained from nmr spectra of the

[Structures [3] and [4] shown]

complexes generated in DMSO medium (Crampton and Gold, 1964). The equilibrium constant for formation of these complexes in DMSO is greatly enhanced relative to a protic solvent. This property of DMSO was utilized over the next decade for the isolation and characterization of a host of complexes of different structural types, e.g. [5] and [6]. A number of reviews highlighting such complexes

[Structures [5] and [6] shown]

are available (Foster and Fyfe, 1966; Buncel et al., 1968; Buck, 1969; Crampton, 1969; Strauss, 1970; Hall and Poranski, 1970; Strauss, 1974).

The existence of such complexes does not prove that reaction (26) proceeds via a discrete intermediate I, rather than by a synchronous displacement mechanism. However, studies of aminolysis reactions have given kinetic evidence for the existence of an intermediate on the reaction path and also revealed mechanistic complexities which are not apparent from (26). Thus it has been shown that decomposition of the zwitterionic intermediate [I'] can proceed by both the catalysed and uncatalysed processes shown in (27). The techniques

$$\text{ArX(NO}_2)_2 + R_2NH \underset{k_{-1}}{\overset{k_1}{\rightleftharpoons}} [\text{I}'] \underset{k_3[B]}{\overset{k_2}{\rightleftharpoons}} \text{ArNR}_2(NO_2)_2 + H^+ + X^-$$

(27)

used to detect this catalysis, the factors favouring it, and the mechanism of the transformation of [I'] to product are of considerable interest (Pietra, 1969; Ross, 1972; Bernasconi, 1973). Recent work on this problem indicates, moreover, that even (27) may not provide an adequate description of the operative catalytic pathways.

The first study of a nucleophilic aromatic substitution in which formation of a Meisenheimer-type complex and its subsequent decomposition were *separately observable* was reported by Orvik and Bunnett (1970). The study involved the reaction of 2,4-dinitro-1-naphthyl ethyl ether [7] with n-butyl- and t-butylamine in DMSO. The use of DMSO in this kinetic study enabled the rate behaviour to be unambiguously interpreted by avoiding complications due to aggregation phenomena, while stabilizing any σ-complexes which are formed. The reaction sequence is given in equation (28). In this

(28)

scheme it is proposed that there is a fast acid-base equilibrium between the zwitterionic intermediate [8] and its conjugate base [9] [a step which is not specifically given in (27)] followed by a rate-limiting general acid catalysed expulsion of the ethoxy group. The transition state for this rate determining step is represented in [12]. The final stage is a rapid deprotonation of [10] with formation of the coloured, delocalized, anion [11]. This mechanistic scheme for the decomposition of zwitterion [8] has been termed the SB-GA (specific base-general acid) mechanism (Bernasconi, 1973). Certain other suggested pathways for the base catalysed step, such as an E2-type process, are inconsistent with the results of this study.

The proposal of the general-acid catalysed expulsion of the leaving group (28) raises the question of possible catalysis by other electrophilic species. No definitive case of catalysis by metal ions has been reported (Lam and Miller, 1966; Giles and Parker, 1970), which may be a consequence of competition from the uncatalysed step which is believed to occur via the entropically favoured transition states [13] or [14] (Kirby and Jencks, 1965; Bernasconi, 1973).

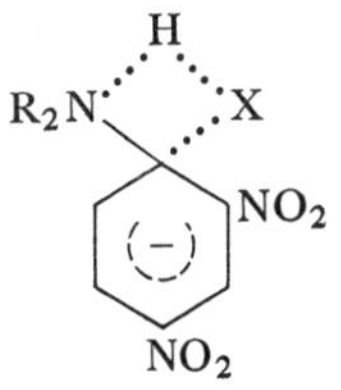

[13]

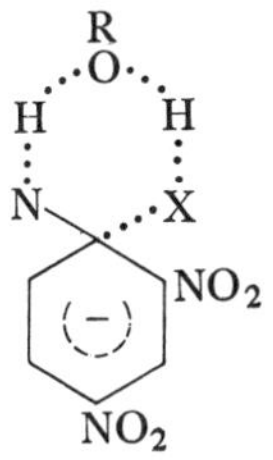

[14]

The search for electrophilic catalysis could in principle be aided by use of tertiary amines as nucleophiles, in which case an intramolecular or solvent assisted proton shift, formulated in [13] and [14], is eliminated. Thus Ayediran, Bamkole and Hirst (1974) studied salt effects on the reactions of 4-fluoro- and 4-chloronitrobenzene with trimethylamine in DMSO, which proceed according to equations (29) and (30). It had been noted (Suhr, 1967) that the

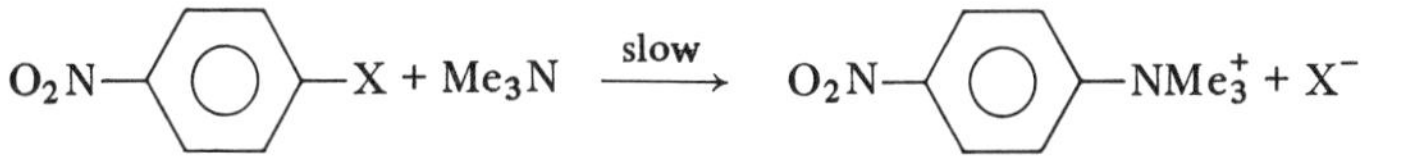

$$O_2N-C_6H_4-X + Me_3N \xrightarrow{slow} O_2N-C_6H_4-NMe_3^+ + X^- \quad (29)$$

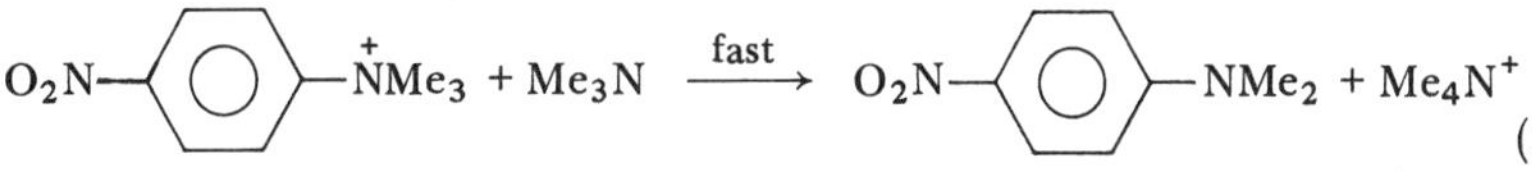

$$O_2N-C_6H_4-\overset{+}{N}Me_3 + Me_3N \xrightarrow{fast} O_2N-C_6H_4-NMe_2 + Me_4N^+ \quad (30)$$

fluoro derivative reacts more slowly than its chloro analog, from which it was inferred that, for the fluoro compound at least, the decomposition of the zwitterionic intermediate is kinetically significant.

The results obtained by Ayediran et al. (Table 19) show that the second-order rate constants for reaction of both substrates are

TABLE 19[a]

Rate Constants (l. mol^{-1} s^{-1}) for the Reactions of 4-X-Nitrobenzenes with Trimethylamine in DMSO at 120°C

X	10^2 [amine] /M	7·69	10·8	14·0	16·2	17·9	21·2	27·0	30·0	40·0
Cl	10$^5 k_A$:	9·90	10·1	10·3	10·3		10·0		10·1	
F	10$^5 k_A$:	7·60				7·70		7·70		7·52

[a] Ayediran et al., 1974.

independent of [Me$_3$N]. However, it was found that added salts have a markedly different effect on the rates for the two substrates, as shown in Fig. 12. While the chloro compound is subject to a small positive salt effect, the fluoro derivative exhibits a curvilinear

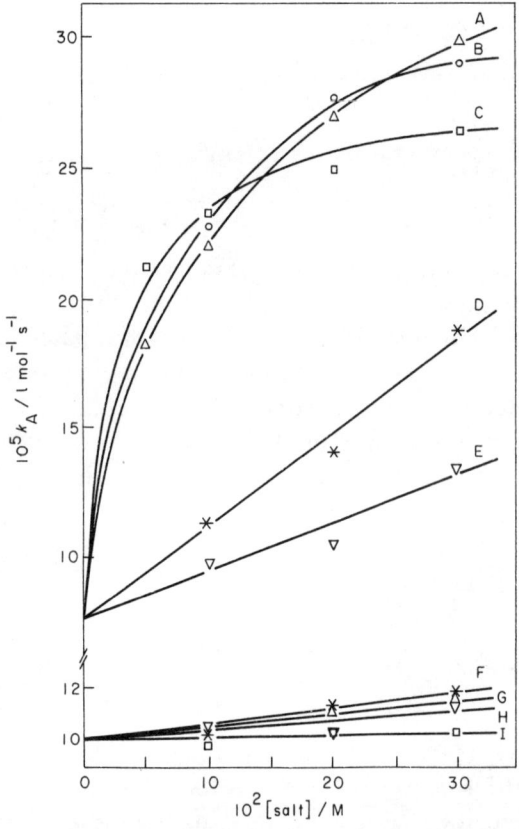

Figure 12. The influence of added salts on the reactions of trimethylamine with 4-X-nitrobenzenes in dimethyl sulphoxide at 120°C; curves A–E, X = F; curves F–I, X = Cl; A and G, Me$_3$NHCl; C and I, LiClO$_4$; D and F, KClO$_4$; E and H, Bu$_4^n$NClO$_4$; B, LiCl (Ayediran et al., 1974).

dependence of the observed second-order rate constant on salt concentration. Such dependence can be explained on the basis of electrophilic catalysis of the decomposition of the intermediate (31).

$$\underset{NO_2}{\underset{|}{\bigcirc}}\!\!\!\!\!-F + Me_3N \underset{k_{-1}}{\overset{k_1}{\rightleftarrows}} \underset{NO_2^-}{\underset{\|}{\bigcirc}}\!\!\!\!\!\overset{Me_3\overset{+}{N}\diagup F}{\diagdown} \overset{k_2}{\underset{k_3'[E]}{\diagdown\diagup}} \text{Products} \quad (31)$$

For this mechanism the rate eqn (32) can be derived and predict the curvilinear dependence of k_{obs} on the concentration of the electrophilic species.

$$k_{obs} = \frac{k_1 k_2 + k_1 k_3' [E]}{k_1 + k_2 + k_3' [E]} \quad (32)$$

Kinetic analysis of this system enables the data of Table 20 to be derived. The ratios k_3'/k_2 give a measure of the relative catalytic effectiveness of the cations since k_2 is independent of the electrophile. The resulting order, $Li^+ > K^+ > n\text{-}Bu_4N^+$, is similar to that reported for the electrophile-assisted ionizations of p-methoxyneophyl tosylate and the p-nitrobenzoate of spiro-[4,5]-deca-6,9-dien-8-ol in acetone (Winstein et al., 1964). The absence of these effects with the chloro compound is in accord with the observations that aromatic chloro derivatives are not usually subject to electrophilic catalysis of leaving group departure (Bernasconi, 1973).

Another aspect of the SB–GA mechanism (28) concerns proton removal from the initially formed zwitterionic intermediate. It was originally suggested (Bunnett and Randall, 1958) that proton transfer from such an intermediate could be rate-determining, but this was

TABLE 20[a]

Values of k_1 (l mol^{-1} s^{-1}) and k_3'/k_2 Derived from Salt Effects for the Reaction of 4-Fluoronitrobenzene with Trimethylamine in Dimethyl Sulphoxide at 120°C

Salt	$10^4 k_1$	k_3'/k_2
LiClO$_4$	2·8	100
LiCl	3·5	95
Me$_3$NHCl	3·6	48
KClO$_4$		5
Bu$_4^n$NClO$_4$		2

[a] Ayediran et al., 1974.

largely disregarded because of the apparent unlikelihood of observing slow proton transfers between electronegative atoms (Eigen, 1964).

Bernasconi and Gehriger (1974) have presented evidence in a recent study of spiro complex formation (33) which indicates that

[AH$_2^+$] ⇌ [AH] ⇌ [XH] ⇌ [X$^-$]   (33)

proton transfer may indeed be rate-determining in the aqueous medium employed. It is found that the rate of equilibration between the starting material, AH$_2^+$, and the two forms of the complex, [XH] and [X$^-$], shows the expected pH dependence, but in addition the

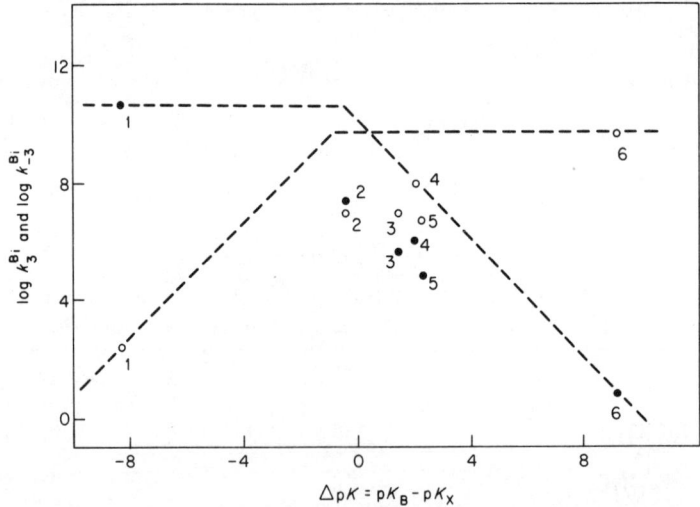

Figure 13. Eigen plot for proton transfer rates: ○, k_3^{Bi} (including $k_3^{H_2O}$ and $k_3^{OH^-}$); ●, k_{-3}^{Bi} (including k_{-3} and $k_{-3}^{H_3O}$); (1) H$_2$O; (2) phosphate; (3) Tris; (4) AH; (5) borate; (6) OH$^-$ (Bernasconi and Gehriger, 1974).

rate increases with buffer concentration in a curvilinear fashion. The data are interpreted in terms of Brønsted acid-base catalysis of the interconversion of [XH] and [X⁻] by the buffer species. The Eigen plot for these processes is shown in Fig. 13.

It therefore appears that in certain instances the observation of general base catalysis in aminolysis may be a consequence of rate-determining deprotonation of the zwitterionic intermediate, rather than of general acid catalysed removal of the leaving group (see also Jencks, 1972). The reaction of 1-fluoro-2,4-dinitrobenzene with N-methylaniline, originally investigated by Bunnett and Randall, may be such a case.

Kinetic study of the ring opening of the two spiro complexes [15] and [16] has also been reported (Crampton and Willison, 1974), and

[15]   [16]

the reaction is found to be subject to general acid catalysis in water. However, comparison with acyclic systems points to significant differences in the rate parameters. This suggests that extrapolation from spiro to acyclic analogs may not always be unambigous.

It may be noted that the above studies refer to an aqueous medium, but the possibility of rate-limiting proton transfer from a zwitterionic intermediate also exists in DMSO solution. This is indicated by the finding that σ-complex formation between trinitrobenzene and aniline in DMSO is catalysed by triethylamine or 1,4-diazabicyclo[2,2,2]octane (Buncel and Leung, 1975). A possible mechanism for the transformation is given in eqns (34) and (35) Buncel et al., 1976).

$$\text{O}_2\text{N-C}_6\text{H}_3(\text{NO}_2)_2 + \text{ArNH}_2 \rightleftharpoons [\text{complex with } \overset{+}{\text{NH}_2}\text{Ar and H}] \quad (34)$$

$$[\text{complex with } \overset{+}{\text{NH}_2}\text{Ar and H}] + \text{R}_3\text{N} \rightleftharpoons [\text{complex with NHAr and H}] + \text{R}_3\overset{+}{\text{NH}} \quad (35)$$

In conclusion, it now appears that the cause of base catalysis in the aminolysis of aromatic substrates may not be of common origin. Further investigations of the mode of decomposition of these zwitterionic intermediates and the analogous σ-complexes are clearly desirable. DMSO will remain central to these studies (Fendler *et al.*, 1975b), due to the possibility of isolation of such complexes in this medium. It now appears that DMSO may also be valuable for studies of electrophilic catalysis, since its use tends to avoid ion pairing effects which complicate kinetic studies in aprotic media of lower dielectric constant. The problem of ion pairing is considered further in the following section.

The Steric Course of Bimolecular Olefin-forming Eliminations

Recent work has revealed that base association phenomena can have a profound effect on the stereochemical outcome of olefin-forming eliminations. Changes in the degree of association can be brought about by solvent variation, and in the studies to be described media as diverse as benzene and DMSO have been utilized. Crown ethers (Pederson and Frensdorff, 1972) provide another means to this end.

It is noteworthy that the use of DMSO in this manner contrasts with its established usage in elimination reactions. DMSO has normally been used because of its ability to promote eliminations by weakly basic anions (Parker, 1971), its effectiveness in enhancing the basicity of alkoxides (Cram *et al.*, 1961), and also because it allows variation of base without the usual variation of solvent (Dolman and Stewart, 1967). Full details of such studies and their importance in the development of an understanding of elimination mechanisms are given in a recent text (Saunders and Cockerill, 1973).[8]

This new use of DMSO has not only yielded information relating to the recently exposed *syn-anti* dichotomy (Sicher, 1972), but has also revealed new facets of the century-old Hoffmann–Saytzeff orientation controversy.

---

[8] For interesting accounts of the "spectrum of transition states" which have been proposed for β-eliminations see also Bunnett (1969), Smith (1976), and Winey and Thornton (1975).

The syn-anti *dichotomy*

Until recently the commonly held view has been that any system having the possibility of free choice will prefer the *anti* over the *syn* elimination path. However, where steric constraints make *anti* elimination unfavourable as in cyclic systems, the normally difficult *syn* eliminations can occur.

In the case of acyclic compounds there is a strong preference for *anti* elimination, at least with short-chain alkyl halides. The first example of departure from this norm was observed (Pánková *et al.*, 1967) in the Hoffmann elimination from *threo*-[^{2}H]-[17] and *erythro*-[^{2}H]-[17] (Fig. 14). Formation of *trans*-[18] and *cis*-[18] [9]

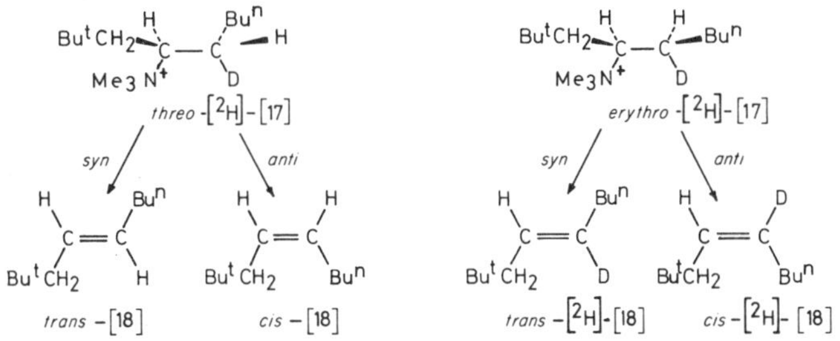

Figure 14. Reaction scheme showing the *syn* and *anti* elimination pathways (Pánková *et al.*, 1967).

from *threo*-[^{2}H]-[17] was accompanied by an appreciable isotope effect (k_H/k_D = 2·3–4·2 and 3·1–4·7 respectively) whereas for *erythro*-[^{2}H]-[17] there is no isotope effect. The results are in agreement with the view that *cis*-[18] arises by *anti* elimination and *trans*-[18] by *syn* elimination.

A bulky leaving group such as NMe_3^+ would tend to favour *syn* elimination by forcing the alkyl chain into a conformation hindering approach to the *anti*-β-hydrogen. It would thus be expected that a smaller leaving group such as fluorine should give much less *syn* elimination. It has also been proposed that ion-paired species would promote *syn* elimination in a neutral substrate whereas with a charged substrate free anions should be more effective, as indicated in the transition state structures [19] and [20]. To test these

---

[9] In addition, there was also obtained in very small amount the product of elimination originating from abstraction of the hydrogen *alpha* to the t-butyl group; this has no bearing on the *syn-anti* problem. The question of positional orientation will be considered in the following section.

hypotheses the stereochemistry of elimination reactions of 3-hexyl fluoride and tosylate were studied in a variety of solvent systems which involve different degrees of base association (Borchardt et al., 1974).

$$\underset{\underset{K^+ \cdots OR}{X\delta- \quad H}}{-C=C-} \qquad \underset{\underset{OR}{Me_3N\overset{\delta+}{}\overset{\delta-}{H}}}{-C=C-}$$

[19]　　　　　　　　　　　　[20]

The results with 3-hexyl fluoride and t-BuOK/t-BuOH (Table 21) show a substantial degree of *syn* elimination, indicating that a large leaving group is not an essential factor. It is seen that for the 3-hexyl tosylates undergoing elimination in DMSO the proportion of *syn* elimination is low; however significant variations do occur. Thus potassium phenoxide gives substantially more *syn* elimination than potassium *p*-nitrophenoxide; ion pairing is more extensive with the former (Zaugg and Schaeffer, 1965). The greater proportion of *syn* elimination with lithium phenoxide, which is even more extensively ion-paired, is a further indication of this effect. It is interesting that addition of Me$_4$NI suppresses *syn* elimination completely. This is in

TABLE 21[a]

Syn Elimination (%) from Erythro and Threo 3-Hexyl-4-d_1 Tosylate and Fluoride[b]

Leaving Group, X	Base	Reactant	d_0, % obsd.	% Syn Uncorr.	% Syn Corr.
OTs	PhOK	er	25·69	23·5	9·3[c]
OTs	p-NO$_2$PhOK	er	35·40	9·2	3·3[c]
OTs	PhOLi	th	13·39	18·1	16·5[d]
OTs	PhOLi + NMe$_4$I	th	9·64	ca. 0[g]	ca. 0[g]
F	t-BuOK	er	11·93	89·0	67·9[e]
		th	24·02	47·9	68·0[f]
F	n-BuOK	er	14·93	38·4	13·9[e]

[a] Borchardt et al., 1974.
[b] The conditions were 95% dimethyl sulfoxide–5% t-butyl alcohol at 60° for 24 hr with tosylates; t-butyl, and n-butyl alcohols, respectively, at 163° for 140 hr with the fluorides.
[c] Corrected for assumed $(k_H/k_D)_{anti}$ of 3·0.
[d] Corrected for assumed $(k_H/k_D)_{syn}$ of 2·2.
[e] Corrected for $(k_H/k_D)_{anti}$ of 3·84, calculated from data on this reaction.
[f] Corrected for $(k_H/k_D)_{syn}$ of 2·37, calculated from data on this reaction.
[g] Apparent slight increases in deuterium content of olefin relative to reactant.

accord with equilibrium (36) being shifted to the right (Borchardt and Saunders, 1972), thereby decreasing the proportion of ion-paired phenoxide.

$$[PhO^-, Li^+] + Me_4N^+ + I^- \rightleftharpoons [Li^+, I^-] + Me_4N^+ + PhO^- \tag{36}$$

Similarly, the proportion of *syn* elimination in the alkyl fluorides is found to be medium dependent. Thus the proportion of *syn* elimination brought about by t-butoxide ion decreases on going from benzene to t-butyl alcohol to DMSO (Pánková *et al.*, 1972).

An interesting demonstration of the importance of base association upon elimination stereochemistry is the β-elimination (37)

$$\underset{[23]}{\text{structure}} \longleftarrow \underset{[21]}{\text{structure}} \longrightarrow \underset{[22]}{\text{structure}} \tag{37}$$

from *exo*-2-norbornyl tosylate [21]. It had previously been reported (Brown and Liu, 1970) that, with the sodium salt of 2-cyclohexylcyclohexanol in triglyme, there is exclusive *syn-exo*-elimination yielding [22]. In the presence of crown ether, however, this stereoselectivity breaks down and a proportion (27%) of [23] is formed (Bartsch and Kayser, 1974).

Hoffmann–Saytzeff orientation

Logically it would be expected that ion pairing phenomena would also affect the product distribution in base-promoted β-elimination from 2-alkyl halides. In such systems one must distinguish between positional orientation, which refers to the relative proportions of 1- and 2-alkenes formed, and geometrical orientation, relating to the *cis* and *trans* 2-alkenes produced.

Thus it is found that the relative amounts of olefinic products which result from reactions of 2-bromobutane with MeOK/MeOH, EtOK/EtOH and t-BuOK/DMSO are insensitive to changes in the base concentration (Bartsch *et al.*, 1973a). In contrast, for eliminations induced by t-BuOK/t-BuOH, both positional and geometrical orientations are base concentration dependent. As the concentration of t-BuOK increases, a relatively higher proportion of 1-butene is

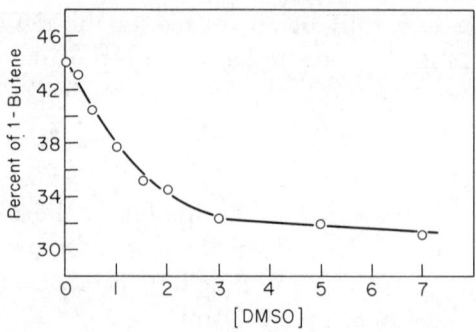

Figure 15. Percent of 1-butene formed in reactions of 0·1 M 2-bromobutane with 0·5 M potassium t-butoxide in *t*-butyl alcohol in the presence of added dimethyl sulfoxide at 50·0° (Bartsch *et al.*, 1973a).

produced, and the ratio of *trans:cis* 2-butene decreases. It is suggested that these changes in orientation result from a greater proportion of the elimination being induced by associated potassium t-butoxide relative to the free ions as the total base concentration is increased. In accord with this hypothesis, it was found that a lower percentage of l-butene and a higher *trans:cis* 2-butene ratio resulted when the proportion of dissociated base was increased, thus reversing the above trend. This effect was accomplished by addition of crown ethers, DMSO, hexamethylphosphortriamide, tetramethylenesulfone, or the use of tetramethylammonium t-butoxide as base.

Figures 15 and 16 illustrate the effect of addition of varying amounts of DMSO upon positional and geometrical orientation in reactions of 2-bromobutane with 0·5 M potassium t-butoxide in

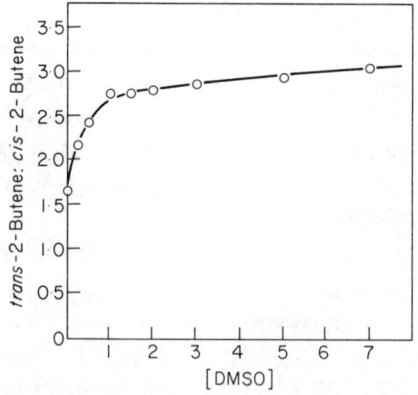

Figure 16. The *trans*-2-butene : *cis*-2-butene ratios for reactions of 0·1 M 2-bromobutane with 0·5 M potassium *t*-butoxide in *t*-butyl alcohol in the presence of added dimethyl sulfoxide at 50·0° (Bartsch *et al.*, 1973a).

t-butyl alcohol. Marked sensitivity of orientation to the amount of DMSO is observed until a concentration of 2-3 M is reached. Further increases in the DMSO concentration produce only very small changes. These results are readily understandable if the effect of the added DMSO is to increase the proportion of dissociated base.

One of the factors influencing orientation is base strength (pK_a). However, in order to evaluate the importance of this factor it is necessary to eliminate problems due to ion pairing. Whereas the extent of ion pairing for common alkoxides in the corresponding alcohols varies greatly, DMSO as solvent causes a levelling of these

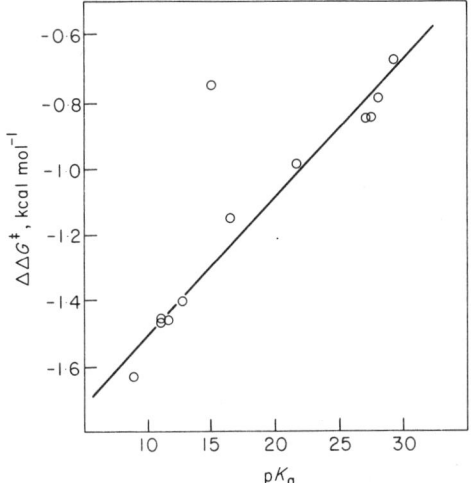

Figure 17. Plot of the free-energy difference for formation of 1-butene and *trans*-2-butene vs. pK_a for the conjugate acid of the base. The point showing marked deviation from the line corresponds to the sterically hindered base [24] (Bartsch *et al.*, 1973b).

effects with but little association (at least for the potassium salts, see below). Moreover, the use of DMSO has the added advantage that pK_a values are known in this medium (Section 2). Accordingly, it has been found that a linear relationship exists between the free energy differences of transition states for formation of terminal and internal olefins and pK_a values, as shown in Fig. 17 (Bartsch *et al.*, 1973b). Significant divergence is observed, however, for the sterically hindered base [24]: other sterically hindered bases, such as [25] have likewise been found to deviate markedly (Brown, 1974).

It is interesting to note that related elimination studies (Sera *et al.*, 1973; Alunni *et al.*, 1974), using different substrates (phenylsulfonyl acetates and 1-phenyl-2-propyl chloride), have revealed contrasting base association effects. Thus the effect of added crown ethers in these systems is to produce an opposite pattern to that discussed above. This is probably a consequence of fundamental differences in mechanism which points to the dangers of overgeneralization.

[24]   [25]

The use of DMSO in recent studies has been largely upon the premise that the problem of base association would be avoided. Some doubt as to the validity of this assumption arises when the results of a conductometric study are considered (Exner and Steiner, 1974). Ion-pairing constants for lithium, sodium, potassium and cesium t-butoxide in DMSO have been evaluated as 10^8, 10^6, 270 and 200 M^{-1} respectively. Not only do these results suggest that there is base association in DMSO but they also imply that base-catalysed reactions involving alkali metal alkoxides in DMSO should be affected by the nature of the cation. If these conclusions are valid and if the possible involvement of the dimsyl anion in these reactions is also taken into consideration, then the choice of DMSO to remove the problem of base association can be a poor one, especially if the base is a lithium or sodium salt of a hindered alkoxide. It is far better to avoid association effects by the use of crown ethers (Bartsch et al., 1973, 1974, 1975). On the other hand, the use of lithium and potassium t-butoxide in DMSO solvent might aid in distinguishing reactivities of free ions and of ion pairs in certain processes.

6. FUTURE DEVELOPMENTS

We have dealt in the previous sections with the use of DMSO in certain selected reactions in some depth. We wish to round off this article with brief consideration of a few systems to illustrate some possible future directions in this field. Thus, reference will be given to micellar catalysis, C versus O alkylation, hydride reactions, and the solvated electron, in the context of studies in DMSO.

The use of micellar systems as models for enzyme catalyzed reactions and hydrophobic interactions in biochemical systems has resulted in a great deal of research on micelle catalysed organic reactions (Fendler and Fendler, 1975). The goal of many of these studies has been to investigate the frequently observed substrate specificity and to determine the factors which influence the rate and hence the magnitude of the catalysis. In this latter regard a rather interesting correspondence has been observed between the results of base-catalysed hydrolysis of *p*-substituted phenyl acetates and N-methylacetanilides carried out in the presence of DMSO and the cationic micelle CTAB [$CH_3(CH_2)_{15}N^+(CH_3)_3Br^-$] respectively (Gani and Lapinte, 1973). DMSO has also been used in a more direct manner to determine the interactions in micellar solutions. thus changes in the chemical shifts of the magnetically discrete protons of hexylammonium propionate have been used to establish aggregation behaviour in DMSO-benzene and DMSO-water solvent systems (Fendler *et al.*, 1975a).

Dipolar aprotic solvents have been found to be of value for the selective alkylation of alkali metal salts of carbonyl compounds and of phenols (House, 1972). Various observations suggest that the reaction of a particular enolate anion with a specific alkylating agent gives the greatest proportion of O-alkylation when conditions (solvent, cation, temperature) are chosen so as to ensure the maximum amount of free enolate ion (as opposed to ion pairs or aggregates). In the case of alkylation of phenoxides, the O-product normally predominates in dipolar aprotic solvents of high dielectric constant such as DMSO (Kornblum *et al.*, 1963). However, the interaction between potassium phenoxide and 1,3,5-trinitrobenzene in DMSO-methanol (80:20) has yielded the product of C-alkylation [26] (Buncel and Webb, 1973). The O-alkylated product, [27], might indeed be formed in a kinetically preferred process, though in a reversible manner. However, formation of [26] from TNB should be irreversible, so that thermodynamic control leads to this product (Buncel et al., 1975). It is intriguing to speculate on the possibility of phenoxide ion acting as an ambident nucleophile in, say, the reaction with picryl chloride.

Reactions in which complex metallic hydrides (LiAlH$_4$, etc.) act as sources of nucleophilic hydride are of considerable synthetic value (Gaylord, 1956; Wiberg and Amberger, 1971). In the case of nitro-aromatics, some interesting hydride displacements [e.g. (38)] have

been observed in the reaction with NaBH$_4$ in DMSO (Lamson et al., 1973). These displacements plausibly proceed via intermediate hydride σ-complexes (e.g. [28]), for which the prototype TNB

σ-complexes (e.g. [30]) are known (Taylor, 1970; Kaplan and Siedle, 1973). It is noteworthy that the major product might have been expected to be the one formed via [29], since it is known that a *p*-nitro group is highly effective in charge stabilization of σ-complexes (Caveng et al., 1967). Possibly one is dealing once more with kinetic and thermodynamic factors in product formation. The parent σ-complex [31], involving just one nitro substituent, is implicated as a reversibly formed intermediate in a kinetic study of isotope exchange between tritium labelled sodium borohydride and nitrobenzene in DMSO solution at 25° (Gold and Nowlan, 1974). This reaction completes the trio of aromatic exchange processes occurring via cyclohexadienyl intermediates, the other two being cationic (Gold, 1964; Kresge and Chiang, 1967) and free radical (Adsetts and Gold, 1969).

[31]

Finally, we come to the solvated electron (e_{aq}^-). As the elementary nucleophilic species, it is not surprising that its properties in DMSO containing media should have come under scrutiny. In fact, evidence from pulse radiolysis studies in aqueous DMSO, including the spectral characteristics and kinetics of the reaction with N_2O in competition with scavengers, has shown that the solvated electron is but weakly trapped in the DMSO-water mixtures (Cooper et al., 1973; Koulkes-Pujo et al., 1974). The results show that above 90 volume % DMSO the electron is extremely reactive and may be considered as "dry". These findings are of significance in the context of the interaction of e_{aq}^- with organic and biological systems (Hart and Anbar, 1970).

7. CONCLUSION

It has been amply demonstrated that DMSO is a valuable solvent in mechanistic investigations. Also, there is every indication that the use of DMSO will continue as an extremely popular mechanistic probe. It may be hoped in this regard that the methods outlined in this article will lead to a more informed usage than has sometimes been the case in the past.

Matthews, Bordwell et al. (1975) have commented that DMSO is destined to achieve an importance as a solvent for quantitative work in organic chemistry comparable to that of water in inorganic chemistry. However, for this prediction to come to fruition, it should be emphasized that there is still a need to garner fundamental thermodynamic data relating to the transfer of solutes to DMSO from the aqueous state. This will be of even greater necessity if the effect of DMSO on biological phenomena is to be understood.

Thus, one can hope that much of our current qualitative discussion of measurements in DMSO media will be replaced by more concrete rationalization, based on direct evidence of solvent-solute interactions at the molecular level.

ACKNOWLEDGEMENTS

Support of our research by the National Research Council of Canada is acknowledged with thanks.

We gratefully acknowledge the permission granted by copyright holders to reproduce the Tables 1-4 and 7-24, and the Figures 1-3 and 6-21.

The authors wish to express their sincere thanks to colleagues for discussion and constructive criticism as well as for providing information prior to publication.

APPENDIX: A COMPILATION OF SELECTED PROPERTIES OF DMSO AND AQUEOUS DMSO SYSTEMS

TABLE 22[a]

Selected Physical Properties of DMSO

Property	Value
Boiling point at 760 mm Hg/°C	189·0
Vapor pressure at 20°C/mm Hg	0·37, 0·417
Specific heat at 25°C/cal g^{-1}	0·4698
Specific heat of vaporization at b.p./cal g^{-1}	175
Molar heat of vaporization at b.p./kcal mol^{-1}	13·67
Entropy of vaporization at b.p./cal deg^{-1} mol^{-1} (Trouton)	29·6
Enthalpy of vaporization at 25°C/kcal mol^{-1}	12·64
Coefficient of expansion/ml g^{-1} deg^{-1}	0·00088
Specific gravity at 25°C/g ml^{-1}	1·0958
Molar volume at 20° C/ml mol^{-1}	71·03
Melting point/°C	18·55
Specific heat of fusion/cal g^{-1}	38·8
Molar heat of fusion/kcal mol^{-1}	3·43
Entropy of fusion/cal deg^{-1} mol^{-1}	10·4
Specific heat of solid at 18·45° C/cal g^{-1}	0·5
Heat of formation (graphite and $S_{rhomb.}$)/kcal mol^{-1}	−47·7
Heat of combustion/cal g^{-1}	6050

Table 22—continued

Property	Value
Flash point, open vessel/°C	95
Molal f.p. depression/deg mol^{-1} kg^{-1}	4·36
Refractive index ($n_D{}^{25}$)	1·4767
Molar refractivity, R_m	20·12
Polarizability/cm^3	7·97 × 10^{-24}
Dielectric constant at 25°C	46·4
Dipole moment 25° C/D	4·11
Viscosity, 25° C/cP	1·99
Surface tension at 20° C/dyne cm^{-1}	46·2
Parachor	184·4
Specific conductance at 20° C/ohm^{-1} cm^{-1}	3 × 10^{-8}

[a] Taken in part from Szmant, 1971.

TABLE 23[a]

Activity Coefficients of Water in H_2O–DMSO Mixtures at 20, 25, 30°C; p$K_{a/a}$, ΔG^0, $T\Delta S^0$ and ΔH^0 Values for the Ionization Process at 25°C

X_{DMSO}	γ_{H_2O} 20°C	γ_{H_2O} 25°C	γ_{H_2O} 30°C	p$K_{a/a}$ 25°C	$\Delta G^0_{25°C}$ kcal mol^{-1}	$T\Delta S^0_{25°C}$ kcal mol^{-1}	$\Delta H^0_{25°C}$ kcal mol^{-1}
0	1	1	1	14·00	19·10	−5·61	13·49
0·1	0·948	0·951	0·955	14·79	20·18	−5·07	15·11
0·2	0·852	0·859	0·867	15·64	21·34	−8·89	12·46
0·3	0·710	0·721	0·731	16·43	22·41	−11·09	11·32
0·4	0·571	0·586	0·602	17·26	23·55	−13·09	10·46
0·5	0·466	0·481	0·498	18·21	24·84	−14·64	10·20
0·6	0·390	0·405	0·421	19·09	26·04	−14·16	11·88
0·7	0·331	0·345	0·359	19·89	27·13	−10·58	16·55
0·8	0·304	0·317	0·330	20·54	28·03	−8·77	19·26

[a] Fiordiponti et al., 1974.

TABLE 24[a]

Free Energies of Transfer of H$^+$ and OH$^-$ Ions from Water to H_2O–DMSO Media (kcal mol^{-1}, 25°)

wt. % DMSO	mol. fraction H_2O	$\Delta G^0_{tr}(H^+)$	$\Delta G^0_{tr}(OH^-)$
20	0·94	−0·21	1·2
40	0·86	−1·69	3·2
60	0·74	−3·52	6·2
80	0·53	−6·20	10·2

[a] Villermaux and Delpuech, 1974.

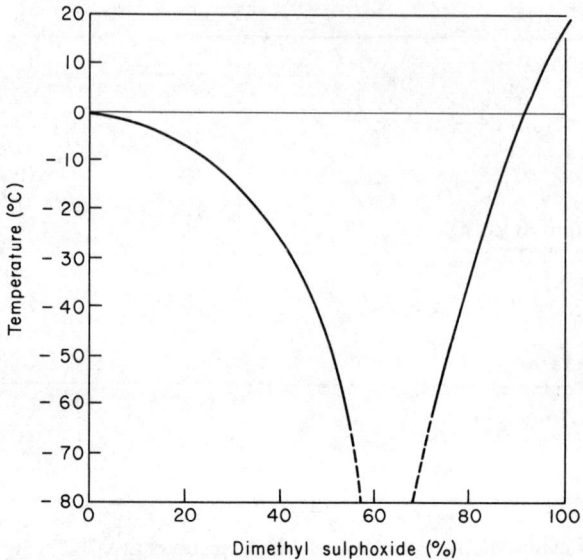

Figure 18. Freezing points of aqueous solutions of dimethyl sulphoxide (Ranky and Nelson, 1961).

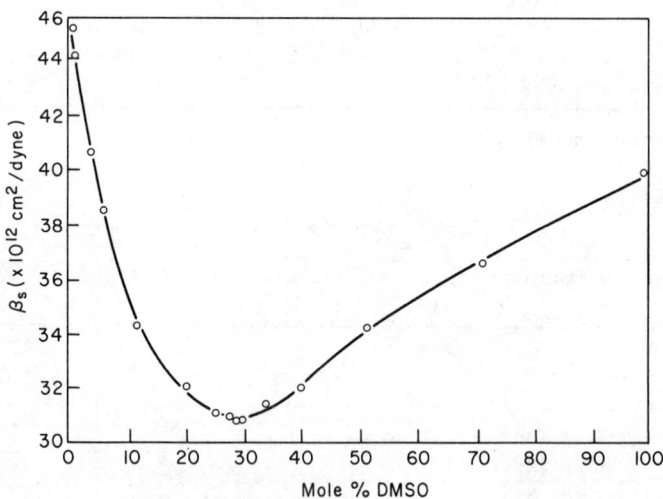

Figure 19. Adiabatic compressibility (β_S) of aqueous solutions of dimethyl sulphoxide at 20° (Bowen et al., 1974).

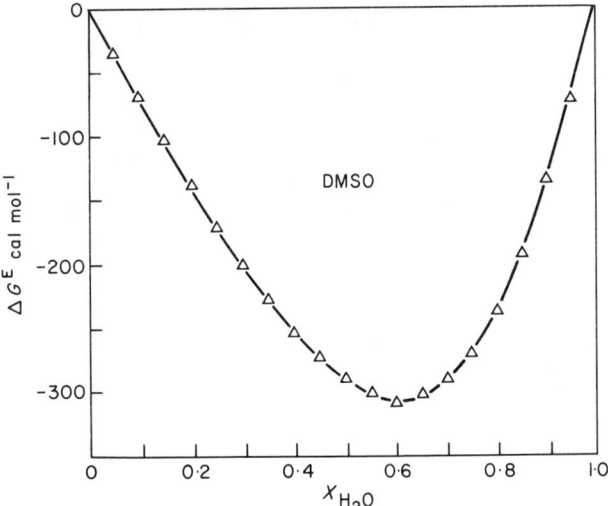

Figure 20. Excess free energies of mixing for the system DMSO–H_2O at 25°C (Lam and Benoit, 1974).

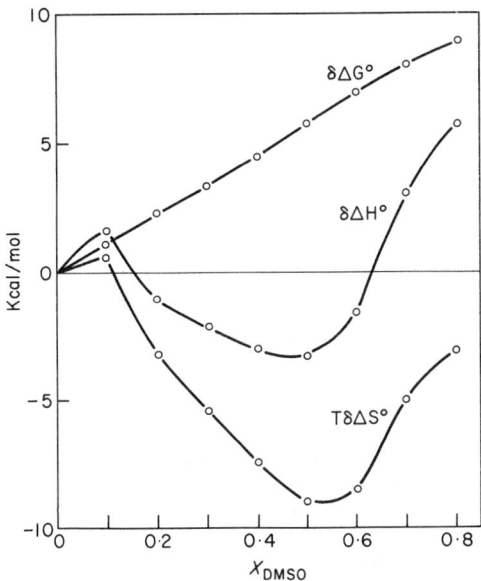

Figure 21. Transfer enthalpy, entropy, and free energy values for the ionization of water in H_2O–DMSO mixtures at 25°C (Fiordiponti et al., 1974).

REFERENCES

Abraham, M. H. (1969). *Chem. Comm.* 1307.
Abraham, M. H. (1973). *J. C. S. Faraday I* 69, 1375.
Abraham, M. H. (1974). *Prog. Phys. Org. Chem.* 11, 2.
Adsetts, J. R., and Gold, V. (1969). *J. Chem. Soc. (B)* 1108.
Albagli, A., Stewart, R., and Jones, J. R. (1970). *J. Chem. Soc. (B)* 1509.
Albagli, A., Buckley. A., Last, A. M., and Stewart, R. (1973). *J. Amer. Chem. Soc.* 95, 4711.
Alexander, R., Ko, E. C. F., Parker, A. J., and Broxton, T. J. (1968). *J. Amer. Chem. Soc.* 90, 5049.
Allerhand, A., and Schleyer, P. von R. (1963). *J. Amer. Chem. Soc.* 85, 371.
Alunni, S., Baciocchi, E., Ruzziconi, R., and Tingoli, M. (1974). *J. Org. Chem.* 39, 3299.
Amis, E. S. (1966). "Solvent Effects on Reaction Rates and Mechanisms", Academic Press, New York.
Arnett, E. M., and Scorrano, G. (1976). *Adv. Phys. Org. Chem.* 13, 83.
Arnett, E. M., Johnston, D. E., and Small, L. E. (1975). *J. Amer. Chem. Soc.* 97, 5598.
Arnett, E. M., Bentrude, W. G., Burke, J. J., and Duggleby, P. McC. (1965). *J. Amer. Chem. Soc.* 87, 1541.
Arnett, E. M., Moriarity, T. C., Small, L. E., Rudolph, P. J., and Quirk, R. P. (1973). *J. Amer. Chem. Soc.* 95, 1492.
Ayediran, D., Bamkole, T. O., and Hirst, J. (1974). *J. C. S. Perkin II*, 1013.
Balakrishnan, M., Venkoba Rao, G., and Venkatasubramanian, N. (1974a). *J.C.S. Perkin II*, 6.
Balakrishnan, M., Venkoba Rao, G., and Venkatasubramanian, N. (1974b). *Aust. J. Chem.* 27, 2325.
Bartsch, R. A. (1975). *Accounts Chem. Res.* 8, 239.
Bartsch, R. A., Pruss, G. M., Cook, D. M., Buswell, R. L., Bushaw, B. A., and Wiegers, K. E. (1973a). *J. Amer. Chem. Soc.* 95, 6745.
Bartsch. R. A., Pruss, G. M., Bushaw, B. A., and Wiegers, K. E. (1973b). *J. Amer. Chem. Soc.* 95, 3405.
Bartsch, R. A., and Kayser, R. H. (1974). *J. Amer. Chem. Soc.* 96, 4346.
Bartsch, R. A., Mintz, E. A., and Parlman, R. M. (1974). *J. Amer. Chem. Soc.* 96, 4249.
Bell, R. P. (1973). "The Proton in Chemistry", 2nd ed., Cornell University Press, Ithaca, New York.
Bell, R. P. (1974). *Chem. Soc. Rev.* 3, 513.
Bell, R. P., and Cox, B. G. (1970). *J. Chem. Soc. (B)* 194.
Bell, R. P., and Cox, B. G. (1971). *J. Chem. Soc. (B)* 783.
Bell, R. P., and Goodall, D. M. (1966). *Proc. Roy. Soc. (London)* A294, 273.
Bell, R. P., and Lidwell, O. M. (1940). *Proc. Roy. Soc. (London)* A176, 88.
Bell, R. P., Sachs, W. H., and Tranter, R. L. (1971). *Trans. Faraday Soc.* 67, 1995.
Bender, M. L. (1971). "Mechanisms of Homogeneous Catalysis from Protons to Proteins", Wiley, New York.
Bernasconi, C. F. (1973). *M.T.P. Intern. Rev. Science, Org. Chem. Ser.* 1, 3, 33.
Bernasconi, C. F., and Gehriger, C. L. (1974). *J. Amer. Chem. Soc.* 96, 1092.
Berson, J. A., Hamlet, Z., and Mueller, W. A. (1962). *J. Amer. Chem. Soc.* 84, 297.
Borchardt, J. K., and Saunders, W. H., Jr. (1972). *Tetrahedron Lett.* 3439.
Borchardt, J. K., Swanson J. C., and Saunders, W. H., Jr. (1974). *J. Amer. Chem. Soc.* 96, 3918.

Bordwell, F. G., and Boyle, W. J., Jr. (1975). *J. Amer. Chem. Soc.* 97, 3447.
Bordwell, F. G., and Matthews, W. S. (1974). *J. Amer. Chem. Soc.* 96, 1214.
Bordwell, F. G., Bartmess, J. E., Drucker, G. E., Margolin, Z., and Matthews, W. S. (1975). *J. Amer. Chem. Soc.* 97, 3226.
Bowden, K. (1966). *Chem. Rev.* 66, 119.
Bowden, K., and Cockerill, A. F. (1970). *J. Chem. Soc. (B)* 173.
Bowden, K., and Cook, R. S. (1968). *J. Chem. Soc. (B)* 1529.
Bowden, K., and Cook, R. S. (1971a). *J. Chem. Soc. (B)* 1765.
Bowden, K., and Cook, R. S. (1971b). *J. Chem. Soc. (B)* 1771.
Bowden, K., and Cook, R. S. (1972). *J. C. S. Perkin II*, 1407.
Bowden, K., and Stewart, R. (1965). *Tetrahedron* 21, 261.
Bowen, D. E., Priesand, M. A., and Eastman, M. P. (1974). *J. Phys. Chem.* 78, 2611.
Brauman, J. I., Bryson, J. A., Kahl, D. C., and Nelson, N. J. (1970). *J. Amer. Chem. Soc.* 92, 6679.
Brønsted, J. N., and Pedersen, K. J. (1923). *Zeit. Physik. Chem.* 108, 185.
Brown, C. A. (1974). *Chem. Comm.* 680.
Brown, H. C., and Liu, K.-J. (1970). *J. Amer. Chem. Soc.* 92, 200.
Brownstein, S. (1960). *Can. J. Chem.* 38, 1590.
Bruice, T. C., and Fife, T. H. (1973). *J. Amer. Chem. Soc.* 94, 1962.
Buck, P. (1969). *Angew. Chem. Int. Ed. Engl.* 8, 120.
Buncel, E. (1975). "Carbanions. Mechanistic and Isotopic Aspects", Elsevier, Amsterdam.
Buncel, E., and Webb, J. G. K. (1973). *J. Amer. Chem. Soc.* 95, 8470.
Buncel, E., and Webb, J. G. K. (1974). *Can. J. Chem.* 52, 630.
Buncel, E., and Leung, H. W. (1975). *Chem. Comm.* 19.
Buncel, E., and Symons, E. A. (1976). *J. Amer. Chem. Soc.* 98, 656.
Buncel, E., Symons, E. A., and Zabel, A. W. (1965). *Chem. Comm.* 173.
Buncel, E., Norris, A. R., and Russell, K. E. (1968). *Quart. Rev. (London)* 22, 123.
Buncel, E., Jonczyk, A., and Webb, J. G. K. (1975). *Can. J. Chem.* 53, 376.
Buncel, E., Eggimann, W., and Leung, H. W. (1976). *Chem. Comm.* in press.
Buncel, E., Symons, E. A., Dolman, D., and Stewart, R. (1970). *Can. J. Chem.* 48, 3354.
Bunnett, J. F. (1958). *Quart. Rev. (London)* 12, 1.
Bunnett, J. F. (1969). *Surv. Prog. Chem.* 5, 53.
Bunnett, J. F., and Olsen, F. P. (1966). *Can. J. Chem.* 44, 1899.
Bunnett, J. F., and Randall, J. J. (1958). *J. Amer. Chem. Soc.* 80, 6020.
Bunnett, J. F., and Zahler, R. E. (1951). *Chem. Rev.* 51, 273.
Caldin, E. F. (1964). "Fast Reactions in Solution", Blackwell, Oxford.
Caldin, E. F., and Mateo, S. (1973). *Chem. Comm.* 854.
Caldin, E. F., Burfoot, G. D., and Goodman, H. (1974). *J. C. S. Faraday I* 69, 105.
Caveng, P., Fischer, P. B., Heilbronner, E., Miller, A. L., and Zollinger, H. (1967). *Helv. Chim. Acta* 50, 848.
Chrisment, J., Delpuech, J. J., and Rubini, P. (1974). *Mol. Phys.* 27, 1663.
Clarke, G. A., and Taft, R. W. (1962). *J. Amer. Chem. Soc.* 84, 2295.
Cockerill, A. F., Earls, D. W., Jones, J. R., and Rumney, T. G. (1974). *J. Amer. Chem. Soc.* 96, 575.
Coetzee, J. F., and Ritchie, C. D. (1969). "Solute-Solvent Interactions", Marcel Dekker, New York.
Cohen, A., and Marcus, R. A. (1968). *J. Phys. Chem.* 72, 4249.
Cook, D., Evans, I. P., Ko, E. C. F., and Parker, A. J. (1966). *J. Chem. Soc. (B)* 404.

Cooper, T. K., Walker, D. C., Gillis, H. A., and Klassen, N. V. (1973). *Can. J. Chem.* 39, 2241.
Corey, E. J., and Chaykovsky, M. (1962). *J. Amer. Chem. Soc.* 84, 866.
Cox, B. G. (1973). *Ann. Rep. Chem. Soc. (A)* 70, 249.
Cox, B. G., and Parker, A. J. (1973). *J. Amer. Chem. Soc.* 95, 408.
Cox, B. G., and Gibson, A. (1974). *Chem. Comm.* 639.
Cox, B. G., Hedwig, G. R., Parker, A. J., and Watts, D. W. (1974). *Aust. J. Chem.* 27, 477.
Cox, R. A., and Stewart, R. (1976). *J. Amer. Chem. Soc.* 98, 488.
Cox, R. A., Last, A. M., and Stewart, R. (1974). *J. C. S. Perkin II*, 1678.
Cram, D. J. (1965). "Fundamentals of Carbanion Chemistry", Academic Press, New York.
Cram, D. J., Hauck, F., Kopecky, K. R., and Nielsen, W. D. (1959) *J. Amer. Chem. Soc.* 81, 5767.
Cram, D. J., Rickborn, B., Kingsbury, C. A., and Haberfield, P. (1961). *J. Amer. Chem. Soc.* 83, 3678.
Crampton, M. R. (1969). *Adv. Phys. Org. Chem.* 7, 211.
Crampton, M. R., and Gold, V. (1964). *J. Chem. Soc. (B)* 4293.
Crampton, M. R., and Willison, M. J. (1974). *J. C. S. Perkin II*, 1686.
Delpuech, J. J. (1966). *Bull. Soc. Chim. France* 1624.
Delpuech, J. J., and Nicole, D. (1974). *J. C. S. Perkin II*, 1025.
Delpuech, J. J., and Nicole, D. (1976). *J. C. S. Perkin II* in press.
Delpuech, J. J., Serratrice, G., Strich, A., and Veillard, A. (1972). *Chem. Comm.* 817.
Dimroth, K., Reichardt, C., Siepmann, T., and Bohlmann, F. (1963). *Ann.* 661, 1.
Dolman, D., and Stewart, R. (1967). *Can. J. Chem.* 45, 911.
Dubois, J. E., Goetz, E., and Bienvenue, A. (1964). *Spectrochim. Acta* 20, 1815.
Durst, T. (1969). *Adv. Org. Chem.* 6, 285.
Earls, D. W., Jones, J. R., Rumney, T. G., and Cockerill, A. F. (1974). *J. C. S. Perkin II*, 1806.
Eigen, M. (1964). *Agnew. Chem. Int. Ed. Engl.* 3, 1.
Euranto, E. K. (1969). In "The Chemistry of Carboxylic Acids and Esters" (S. Patai, ed.), Wiley, New York.
Exner, J. H., and Steiner, E. C. (1974). *J. Amer. Chem. Soc.* 96, 1782.
Fendler, E. J., Constien, V. G., and Fendler, J. H. (1975a). *J. Phys. Chem.* 79, 917.
Fendler, J. H., and Fendler, E. J. (1975). "Catalysis in Micellar and Macromolecular Systems", Academic Press, New York.
Fendler, J. H., Hinze, W. L., and Liu, L. J. (1975b). *J. C. S. Perkin II*, 1768.
Fiordiponti, P., Rallo, F., and Rodante, F. (1974). *Z. Phys. Chem. N. F.* 88, 149.
Foster, R., and Fyfe, C. A. (1966). *Rev. Pure Appl. Chem.* 16, 61.
Fuchs, R., and Caputo, A. (1966). *J. Org. Chem.* 31, 1524.
Fuchs, R., Hagan, C. P., and Rodewald, R. F. (1974). *J. Phys. Chem.* 78, 1509.
Gani, V., and Lapinte, C. (1973). *Tetrahedron Lett.* 2775.
Gaylord, N. G. (1956). "Reduction With Complex Metal Hydrides", Interscience, New York.
Giles, D. E., and Parker, A. J. (1970). *Aust. J. Chem.* 23, 1581.
Goitein, R., and Bruice, T. C. (1972). *J. Phys. Chem.* 76, 432.
Gold, V. (1964). In "Friedel-Crafts and Related Reactions" (G. A. Olah, ed.), Interscience, New York, Chapter 29.
Gold, V., and Nowlan, V. (1974). *Chem. Comm.* 482.

Grunwald, E., and Price, E. (1964). *J. Amer. Chem. Soc.* 86, 4517.
Grunwald, E., and Winstein, S. (1948). *J. Amer. Chem. Soc.* 70, 846.
Grunwald, E., Baugham, G., and Kohnstam, G. (1960). *J. Amer. Chem. Soc.* 82, 5801.
Haberfield, P., Clayman, L., and Cooper, J. S. (1969). *J. Amer. Chem. Soc.* 91, 787.
Haberfield, P., Friedman, J., and Pinkston, M. F. (1972). *J. Amer. Chem. Soc.* 94, 71.
Haberfield, P., Nudelman, A., Bloom, A., Romm, R., and Ginsberg, H. (1971). *J. Org. Chem.* 36, 1792.
Hall, T. N., and Poranski, C. F., Jr. (1970). In "The Chemistry of the Nitro and Nitroso Groups", Part 2 (H. Feuer, ed.), Interscience, New York.
Halle, J. C., Terrier, F., and Schaal, R. (1973). *Bull. Soc. Chim. France* 1225.
Hammett, L. P. (1940). "Physical Organic Chemistry", McGraw-Hill, New York, Chapter 9.
Hammett, L. P. (1970). "Physical Organic Chemistry", 2nd ed., McGraw-Hill, New York, Chapter 9.
Hammett, L. P., and Deyrup, A. J. (1932). *J. Amer. Chem. Soc.* 54, 2721.
Hanna, S. B., Jermini, C., Loewenschuss, H., and Zollinger, H. (1974). *J. Amer. Chem. Soc.* 96, 7222.
Hart, E. J., and Anbar, M. (1970). "The Hydrated Electron", Wiley-Interscience, New York.
Hepler, L. G., Stokes, J. M., and Stokes, R. H. (1965). *Trans. Faraday Soc.* 61, 20.
Herbrandson, H. F., and Neufeld, F. R. (1966). *J. Org. Chem.* 31, 1140.
Hojo, M., Utaka, M., and Yoshida, Z. (1966). *Tetrahedron Lett.* 25.
House, H. O. (1972). "Modern Synthetic Reactions", 2nd ed., Benjamin, New York.
Hughes, E. D., and Ingold, C. K. (1935). *J. Chem. Soc.* 244.
Hyne, J. B., Wills, R., and Wonkka, R. E. (1962). *J. Amer. Chem. Soc.* 84, 2914.
Ingold, C. K. (1953). "Structure and Mechanism in Organic Chemistry", Cornell University Press, Ithaca, New York.
Ingold, C. K. (1969). "Structure and Mechanism in Organic Chemistry", 2nd ed., Cornell University Press, Ithaca, New York.
Ivin, K. J., McGarvey, J. J., Simmons, E. L., and Small, R. (1971). *Trans. Faraday Soc.* 67, 104.
Jacob, S. W., Rosenbaum, E. E., and Wood, D. C. (1971). "Dimethyl Sulphoxide", Vol. 1. Marcel Dekker, New York.
Jencks, W. P. (1969). "Catalysis in Chemistry and Enzymology", McGraw-Hill, New York.
Jencks, W. P. (1972). *Chem. Rev.* 72, 705.
Jones, J. R. (1972). *Prog. Phys. Org. Chem.* 9, 241.
Jones, J. R. (1973). "The Ionization of Carbon Acids", Academic Press, London.
Jones, J. R., and Stewart, R. (1967). *J. Chem. Soc. (B)* 1173.
Jones, J. R., Jaiswal, D. K., and Fuchs, R. (1976). *J. C. S. Perkin II*, 102.
Kaplan, L. A., and Siedle, A. R. (1971). *J. Org. Chem.* 36, 937.
Kharasch, N., and Thyagarajan, B. S. (1966). *Quart. Rep. Sulphur Chem.* 1, 1-91.
Kirby, A. J., and Jencks, W. P. (1965). *J. Amer. Chem. Soc.* 87, 3217.
Kollmeyer, W. D., and Cram, D. J. (1968). *J. Amer. Chem. Soc.* 90, 1784.
Kolthoff, I. M., and Chantooni, M. K., Jr. (1972). *J. Phys. Chem.* 76, 2024.
Kolthoff, I. M., and Reddy, T. B. (1962). *Inorg. Chem.* 1, 189.

Kolthoff, I. M., Chantooni, M. K., Jr., and Bhowmik, S. (1968). *J. Amer. Chem. Soc.* **90**, 23.
Koppel, I. A., and Palm, V. A. (1972). In "Advances in Linear Free Energy Relationships" (N. B. Chapman and J. Shorter, eds.), Plenum Press, London, Chapter 5.
Kornblum, N., Berrigan, P. J., and le Noble, W. J. (1963). *J. Amer. Chem. Soc.* **85**, 1141.
Kosower, E. M. (1958). *J. Amer. Chem. Soc.* **80**, 3253.
Kosower, E. M. (1968). "An Introduction to Physical Organic Chemistry", Wiley, New York, p. 293.
Koulkes-Pujo, A. M., Gilles, L., and Sutton, J. (1974). *Chem. Comm.* 912.
Kreevoy, M. M., and Baughman, E. H. (1973). *J. Amer. Chem. Soc.* **95**, 8178.
Kresge, A. J. (1973). *Chem. Soc. Rev.* **2**, 475.
Kresge, A. J. (1975). In "Proton Transfer Reactions" (E. F. Caldin and V. Gold, eds.), Chapman and Hall, London.
Kresge, A. J., and Chiang, Y. (1967). *J. Amer. Chem. Soc.* **89**, 4411.
Kroeger, D. J., and Stewart, R. (1970). *J. Chem. Soc. (B)* 217.
Krygowski, T. M., and Fawcett, W. R. (1975). *J. Amer. Chem. Soc.* **97**, 2143.
Kurz, J. L., and Kurz, L. C. (1972). *J. Amer. Chem. Soc.* **94**, 4451.
Laidler, K. J., and Eyring, H. (1940). *Ann. N.Y. Acad. Sci.* **39**, 303.
Lam, K. B., and Miller, J. (1966). *Chem. Comm.* 642.
Lam, S. Y., and Benoit, R. L. (1974). *Can. J. Chem.* **52**, 718.
Lamson, D. W., Ulrich, P., and Hutchins, R. O. (1973). *J. Org. Chem.* **38**, 2928.
Leffek, K. T. (1976). In "Isotopes in Organic Chemistry", Vol. 2 (E. Buncel and C. C. Lee, eds.), Elsevier, Amsterdam.
Levi, A., Modena, G., and Scorrano, G. (1974). *J. Amer. Chem. Soc.* **96**, 6585.
Lewis, E. S. (1975). In "Proton Transfer Reactions" (E. F. Caldin and V. Gold, eds.), Chapman and Hall, London.
Lewis, E. S., and Funderburk, L. H. (1967). *J. Amer. Chem. Soc.* **89**, 2322.
Liotta, C. L., Abidaud, A., and Hopkins, H. P., Jr. (1972). *J. Amer. Chem. Soc.* **94**, 8624.
MacDonald, D. D., and Hyne, J. B. (1971). *Can. J. Chem.* **49**, 611.
MacGregor, W. S. (1967). *Ann. N.Y. Acad. Sci.* **141**, 3.
MacGregor, W. S. (1968). *Quart. Rep. Sulphur Chem.* **3**, 149.
Marcus, R. A. (1968). *J. Phys. Chem.* **72**, 891.
Marziano, N. C., Cimino, G. M., and Passerini, R. C. (1973). *J. C. S. Perkin II*, 1915.
Matthews, W. S., Bares, J. E., Bartmess, J. E., Bordwell, F. G., Cornforth, F. J., Drucker, G. E., Margolin, Z., McCallum, R. J., McCollum, G. J., and Vanier, N. R. (1975). *J. Amer. Chem. Soc.* **97**, 7006.
Meisenheimer, J. (1902). *Ann.* **313**, 242.
Miller, J. (1951). *Rev. Pure and Appl. Chem.* **1**, 171.
Miller, J. (1968). "Aromatic Nucleophilic Substitution", Elsevier, Amsterdam.
Miller, J., and Parker, A. J. (1961). *J. Amer. Chem. Soc.* **83**, 117.
More O'Ferrall, R. A. (1972). *J. C. S. Perkin II*, 976.
More O'Ferrall, R. A. (1975). In "Proton Transfer Reactions", (E. S. Caldin and V. Gold eds.), Chapman and Hall, London.
Murdoch, J. R. (1972). *J. Amer. Chem. Soc.* **94**, 4410.
Noyes, R. M. (1961). *Prog. Reaction Kinetics* **1**, 129.
Orvik, J. A., and Bunnett, J. F. (1970). *J. Amer. Chem. Soc.* **92**, 2417.
Pánková, M., Sicher, J., and Závada, J. (1967). *Chem. Comm.* 394.
Pánková, M., Svoboda, M., and Závada, J. (1972). *Tetrahedron Lett.* 2465.
Parker, A. J. (1962). *Quart. Rev. (London)* **16**, 163.
Parker, A. J. (1967). *Adv. Phys. Org. Chem.* **5**, 173.

Parker, A. J. (1969). *Chem. Rev.* **69**, 1.
Parker, A. J. (1971). *Chem. Tech.* 297.
Pietra, F. (1969). *Quart. Rev. (London)* **23**, 504.
Pederson, C. H., and Frensdorff, H. K. (1972). *Angew. Chem. Int. Ed. Engl.* **11**, 16.
Price, G. G., and Whiting, M. C. (1963). *Chem. Ind. (London)* 775.
Rammler, D. H. (1971). In "Dimethyl Sulphoxide", Vol. 1, (S. W. Jacob, E. E. Rosenbaum and D. C. Wood, eds.), Marcel Dekker, New York.
Ranky, W. O., and Nelson, D. C. (1961). "Organic Sulphur Compounds" (N. Kharasch, ed.), Vol. 1. Pergamon, New York, Chapter 17.
Reichardt, C. (1965). *Angew. Chem. Int. Ed. Engl.* **4**, 29.
Ritchie, C. D. (1969). In "Solute-Solvent Interactions" (J. F. Coetzee and C. D. Ritchie, eds.), Marcel Dekker, New York, Chapter 4.
Ritchie, C. D., and King, H. F. (1968). *J. Amer. Chem. Soc.* **90**, 833.
Ritchie, C. D., and Megerle, G. H. (1967). *J. Amer. Chem. Soc.* **89**, 1447.
Ritchie, C. D., and Uschold, R. E. (1967a). *J. Amer. Chem. Soc.* **89**, 1721.
Ritchie, C. D., and Uschold, R. E. (1967b). *J. Amer. Chem. Soc.* **89**, 2752.
Ritchie, C. D., and Uschold, R. E. (1968). *J. Amer. Chem. Soc.* **90**, 3415.
Roberts, D. D. (1966). *J. Org. Chem.* **31**, 4037.
Robertson, R. E., Heppolette, R. L., and Scott, J. M. W. (1959). *Can. J. Chem.* **37**, 803.
Rochester, C. H. (1966). *Quart. Rev. (London)* **20**, 511.
Rochester, C. H. (1970). "Acidity Functions", Academic Press, London.
Ross, S. D. (1972). In "Comprehensive Chemical Kinetics", Vol. 13 (C. H. Bamford and C. F. H. Tipper, eds.), Elsevier, Amsterdam.
Saunders, W. H., Jr., and Cockerill, A. F. (1973). "Mechanisms of Elimination Reactions", Wiley-Interscience, New York.
Scatchard, G. (1940). *Ann. N.Y. Acad. Sci.* **39**, 341.
Sera, A., Mano, H., and Maruyama, K. (1973). *Bull. Chem. Soc. Japan* **47**, 1754.
Sicher, J. (1972). *Angew. Chem. Int. Ed. Engl.* **11**, 200.
Smith, P. J. (1976). In "Isotopes in Organic Chemistry", Vol. 2 (E. Buncel and C. C. Lee, eds.) Elsevier, Amsterdam.
Smith, S. G., Fainberg, A. H., and Winstein, S. (1961). *J. Amer. Chem. Soc.* **83**, 618.
Steiner, E. C., and Gilbert, J. M. (1965). *J. Amer. Chem. Soc.* **87**, 382.
Steiner, E. C., and Starkey, J. D. (1967). *J. Am. Chem. Soc.* **89**, 2751.
Stewart, R., and O'Donnell, J. P. (1964). *Can. J. Chem.* **42**, 1681, 1694.
Strauss, M. J. (1970). *Chem. Rev.* **70**, 667.
Strauss, M. J. (1974). *Accounts Chem. Res.* **7**, 181.
Streitwieser, A., Jr., and Hammons, J. H. (1965). *Prog. Phys. Org. Chem.* **3**, 41.
Streitwieser, A., Jr., and Taylor, D. R. (1970). *Chem. Comm.* 1248.
Streitwieser, A., Jr., Hammons, J. H., Ciuffarin, E., and Brauman, J. I. (1967a). *J. Amer. Chem. Soc.* **89**, 59.
Streitwieser, A., Jr., Ciuffarin, E., and Hammons, J. H. (1967b). *J. Amer. Chem. Soc.* **89**, 63.
Streitwieser, A., Jr., Chang, C. J., and Young, A. T. (1972). *J. Amer. Chem. Soc.* **94**, 4888.
Suhr, H. (1967). *Ann.* **701**, 101.
Symons, E. A. (1971). *Can. J. Chem.* **49**, 3940.
Symons, E. A., and Buncel, E. (1972). *J. Amer. Chem. Soc.* **94**, 3641.
Symons, E. A., and Buncel, E. (1973). *Can. J. Chem.* **51**, 1673.
Szmant, H. H. (1971). In "Dimethyl Sulphoxide", Vol. 1 (S. W. Jacob, E. E. Rosenbaum and D. C. Wood, eds.), Marcel Dekker, New York, Chapter 1.

Szwarc, M., Streitwieser, A., Jr., and Mowery, P. C. (1974). In "Ions and Ion Pairs in Organic Reactions", Vol. 2 (M. Szwarc, ed.), Interscience, New York.
Taylor, R. P. (1970). *Chem. Comm.* 1463.
Terrier, F., and Schaal, R. (1967). *Compt. Rend.* **264**, 465.
Terrier, F., Millot, F., and Schaal, R. (1969). *Bull. Soc. Chim. France* 3002.
Tommila, E., and Murto, M. J. (1963). *Acta Chem. Scand.* **17**, 1947.
Villermaux, S., and Delpuech, J. J. (1974). *Bull. Soc. Chim. France* 2534.
Westheimer, F. H. (1961). *Chem. Rev.* **61**, 265.
Wiberg, E., and Amberger, E. (1971). "Hydrides of the Elements of Main Groups", Elsevier, Amsterdam.
Wiberg, K. B. (1955). *Chem. Rev.* **55**, 713.
Wilmarth, W. K., Dayton, J. C., and Flournoy, J. M. (1953). *J. Amer. Chem. Soc.* **75**, 4549.
Winey, D. A., and Thornton, E. R. (1975). *J. Amer. Chem. Soc.* **97**, 3102.
Winstein, S., and Fainberg, A. H. (1956). *J. Amer. Chem. Soc.* **78**, 2270.
Winstein, S., and Fainberg, A. H. (1957). *J. Amer. Chem. Soc.* **79**, 5937.
Winstein, S., Frederick, E. C., and Smith, S. (1964). *J. Amer. Chem. Soc.* **86**, 305.
Wong, S. M., Fischer, H. P., and Cram, D. J. (1971). *J. Amer. Chem. Soc.* **93**, 2235.
Wooley, E. M., and Hepler, L. G. (1972). *Anal. Chem.* **44**, 1520.
Yates, K. (1971). *Accounts Chem. Res.* **4**, 136.
Zaugg, H. E., and Schaefer, A. D. (1965). *J. Amer. Chem. Soc.* **87**, 1857.

Kinetics of Organic Reactions in Water and Aqueous Mixtures

MICHAEL J. BLANDAMER

Department of Chemistry, The University, Leicester LE1 7RH, United Kingdom

1.	Introduction	204
	Scope of the Review	204
	Background to the Present Review	205
	Background to Aqueous Solution Chemistry . . .	206
2.	Scope of the Problem	211
	Conventions	211
	Rate Constants	211
	Activation Parameters	213
	Analysis of Activation Parameters	215
	Analysis of Thermodynamic Data	216
3.	Water Molecules and Hydrogen Bonding . . .	219
	A Water Molecule	219
	Hydrogen Bonding	220
4.	Water in the Solid State	223
	Ice	223
	Clathrate Hydrates	225
5.	Water	229
	Properties of Water	229
	Structure for Water	236
6.	Solutes in Water	237
	Hydration	237
	Real Solutions	239
	Solvent Co-spheres and Kinetics of Reaction . .	246
7.	Apolar Solutes in Water	248
	Structure and Properties of Apolar Solutes in Water .	248
	Monofunctional Solutes	251
	Hydrophobic Association	254
	Kinetics of Reactions Involving Apolar Solutes in Water .	256
8.	Hydrophilic Solutes	259
	Small Molecules	259
	Urea	260
	Large Molecules	261

9.	Ionic Solutions	263
	Structure and Properties of Aqueous Salt Solutions	263
	Apolar Solutes in Aqueous Salt Solutions	272
	Ions in Aqueous Salt Solutions	276
	Kinetics of Reactions in Salt Solutions	276
10.	Aqueous Mixtures	280
	Thermodynamic Properties	280
	Solutes in Aqueous Mixtures	285
	Activation Parameters and Thermodynamic Properties of Mixtures	289
11.	Typically Aqueous Mixtures	290
	Properties of the Mixtures	290
	Solutes in Typically Aqueous Mixtures	303
	Equilibria in Typically Aqueous Mixtures	314
	Kinetics of Reactions in Typically Aqueous Mixtures	316
12.	Typically Non-Aqueous Mixtures with G^E Negative (TNAN Mixtures)	325
	Properties of the Mixtures	325
	Solutes in TNAN Aqueous Mixtures	327
	Kinetics of Reactions in TNAN Aqueous Mixtures	331
13.	Typically Non-Aqueous Mixtures with G^E Negative (TNAP Mixtures)	
	Properties of the Mixtures	333
	Solutes in TNAP Mixtures	334
	Kinetics of Reactions in TNAP Mixtures	335
	References	336

1. INTRODUCTION

Scope of Review

Many reasons can be advanced for the wide interest in the kinetics of reactions in aqueous media, not the least being the importance of understanding the role of water, "the matrix of life" (Szent-Gyorgi, 1957), in biological systems. Moreover the extensive information now available concerning the properties of water and aqueous solutions should lead to a detailed understanding of the ways in which water controls the kinetics of chemical reactions. In this review, some kinetic results will be woven into a review of aqueous systems in order to illustrate how information concerning the properties of these systems can be used in the analysis of kinetic data. Thus our aim is to sketch, in broad outline, some important aspects of aqueous chemistry. Within this general approach we will omit two important areas of interest which have been reviewed in some detail by others. These topics are kinetics in micellar systems (Fendler and Fendler, 1970, 1975) and kinetic solvent isotope effects (Laughton and Robertson, 1969; Gold, 1969).

Background to this Review

From the start of physical organic chemistry, explanations have been sought for the observed dependence of rate constants on solvent. Closely allied to the explanations have been arguments concerning the mechanism of reaction. The pioneering work of many workers in the 1930's and 1940's (e.g., Polanyi and Ogg, 1935; Evans, 1946) has been reviewed (Streitwieser, 1956). These early analyses identified the importance of the solvent in determining the mechanism of reaction, indicating how solvation could tip the balance in favour of heterolytic bond fission as the preferred reaction pathway in a polar solvent rather than homolytic fission. Thus in the S_N1 hydrolysis of an alkyl halide, RX, the heterolytic fission is favoured as a result of the solvation energies of the carbonium and halide ions. A variable in these calculations is a quantity, α, which we use to express the extent of charge development in the transition state, α being taken as 1 for the above S_N1 reaction. A similar calculation for the S_N2 hydrolysis of methyl halides in water has been described where $\alpha < 1$ (Robertson et al., 1959). The detailed calculations of the activation parameters for the S_N1 hydrolysis of t-butyl chloride as reported by Franklin (1952) are good examples of the way in which information from many sources can be brought together in the analysis of kinetic data. In general, however, whether the mechanism of hydrolysis of an alkyl halide is S_N1 or S_N2, extensive charge development in the transition state is seen as a prerequisite for reaction so that the attendant solvation energies can offset the energy required for the related gas-phase reaction. We have noted this emphasis on the energetics of the transition state in order to contrast it with another view mentioned later in this review. More sophisticated approaches to the analysis of activation parameters are possible. Thus the derived activation parameters can be treated as the sum of separate contributions characterising solute and solvation components in initial and transition states. While it is easy to criticize this approach, these calculations do point to the importance of the initial state properties in the context of the heat capacity of activation (Robertson, 1960) a point we shall return to later.

Nevertheless the ideas outlined above provided the basis for the generalization of solvent effects on rates of chemical reactions summarized in the well-known table drawn up by Ingold (1953). Here the underlying theme is the extent to which the polarity of the transition state differs from that of the initial state. Thus, in the $S_N 1$ hydrolysis of an alkyl halide, it is argued that, because the transition state is more polar than the initial state, the solvent has a marked effect on the energy of the transition state. Further, on changing the solvent, changes in the kinetic activation parameters stem from changes in the properties of the transition state.

For many years "solvation" (including hydration) was seen as involving electrostatic interactions between solute and solvent. It was a natural development therefore to attempt to relate quantitatively the kinetic parameters with the solvent permittivity[1] (Amis and Hinton, 1973). This approach has only limited success. Even where agreement between theory and experiment is good, there still remains the difficult problem of linking the solvent permittivity with the molecular properties of the system. Where chemical reaction involves a bimolecular rate determining step, rate constants can sometimes be related to the viscosity of the solvent (Moelwyn-Hughes, 1971; Caldin and Hasinoff, 1975). Correlations involving solvent viscosity have been used, for example, in the analysis of inorganic substitution reactions (Bennetto and Caldin, 1971).

Water has provided a convenient solvent in which to study the kinetics of reactions involving ions, but the low solubility of neutral organic solutes often leads to difficulties. In such cases, aqueous mixtures have been extensively used because, by judicious choice of the non-aqueous component and the composition, the solubilities of both ionic and neutral solutes are sufficiently large to overcome these problems. Further, in testing correlations between kinetic parameters and solvent permittivity, the latter can be varied, by changing the composition of an aqueous mixture, over a considerable range, e.g. $2 < \epsilon_r < 80$ for dioxan + water at 298 K.

Background to Aqueous Solution Chemistry

Over the years in which the developments outlined above attracted wide interest, the chemistry of water and aqueous solutions has

[1] Relative permittivity (ϵ_r) = dielectric constant (D).

undergone striking changes which have had important consequences on our understanding of kinetics in aqueous mixtures. The early history of the chemistry of water is described in a fascinating book written by Dorsey (1940), but for the purposes of this review it is convenient to survey briefly the subject starting with the seminal paper by Bernal and Fowler (1933). These authors examined the properties of ice, water and aqueous salt solutions, proposing models for water and ionic hydration. They introduced the now widely used terms "structure-forming" and "structure breaking" to describe the effect which an ion could have on the structure of water. By these terms they meant that, when a structure-breaker was added to water, the effect on the water was similar to an increase in temperature, while a structure former produced an effect like that of a decrease in temperature. These terms have attracted considerable attention, being currently used to describe the effects of ionic and neutral solutes. It is noteworthy that in this important paper the term "hydrogen-bond" was not used.

In the following year, Moelwyn–Hughes (1934) reported that the activation energy E_A for the inversion reaction of cane-sugar was dependent on temperature, i.e. $dE_A/dT < 0$ (see also LaMer, 1933). A similar observation (Moelwyn–Hughes, 1938) for a simpler reaction, the hydrolysis of methyl halides, prompted the proposal that the solvent determined the temperature dependence of E_A (Moelwyn–Hughes, 1936). Interpretation of these and related observations required a clearer understanding of the hydration of neutral solutes in water. This topic was discussed by several authors, the analysis given by Eley (1939) being notable, but it was not until Frank and Evans (1945) examined the thermodynamic properties of ions and neutral solutes in water and proposed structural models for their hydration that the subject made another significant advance. Building on the model of Bernal and Fowler, they established the idea that a solute, ionic or neutral, could affect the structural order of water at some distance from the solute. Robinson and Stokes (1959) drew attention to these proposals. Gurney (1953) showed how B-viscosity coefficients for salts in water could be analysed to obtain ionic contributions (Kaminsky, 1957) and then that large ions (e.g. Cs^+, I^- and Br^-) in water have negative coefficients which can be attributed to disorder in a co-sphere of solvent around these ions. The concept of a solvent co-sphere surrounding a solute where the structure of the solvent differs from that of the bulk solvent has been a valuable one.

In the context of the kinetics of reactions in water, the paper by Glew and Moelwyn-Hughes (1952) is important. These authors noted that the activation energy calculated using the model of Ogg and Polanyi for the reaction between ions and molecules was more sensitive to the reactants than was actually observed. Instead, they proposed that the activation energy for the reaction between OH^- and RX reflected in large measure the need to reorganise solvent molecules around the ion. With reference to the spontaneous hydrolysis of alkyl halides, it was suggested that the activation energy showed the necessity of reorganizing water molecules around the RX molecule and producing OH^- by ionization of water. Although the details of this mechanism have not been pursued, this paper prompted a more thorough consideration of the changes which occur in water when a chemical reaction occurs. To take one further example, it might be anticipated that quite striking changes would be observed in the enthalpies of activation, $\Delta H^{\ddagger}$, for hydrolysis of alkyl halides as the halogen atom is changed. Thus for the hydrolysis of CH_3X (where X = F, Cl, Br, and I), it was suggested on p. 205 that an important contribution to $\Delta H^{\ddagger}$ stems from the hydration of the anion formed by heterolytic fission of the C—X bond. Consequently one might expect a trend in $\Delta H^{\ddagger}$-values that mirrors the solvation enthalpies of these anions. However, enthalpies of hydration [474 (F^-), 340 (Cl^-), 326 (Br^-) and 257 (I^-) kJ mol^{-1} (Friedman and Krishnan, 1973b)] fail to correlate with the $\Delta H^{\ddagger}$ values [90 (CH_3F), 100 (CH_3Cl), 96 (CH_3Br) and 103 (CH_3I) kJ mol^{-1} (Robertson, 1967)]. The lack of correlation indicates that the role of the solvent is not simply one of stabilizing the transition state but must include a significant contribution from the need to re-organize the solvent around the solute. Indeed Moelwyn-Hughes (1953) argued that the dependence of E_A on temperature reflects the dependence on temperature of the enthalpy of the initial state.

It followed that if further progress was to be made, structural models were required for the solvent surrounding a solute in aqueous solution. Some information was available from the structures of clathrate hydrates (Jeffrey and McMullan, 1967) which show how water molecules can be arranged around a hydrophobic organic molecule. There was some natural hesitation in arguing that similar structures might be present in solution. However, Glew (1962) showed that there was a marked similarity between the enthalpy change for the transfer of methane from the gas phase to water at 273 K and that for the formation of methane clathrate hydrate from ice and methane gas (Glew and Moelwyn-Hughes, 1953). This indicated that the organization of water molecules surrounding a methane molecule in water resembles that in a gas hydrate, having extensive hydrogen bonding between water molecules but only a weak interaction between solute and solvent (Pauling, 1959). A similar model has been discussed more recently for cyclopropane in water (Hafemann and Miller, 1969). If these ideas can be extended to describe, for example, the hydration of an alkyl halide, the extent of the changes which must occur in the solvent on forming the hydrolysis transition state can be readily appreciated. Thus even where the solute is too large to form a clathrate (e.g. butane), it may be surrounded by partial cages of water molecules (Frank and Quist, 1961).

The possibility that water molecules organize themselves around a solute without being bonded to it was accepted as an example of the propensity for water molecules to hydrogen bond with each other, the stability of the solvent shell reflecting the co-operative nature of the hydrogen bonding (Frank, 1958) (see p. 222). This co-operative character was originally proposed (Frank and Wen, 1957) as an underlying principle for a new molecular model of liquid water. A two-state mixture model was described in which water molecules were either part of low-density hydrogen-bonded clusters or in a non-hydrogen bonded dense state (p. 236), the lifetime of each state being very short (ca. 10^{-11} s). A major achievement was the work of Nemethy and Scheraga (1962a) who showed how qualitative ideas concerning the structure of water can be set down in a statistical mechanical framework and, using computer based calculations, how this approach provided further insight into the properties of water and deuterium oxide (Nemethy and Scheraga, 1964). Although the analysis was not flawless (Levine and Perram, 1968), it paved the way for more detailed calculations (Section 5).

Thus gradually it became clear that the role of water in chemical reactions is not straightforward, but the problem of understanding this role was and is extremely difficult. One way of testing hypotheses is to monitor the effects of added salts, neutral solutes and co-solvents on the kinetics of a given reaction. An extensive literature of the effect of added solvents on reactions in water has been built up, particularly by Tommila (1967) and co-workers. In addition, renewed interest was shown in the thermodynamic properties of binary aqueous mixtures, the review by Franks and Ives (1966) of the properties of alcohol + water mixtures being noteworthy. Application of spectroscopic (Blandamer and Symons, 1968; Blandamer and Fox, 1973; Zeidler, 1973) and other techniques, e.g. ultrasonic absorption (Blandamer, 1973), to the study of these systems provided a foundation for a better understanding of the kinetics of reactions in these mixtures. It was already well known that plots of activation energies and other derived parameters against solvent composition could have quite complicated shapes ["roller coaster behaviour" (Hyne, 1968)]; the example of t-butyl chloride hydrolysis in ethyl alcohol + water mixtures reported by Winstein and Fainberg (1957) is well known. Nevertheless, further experimental information was necessary before the dependence of, for example, the enthalpy of activation on solvent composition could be assigned to a variation in the enthalpies of either initial or transition states. However, these complex patterns demonstrated that calculations of solvent polarity; e.g. Y-values (Grunwald and Winstein, 1948) have limited scope. At the same time there was evidence for a division into two camps, those who attributed the dependence of kinetic parameters on solvent composition solely to changes in the transition state and those who favoured changes in the initial state. It is now clear that it is dangerous to generalize. Nevertheless the case for serious consideration of changes in initial state properties was made by Arnett (1967) and co-workers in the particular example of t-butyl chloride solvolysis in water + ethyl alcohol mixtures. It was shown that the dependence of $\Delta H^{\neq}$ on composition in the water-rich mixtures could be accounted for almost completely in terms of the effect on the molar enthalpy of the initial state. Also the complex patterns in activation parameters reflect the effects which the organic co-solvent has on water–water interactions, i.e., on the water structure. These findings prompted the currently renewed interest in the kinetics in water and aqueous mixtures of inorganic as well as organic reactions (Blandamer and Burgess, 1974; Caldin and Bennetto, 1973; Caldin and Godfrey, 1974). At the same time, interest has grown in the properties of salts in aqueous mixtures (Feakins, 1967).

The above has been an attempt, obviously subjective, to identify a chronological pattern to the subject under review. In the following discussion, less emphasis will be placed on the sequence of events than on classification of systems and the problems of interpretation.

2. SCOPE OF THE PROBLEM

Conventions

As a starting point, it will be assumed that the rate constant, k, refers to a given reaction in a solvent at specified fixed temperature and pressure. The solvent is either water or an aqueous solution. Water is designated component 1, and the second component of the solvent (salt, solute or co-solvent) as component 2. Thus in a binary mixture, water + co-solvent, x_1 is the mole fraction of water and x_2 ($=1 - x_1$) is the mole fraction of co-solvent. The composition of a solvent mixture can also be expressed in terms of weight per cent, w_2, which is easily converted into mole fractions. The practice of using volume per cent appears to be dying out, which is welcomed because the value depends upon the pressure and temperature. The reactants will be described as components, 3, 4, etc., and the transition state represented by $\ddagger$.

Rate Constants

An important assumption in our discussion is that the rate constant describes a simple reaction, reactants → products. Where this assumption is not justified, the role of the solvent in the kinetics is rarely amenable to detailed analysis, because the rate constant then represents a complex term in rate constants and equilibrium constants which describe different stages in the reaction. For this reason, solvent effects on acid catalysed hydrolysis of esters are difficult to analyse. Attention is also drawn to recent criticisms of the steady state hypothesis in complex reaction schemes (Farrow and Edelson, 1974). The development of numerical integration techniques, especially the method due to Gear (1971), could herald a new way of examining such schemes.

The roller coaster patterns mentioned on p. 210 are normally observed in plots of activation parameters against x_2, calculated from the dependence of rate constant on either temperature or pressure, rather than in the dependence of k on x_2. Thus a careful examination is usually made of the dependence of k on T at fixed p and x_2, and on p at fixed T and x_2. The analysis of these separate dependences has been discussed by Kohnstam (1967) who points out the various pitfalls which await the unsuspecting. Indeed, a growing literature points to the importance of critical analysis of the kinetic data (Koren and Perlmutter-Hayman, 1971). For example, the measured rate constant may describe a reaction scheme involving two simultaneous reactions or a rapid pre-equilibrium, giving rise to a measurable heat capacity of activation, $\Delta C_p^\ddagger$. Thus Albery and Robinson (1969) argue that the large negative $\Delta C_p^\ddagger$ for the hydrolysis of t-butyl chloride is a consequence of an ion-pair mechanism, $RCl \rightleftharpoons R^+Cl^- \rightarrow$ Products, so that $\Delta H^\ddagger$ is a function of the values for the three individual steps. Alternatively Moelwyn-Hughes et al. (1965) discuss $\Delta C_p^\ddagger$ in terms of a single activation process, the negative value reflecting the properties of the initial state solvation. The latter approach has been supported in an analysis of activation parameters for different reaction schemes (Scott and Robertson, 1972). It is also noteworthy that $\Delta C_p^\ominus$ for the dissociation of carboxylic acids in water is negative, e.g. -155 J mol^{-1} K^{-1} for acetic acid (Larson and Hepler, 1969). Here a negative heat capacity term is unambiguously related to the partial molar heat capacities of the dissociated and undissociated acid (cf. transition state and initial state of $S_N 1$ hydrolysis of t-butyl chloride).

In the analysis of the dependence of measured parameters, e.g. rate constants, equilibrium constants and gas solubilities, on temperature, increasing emphasis is being placed on the statistical significance of derived quantities. The availability of computers and packaged numerical analysis routines has removed much of the heartache associated with the arithmetic. While cases for particular treatments have been advanced, e.g. orthogonal polynomials (Ives and Marsden, 1965) and polynomial spline functions (Wold, 1970, 1972), the following method proposed by Clarke and Glew (1966) for the analysis of the dependence of equilibrium constants on temperature has been widely used and extended to include solubility (Clarke and Glew, 1971; Wauchope and Haque, 1972) and kinetic data. The dependence on temperature of equilibrium constants is fitted about a reference temperature, usually chosen near the middle of the measured range, using a series of Taylor expansions of the thermodynamic functions. This method (Clarke and Glew, 1966) offers considerable flexibility with regard to the phenomena being considered (Bolton, 1970; Bezboruah et al., 1973; Bolton et al., 1972). In a recent review of the different methods of analysing the dependence of dissociation constants on temperature, some doubt has been expressed on the validity of calculated $\Delta C_p^\ominus$-values (Timimi, 1974). Similar uncertainties seem to surround $\Delta C_p^\ddagger$-values obtained from kinetic data, and reported ΔC_p-values ought always to be accompanied by clear details of the mode of analysis employed (Gaboriaud, 1971).

Analogous uncertainties apply to the analysis of the dependence of rate constant on pressure. Golinkin et al. (1966) discuss the various equations used to describe this dependence in the context of the hydrolysis of benzyl chloride in aqueous mixtures.

Activation Parameters

On applying transition state theory, eqn (1) expresses at fixed temperature and pressure a rate constant, which describes an irre-

$$k = (kT/h) \exp(-\Delta G^{\ddagger}/RT) \tag{1}$$

versible process, in terms of a quantity, $\Delta G^{\ddagger}$, which can be treated using the thermodynamic theory developed for reversible processes. Thus $\Delta G^{\ddagger}$ is the difference (2) between the chemical potentials of

$$\Delta G^{\ddagger} = \mu^{\ddagger} - \sum_{i=3}^{n} \mu_i^{\ominus} \tag{2}$$

the transition state and reactants in their solution standard states. For example, eqn (3) describes a first order and eqn (4) a second

$$\Delta G^{\ddagger} = \mu^{\ddagger} - \mu_3^{\ominus} \tag{3}$$

$$\Delta G^{\ddagger} = \mu^{\ddagger} - (\mu_3^{\ominus} + \mu_4^{\ominus}) \tag{4}$$

order process. The rate constant therefore yields a difference quantity so that, for example, in (3) $\mu^{\ddagger}$ and $\mu_3^{\ominus}$ may change when the reactant or solvent is changed, but in such a way that $\Delta G^{\ddagger}$ does not. This is rare but it is clearly important to seek evidence from, say, non-kinetic experiments of individual changes in $\mu^{\ddagger}$ or $\mu_3^{\ominus}$ in order to understand patterns in $\Delta G^{\ddagger}$. In this type of analysis, the definition of solution standard states is an important consideration. Three solution standard states are commonly used; (i) the hypothetical solution in which the molar concentration of solute, c_3, is unity and the activity coefficient, $y_3 = 1$ ($y_3 \rightarrow 1$ as $c_3 \rightarrow 0$), (ii) the hypothetical solution in which the mole fraction of solute, x_3, is unity and the asymmetric activity coefficient $f_3^* = 1$ ($f_3^* \rightarrow 1$ as $x_3 \rightarrow 0$), and (iii) the hypothetical solution in which the molality of the solute, m_3, is unity and the activity coefficient, $\gamma_3 = 1$ ($\gamma_3 \rightarrow 1$ as $m_3 \rightarrow 0$). The chemical potentials of the solute in these three standard states are $\mu_3^{\ominus}(c)$, $\mu_3^{\ominus}(x)$ and $\mu_3^{\ominus}(m)$ respectively and are related through the relative molecular mass and density of the solvent (Robinson and Stokes, 1959). Consequently, if a second-order rate constant is expressed in units of dm³ mol⁻¹ s⁻¹, $\Delta G^{\ddagger}$ represents the difference in chemical potentials on the concentration scale, $\Delta G^{\ddagger}(c)$ which will differ from $\Delta G^{\ddagger}(x)$. However, if the reaction is first-order, then $\Delta G^{\ddagger}(x) = \Delta G^{\ddagger}(m) = \Delta G^{\ddagger}(c)$. Although the choice of standard states is often said to be simply a matter of convenience, there are important consequences (see Guggenheim, 1937b). Thus on the mole fraction scale, propane is more soluble in D_2O than in H_2O, i.e. $\Delta G^{\ominus}(x) = \mu_3^{\ominus}$ (propane in D_2O) $- \mu_3^{\ominus}$ (propane in water) < 0, whereas on the molality scale, the reverse holds, propane being more soluble in H_2O than D_2O, $\Delta G^{\ominus}(m) > 0$ (Arnett and McKelvey, 1969).

Standard thermodynamic operations (Prigogine and Defay, 1954) on the Gibbs function, $\Delta G^{\ddagger}$, yield expressions for related thermodynamic activation parameters. Thus the dependence of k on T can be used to calculate the enthalpy of activation, $\Delta H_p^{\ddagger}$, for processes at constant pressure or the thermodynamic energy of activation, $\Delta U_v^{\ddagger}$, for processes at constant volume, which in turn lead to the related entropies of activation, $\Delta S_p^{\ddagger}$ and $\Delta S_v^{\ddagger}$ respectively. The dependence of k on pressure can be used to calculate the volume of activation, $\Delta V^{\ddagger}$ which is related to $\Delta H_p^{\ddagger}$ by eqn (5) where α is the thermal

$$\Delta H_p^{\ddagger} = \Delta U_v^{\ddagger} + (T \cdot \alpha \Delta V^{\ddagger}/\kappa) \qquad (5)$$

expansivity and κ the isothermal compressibility of the reacting system. The temperature dependence of $\Delta H_p^{\ddagger}$ and $\Delta U_v^{\ddagger}$ yields the heat capacities of activation, $\Delta C_p^{\ddagger}$ and $\Delta C_v^{\ddagger}$ respectively. In the following discussion we will write the constant pressure parameters $\Delta H^{\ddagger}$ and $\Delta S^{\ddagger}$ without the subscript p.

The derived parameters can be linked through Maxwell's equations. For example, the dependence of $\Delta V^{\ddagger}$ on temperature is related to the dependence of $\Delta S^{\ddagger}$ on pressure through eqn (6).

$$-(\partial \Delta V^{\ddagger}/\partial T)_p = (\partial \Delta S^{\ddagger}/\partial p)_T \qquad (6)$$

Although kinetic data must be precise before these quantities can be realistically calculated, agreement between, for example, the two separately calculated sides of (6) provides reassurance concerning the analysis of the dependence of k on p and on T. Laidler (1956) discussed application of Maxwell's equations to a range of reaction types in water while Dickson and Hyne (1971) examined, from this

TABLE 1

Hydrolysis of Benzyl Chloride[a, b]

$T = 323 \cdot 2$ K	$p = 101325$ N m^{-2} (1 atm)
$\Delta H^{\ddagger} = 84 \cdot 3$ kJ mol^{-1}	$\Delta V^{\ddagger} = -9 \cdot 8 \times 10^{-6}$ m^3 mol^{-1}

$$d\Delta H^{\ddagger}/dp = 920 \times 10^{-8} \text{ m}^3 \text{ mol}^{-1}$$
$$\Delta V^{\ddagger} - T(d\Delta V^{\ddagger}/dT) = 890 \times 10^{-8} \text{ m}^3 \text{ mol}^{-1}$$

By application of Maxwell's Equations
$$(\partial \Delta H^{\ddagger}/\partial p)_T = \Delta V^{\ddagger} - T(\partial \Delta V^{\ddagger}/\partial T)_p$$

[a] data from Dickson and Hyne, 1971.
[b] analysis by E. Godfrey and M. J. Blandamer (unpublished).

standpoint, the activation parameters for the hydrolysis of benzyl chloride in water (Table 1). Of course, these analyses do not test Maxwell's equations which are thermodynamically rigorous.

Analysis of Activation Parameters

The derivation of the various thermodynamic activation parameters is not an end in it itself; these quantities organize the information in a way amenable to further analysis. It is assumed in this section that, for a given reaction under a specified set of conditions (T, p, x_2), the data have been analysed to obtain the activation parameters $\Delta X^{\ddagger}$, where $X = G, H, S, U, V, C_p \ldots$ and where $\Delta X^{\ddagger} = X^{\ddagger} - X_3^{\ominus}$ [cf. eqn (3) for a first order reaction].

In one class of experiments, the values of $\Delta X^{\ddagger}$ are examined for a series of related reactants. Here one member of the series is chosen as reference and the change in $\Delta X^{\ddagger}$ on going from, say, RX to R'X is described by the operator δ_R, eqn (7) (Leffler and Grunwald, 1963).

$$\delta_R \Delta X^{\ddagger} = \Delta X^{\ddagger}(R'X) - \Delta X^{\ddagger}(RX) \quad (7)$$

If, for example, $X \equiv G$ then $\Delta G^{\ddagger}$ for each species can be represented as in eqn (3) and in analogous fashion, (8) is obtained. This equation

$$\delta_R \Delta G^{\ddagger} = \delta_R \mu^{\ddagger} - \delta_R \mu_3^{\ominus} \quad (8)$$

simply formalizes the idea that the change in rate constant can result from changes in either initial or transition states.

If the reactant is fixed but the solvent is changed, then the variation in activation parameters can be described using a medium operator, δ_m. If the reference solvent is water then the change in $\Delta X^{\ddagger}$ on going to a solvent mixture can be expressed as in (9). We

$$\delta_m \Delta X^{\ddagger} = \Delta X^{\ddagger}(x_2) - \Delta X^{\ddagger}(x_2 = 0) \tag{9}$$

shall adopt eqn (9) as the definition of δ_m functions in this review, i.e. water will be taken as the reference solvent. In the particular case where $X \equiv G$ and the reaction is of first order, it follows that the change in $\Delta G^{\ddagger}$ can stem from a variation either in the initial state, $\delta_m \mu_3^{\ominus}$, or in the transition state, $\delta_m \mu^{\ddagger}$, as shown in eqn (10). As an

$$\delta_m \Delta G^{\ddagger} = \delta_m \mu^{\ddagger} - \delta_m \mu_3^{\ominus} \tag{10}$$

indication of the magnitude of these effects it follows from eqn (1) that if $\delta_m \Delta G^{\ddagger}$ is (i) 1·0 kJ mol^{-1}, the rate constant changes by a factor of 1·5, (ii) 2 kJ mol^{-1}, a factor of 2·2, (iii) 5 kJ mol^{-1}, a factor of 7·5, and (iv) 10 kJ mol^{-1}, a factor of 56.

Analysis of Thermodynamic Data

One way of probing trends in either $\delta_R \Delta X^{\ddagger}$ or $\delta_m \Delta X^{\ddagger}$ is to examine the corresponding effects in either initial or transition states. Although it is not possible to examine the properties of transition states directly, these can often be estimated by measuring the properties of solutes which resemble postulated transition states.

Where the solute is volatile, solubility data can be used to obtain $\Delta G_3^{\ominus}$ for the transfer of the solute from the gas phase to solution using (11).

$$\Delta G_3^{\ominus} = \mu_3^{\ominus}(\text{solution}) - \mu_3^{\ominus}(\text{gas phase}) \tag{11}$$

The units used to express solubilities of gases, e.g. Henry's law coefficients, Ostwald coefficients and Bunsen coefficients, have to be converted to the relevant solution standard state (p. 213). Such solubilities (Battino and Clever, 1966; Wilhelm and Battino, 1973) are valuable in the analysis of kinetic data. For example, the solubility of a neutral solute in a range of aqueous mixtures can provide some indication of the variation of the chemical potential of a neutral reactant because, from eqn (11), $\delta_m \Delta G^{\ominus} = \delta_m \mu_3^{\ominus}$. Where the pure solute is a liquid or solid, it is often convenient to chose the pure solute as a standard state, represented by the symbol, $^{\circ}$ in eqn (12). Similar comments apply to the related thermodynamic quan-

$$\Delta G_3^{\ominus} = \mu_3^{\ominus}(\text{solution}) - \mu_3^{\circ} \tag{12}$$

tities. Thus it is possible to calculate from density data the change in molar volume on going from pure alcohol, V_3^o, to the solution standard state, $V_3^\ominus$. Franks and Reid (1973) discuss the problems associated with the choice of standard states. Thus $V_3^\ominus$ for an alcohol in water reflects the special hydration properties of solutes in water, $V_3^\ominus$ being less than V_3^o. However, the two states, pure alcohol and aqueous solution, are very different and it would possibly be more relevant to compare $V_3^\ominus$ for a solute in water with that in a non-aqueous solvent if such data were available.

Calculation of $\Delta H^\ominus$-quantities from the dependence of $\Delta G^\ominus$ on temperature is less reliable than direct calorimetric measurements (Franks and Reid, 1973; Frank, 1973; Reid et al., 1969). However, disagreement between published ΔH-functions for apolar solutes in aqueous solutions may also stem from practical problems associated with low solubilities (Gill et al., 1975). Calorimetric data have the advantage that, as theory shows, the standard partial molar enthalpy $H_3^\ominus$ for a solute in solution is equal to the partial molar enthalpy in the infinitely dilute solution, i.e. $x_3 \to 0$. A similar identity between $X_3^\ominus$ and X_3 ($x_3 \to 0$) occurs for the volumes and heat capacities but not for the chemical potentials and entropies. The design of a flow system for the measurement of the heat capacity of solutions (Picker et al., 1971) has provided valuable information on aqueous solutions.

If the solute is a salt, then the extrapolation to obtain, say, $V_3^\ominus$ can be based on the Debye-Hückel limiting law (DHLL) or some variant of this equation. However, where non-polar solutes are concerned, there is no simple theory. It is generally assumed that the partial molar volume, V_3 is a linear function of x_3, and $V_3^\ominus$ is obtained by extrapolation to the value of V_3 when $x_3 = 0$ (Franks and Smith, 1968).

A serious problem in the analysis of kinetic data is encountered if one or more of the reactants is ionic. Suppose, for example, the reaction can be represented by the general eqn (13). Analysis of the

$$Z^- + RY \to \text{products} \qquad (13)$$

changes in $\Delta G^\ddagger$ as Z is varied through a series of anions requires information concerning the variation of the standard state chemical potentials, $\delta_Z \mu^\ominus$. Similarly, for one such reaction, the dependence of $\Delta G^\ddagger$ on solvent will reflect the medium dependence of $\mu^\ominus$ (Z^-), i.e. $\delta_m \mu^\ominus$ (Z^-). Calculation of these single ion properties is beset with difficulties as we now explain. In order to minimize the algebra, we confine our attention to 1 : 1 salts.

The chemical potential of a 1 : 1 salt in water can be expressed on the molality scale using eqn (14), where the mean ionic activity coefficient $\gamma_\pm \to 1$ as $m_3 \to 0$ and $\mu_3^\ominus$ is the chemical potential of the salt in the hypothetical solution where $m_3 = 1$ and $\gamma_\pm = 1$ at the same T and p.

$$\mu_3 = \mu_3^\ominus + 2RT \ln m_3 \gamma_\pm \tag{14}$$

The Gibbs function for solvation, $\Delta G_s^\ominus$ is given by $\mu_3^\ominus - \mu_3^\ominus$ (gas phase) and measures the contributions from both the anion and cation. Thus $\mu_3^\ominus = \mu_+^\ominus + \mu_-^\ominus$, and, for example, $V_3^\ominus = V_+^\ominus + V_-^\ominus$. In the context of thermodynamic data, it is not possible to calculate such single ion quantities in an unambiguous and rigorous fashion. Calculation of single ion properties has been discussed extensively (Conway et al., 1965). In certain circumstances single ion properties can, of course, be readily determined. For example, ionic mobilities can be calculated from a knowledge of molar conductances and transport numbers at infinite dilution. Similarly, spectroscopic properties of single ions can be used to study ion-solvent interactions, e.g. charge-transfer-to-solvent spectra of anions (Blandamer and Fox, 1970). Ionic vibration potentials can be used to calculate ionic partial molar volumes (Zana and Yeager, 1966) and solvation numbers (Bockris and Saluja, 1972). In most cases, however, some extrathermodynamic assumption is required before the properties of single ions can be calculated from the properties of salts.

Partial molar entropies of ions can, for example, be calculated assuming $S^\ominus(H^+) = 0$. Alternatively, because K^+ and Cl^- ions are isoelectronic and have similar radii, the ionic properties of these ions in solution can be equated, e.g. analysis of B-viscosity coefficients (Gurney, 1953). In other cases, a particular theoretical treatment which relates solvation parameters to ionic radii indicates how the subdivision could be made. For example, the Born equation requires that $\Delta G_s^\ominus$ (ion) be proportional to the reciprocal of the ionic radius (Friedman and Krishnan, 1973b). However, this approach involves new problems associated with the definition of ionic radius (Stern and Amis, 1959). In another approach to this problem, the properties of a series of salts in solution are plotted in such a way that the value for a common ion is obtained as the intercept. For example, when the partial molar volumes of some alkylammonium iodides, $V^\ominus(R_4N^+I^-)$ in water (Millero, 1971) are plotted against the relative molecular mass of the cation, M_+, the intercept at $M_+ = 0$ is equated to $V^\ominus(I^-)$ (Conway et al., 1966). This procedure has been used to examine the dependence of $V^{\ominus}$ (ion) on x_2 in ethyl alcohol + water mixtures (Lee and Hyne, 1968).

Returning to eqn (13), it is clear that if the ion Z⁻ is changed in a given solvent at the same temperature and pressure, analysis of the kinetic data requires, as a minimum, a knowledge of the properties of the salts with a common counter ion in this solvent. Equation (15)

$$\delta_z C_p^{\ominus} = C_p^{\ominus}(\text{Br}^-) - C_p^{\ominus}(\text{Cl}^-) = C_p^{\ominus}(\text{KBr}) - C_p^{\ominus}(\text{KCl}) \quad (15)$$

illustrates the point with reference to heat capacity data. However, calculation of the corresponding medium effects on the properties of a given ion is not so straightforward. A number of extrathermodynamic or empirical approaches to the problem have been proposed.

Currently, a considerable amount of information is available concerning thermodynamic functions characterizing the transfer of salts from one solvent to another, $\delta_m X_2^{\ominus}$. Most of this information has been obtained from emf studies of cell reactions. A method based on analysis of nmr data also seems promising (Covington et al., 1973), the results agreeing closely with emf data (Covington et al., 1972). However, calculation of the related transfer functions for ions remains a subject for considerable debate. For example, we might assume that for transfer between two solvents, $\delta_m X^{\ominus}(\text{Na}^+)$ = 0 and thus $\delta_m X^{\ominus}(\text{Cl}^-) = \delta_m X^{\ominus}(\text{NaCl})$. Thus, by selecting a value for one ion, the values for the other ions can be readily calculated. In a similar way Abraham (1973) assumes $\delta_m X^{\ominus}(\text{Me}_4\text{N}^+) = 0$ where $X = G, H$ and S. Alternatively it might be assumed that both ions in a given salt contribute equal amounts to the overall transfer function, e.g. $\delta_m X^{\ominus}(\text{Ph}_4\text{As}^+) = \delta_m X^{\ominus}(\text{Ph}_4\text{B}^-)$, (Parker, 1969; Cox et al., 1974). Here the assignment can be justified in part on the grounds that in both anion and cation, the same groups are in contact with the solvent.

3. WATER MOLECULES AND HYDROGEN BONDING

A Water Molecule

A water molecule is triangular in shape, the HOH angle being 104·523° and the O—H bond length being 0.9571×10^{-8} cm in the equilibrium state (Eisenberg and Kauzmann, 1969). The molecular dimensions are precisely known following analysis of vibration–rotation spectra (Darling and Denison, 1940; Izatt et al., 1969). The three normal modes of vibration are: symmetric stretch, ν_1 (3656 cm^{-1}), asymmetric stretch, ν_3 (3755 cm^{-1}) and bond bend, ν_2 (1594 cm^{-1}). The OH bonds are strong, having a bond energy of 460 kJ mol^{-1}. A water molecule has a dipole moment, 1·84 D, usually written with the negative end at the oxygen atom.

A water molecule is isoelectronic with a neon atom, i.e. 10 electrons. There have been numerous attempts to describe the bonding in an isolated water molecule because, as a simple non-linear molecule, it provides a convenient test for bonding theories (Kern and Karplus, 1973). For present purposes a simple model (Pople, 1950; Duncan and Pople, 1953) will suffice. Of the four sp^3-hybrid orbitals on oxygen, two are involved in σ-covalent bonds with hydrogen atoms, and two are doubly filled, the four orbitals pointing towards the corner of a tetrahedron with the oxygen atom at the centre. Indeed Duncan and Pople, using Slater atomic orbitals in a molecular orbital treatment, concluded that the two lone pairs are almost exactly sp^3 in character, with the angle between the lone pairs approximately 120°.

Hydrogen Bonding

The phenomenon of hydrogen bonding is obviously of paramount importance to aqueous solutions. The theory of hydrogen bonding has seen remarkable progress (Rao, 1973; Kollman and Allen, 1972) although important problems still remain where more than two water molecules are involved.

The most satisfactory treatments of hydrogen bonding use either *ab initio* or semi-empirical molecular orbital theory. The most stable form of the dimer has the configuration shown in Fig. 1 (Shipman *et al.*, 1974; Stillinger and Lemberg, 1975), It is generally assumed, in the absence of information to the contrary, that in water, ice and hydrates one H-atom is collinear with the two oxygen atoms.

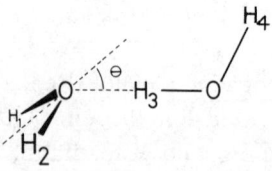

Figure 1. Water Dimer; H$_1$ and H$_2$ are in the *xy*-plane and H$_3$ and H$_4$ are in the *xz*-plane.

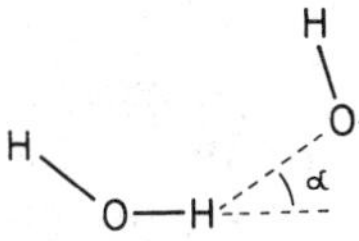

Figure 2. Water Dimer with bent hydrogen-bond.

Estimates of the equilibrium O–O distance in the dimer vary in the range $2 \cdot 53 - 3 \cdot 00 \times 10^{-8}$ cm. *Ab initio* calculations show that the OH group approaches the lone pair of the second so that the angle θ in Fig. 1 is $57°$ (Morokuma and Winick, 1970; Del Bene and Pople, 1970). Estimates of the dimerization energy differ but a value of ca. 20 kJ mol^{-1} is obtained by a number of theoretical studies (Kollman and Allen, 1972) and agrees with that calculated from second virial coefficients for water (Rowlinson, 1951). The effect of H-bond bending on the energy of the dimer has been examined as a function of the angle α (Fig. 2) using CNDO/2 methods (Goel *et al.*, 1971). The energy is relatively insensitive to bending until $\alpha > 25°$, when the energy increases rapidly. Rao (1973) suggests that this is one reason why in hydrates, hydrogen bonds can bend without significant loss of stability. The extent to which a hydrogen bond in liquid water can bend without breaking has been a topic of considerable debate.

Perhaps the most exciting development in the theory of hydrogen bonding has concerned the analysis of the stability of water polymers. The three structures for the trimer shown in Fig. 3 differ in the way in which the "central" water molecule of the trimer takes part in the hydrogen bonding, and it is possible to examine theoretically the stabilization energy per H-bond in each case. There is a non-

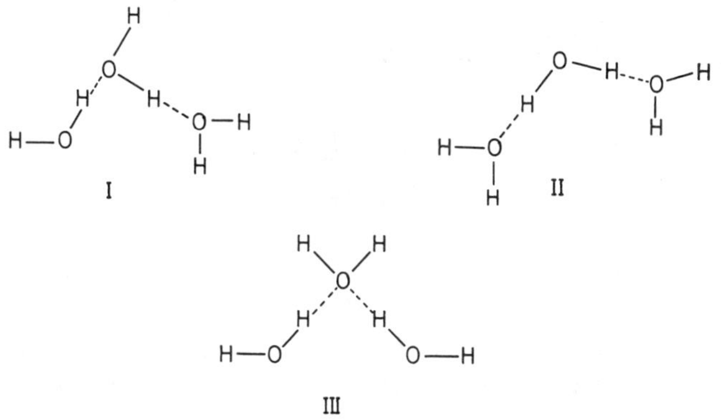

Figure 3. Water trimers showing three possible arrangements where two water molecules are hydrogen bonded to a third water molecule; I is asymmetric while II and III are symmetric.

additivity in hydrogen bond energies (Del Bene and Pople, 1973; Hankins *et al.*, 1970). The assymmetric arrangement I (Fig. 3) is more stable and, unlike structures II and III, has a stability per H-bond which is larger than that for the dimer.

These calculations provide a quantitative basis for the qualitative model proposed by Frank and Wen (1957). These authors used the term "co-operative" to describe the tendency for water molecules to hydrogen-bond together and suggested that, once two water molecules had linked together, further association would readily occur.

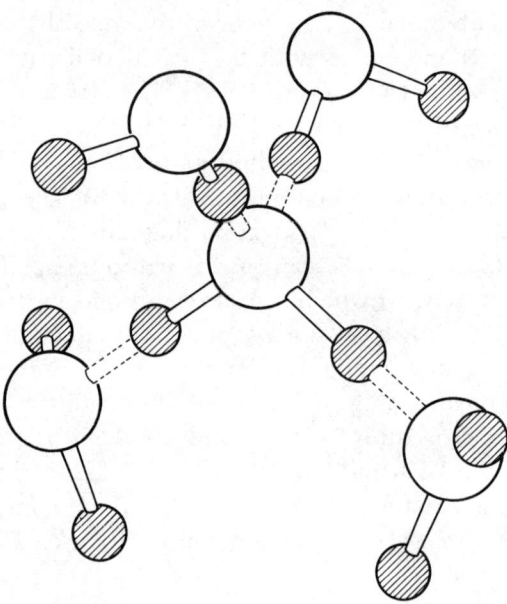

Figure 4. Water pentamer with four water molecules hydrogen bonded to a central water molecule (redrawn from Walrafen's (1973) diagram).

This notion of co-operative build-up of a cluster of hydrogen-bonded water molecules is central to Frank and Wen's mixture model for water (p. 236).

A key configuration in these water clusters, in ice and in clathrate hydrates is the pentamer (Fig. 4) in which one water molecule at the centre of a tetrahedron is hydrogen-bonded to four other water molecules at the vertices. Here two hydrogen bonds are formed by "H-transfer" and two by "electron transfer". The calculated average bond energy is similar to that for the dimer (Kollmann and Allen, 1970; Hoyland and Kier, 1969). More sophisticated calculations on the pentameric unit may reveal otherwise but the calculations involved are awesome.

Nevertheless, the facility of water molecules to form such a pentamer explains in part the complexity of aqueous solutions. Thus it is possible to arrange a series of linked pentameric units in space such that each water molecule lies at the centre of a tetrahedron and is hydrogen bonded to four water molecules at the vertices. This is a unique feature which sets water apart from other molecules (Symons, 1975). The situation can be compared with that in, say, methyl alcohol each molecule of which can form up to two hydrogen bonds by electron-transfer but only one by H-atom transfer. The resulting unit lacks the uniformity for growth in three dimensions.

4. WATER IN THE SOLID STATE

Ice

The general term *ice* describes a wide range of solid phases formed by water (Franks, 1973; Kamb, 1968). Preparation of a pure polymorph is not straightforward. Ice-Ih is the most widely studied polymorph but considerable information is available concerning many others. Eight solid phases, Ih, II, III, V, VI, VII, VIII and IX, are shown in the phase diagram for ice (Franks, 1973). In addition, two more polymorphs, ice-Ic and IV, are thermodynamically unstable.

The polymorph formed when water freezes at ambient pressure is called ice-Ih, the letter "h" indicating that the unit cell is hexagonal. In the unit cell (Fig. 5) each O-atom is surrounded tetrahedrally by four other O-atoms at a distance of 2.78×10^{-8} cm near the melting point. The structure is open, producing a low density/high volume system. Along each O—O link, there is one hydrogen atom. At a given instant, each oxygen atom has two hydrogen atoms at a distance 0.97×10^{-8} cm and two more at 1.79×10^{-8} cm. Neutron diffraction studies (Peterson and Levy, 1975) confirmed Pauling's suggestion (1960) that the hydrogen atoms are disordered.

The infrared and Raman spectra of ice is complicated by inter- and intramolecular coupling of vibrations (Bertie, 1968). However, the problem of spectral analysis is simplified by measuring the spectra of, for example, dilute H_2O in D_2O ice (Hornig et al., 1958). In this case, the O—H stretching vibration is uncoupled, having a broad absorption band at 3277 cm^{-1} with a band width of 50 cm^{-1} (Bertie and Whalley, 1964). The shift of the band from that produced by water vapour is consistent with hydrogen bond formation while the band-width reflects the disordered proton configuration.

The static permittivity, ϵ', of ice at 273 K is higher than that of water (Whalley et al., 1966; Wilson et al., 1965). However, ϵ'-values for two polymorphs, ice-II and ice-VIII, stand out as being very small and in the range expected for non-polar systems. These polymorphs

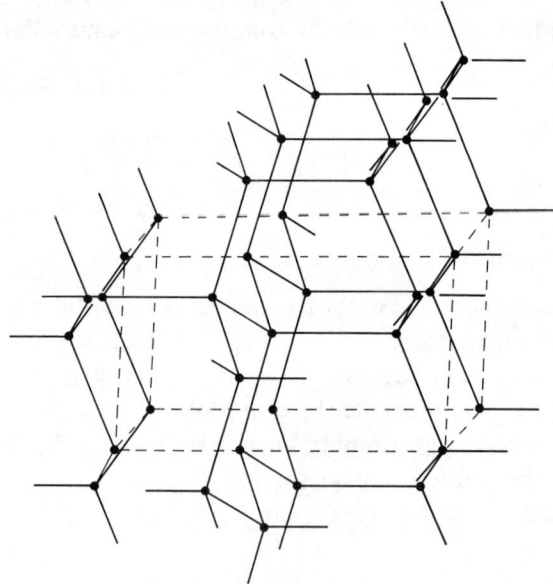

Figure 5. Structure of ice-lh; oxygen atoms are located at the vertices.

comprise discrete water molecules which do not exchange H-atoms between positions along O—O links.

In all the polymorphs, each water molecule is hydrogen bonded to four other water molecules. Polymorphs VI has a remarkable structure which comprises two discrete interpenetrating but not interconnected frameworks of H-bonded water molecules (Kamb, 1965). Similarly ice-VII and VIII comprise two interpenetrating frameworks of water molecules (Fig. 6) each framework having the same structure as the thermodynamically unstable polymorph ice-Ic. In ice-VII the hydrogen atoms are disordered, but they are ordered in ice-VIII.

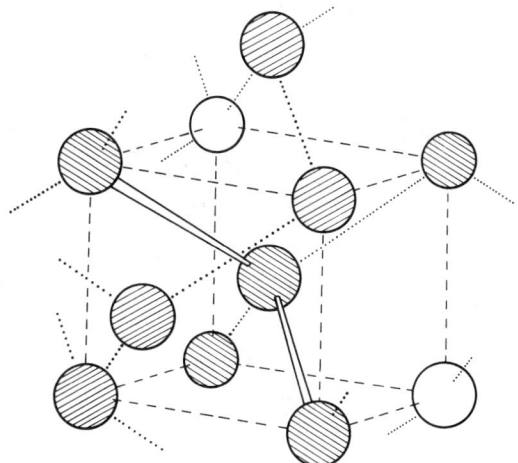

Figure 6. Structure of ice-VII showing oxygen positions (redrawn from Kamb and Davis, 1964).

Finally, we draw attention to freezing potential phenomena, the Workman-Reynolds effect (Gross et al., 1975; Gross, 1968; Drost-Hansen and Curry, 1970). A freezing potential is generated between an "ice" layer and aqueous solution by virtue of solute redistribution at an advancing ice/water phase boundary. Understanding such phenomena is important in the study of, for example, thunderstorms and cloud condensation. In another field of study, there has been speculation concerning the catalytic participation of ice structure in the kinetics of chemical reactions but the outcome of the discussion is not clear cut (Grant and Alburn, 1967).

Clathrate Hydrates

Solids formed by gases and water are called gas hydrates when the molecules of the volatile component (guest) are trapped inside a lattice (host) formed by hydrogen bonded water molecules (Davidson, 1973; Jeffrey and McMullan, 1967). The structures of these clathrate hydrates are reviewed from the standpoint of idealised structures of the water host framework, recognizing that in real clathrates, this framework describes one component of an at least two component system.

A basic building block in many clathrate hydrates is a pentagonal dodecahedron of water molecules (Fig. 7). This structure is found in gas hydrates of structure I (12 Å) and structure II (17 Å) types

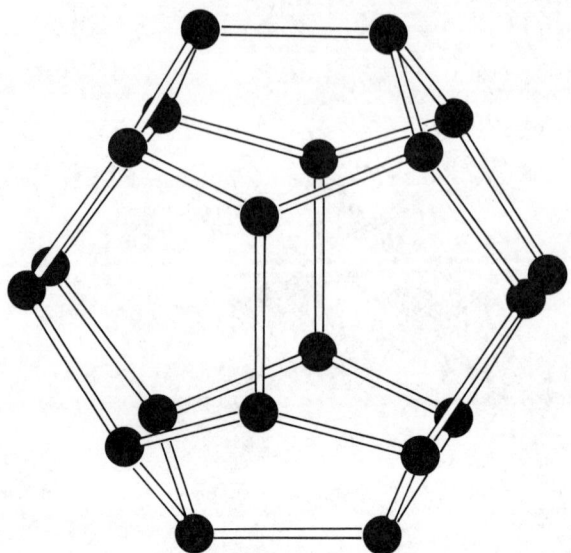

Figure 7. Pentagonal dodecahedron.

(Claussen, 1951). Molecules having radii less than ca. 5×10^{-8} cm, e.g. H_2S and CH_4 form structure I hydrates, while molecules with radii between 5·5 and $6·6 \times 10^{-8}$ cm, e.g. CH_3I, $(CH_3)_3CF$, C_2H_5Br and tetrahydrofuran, form structure II hydrates.

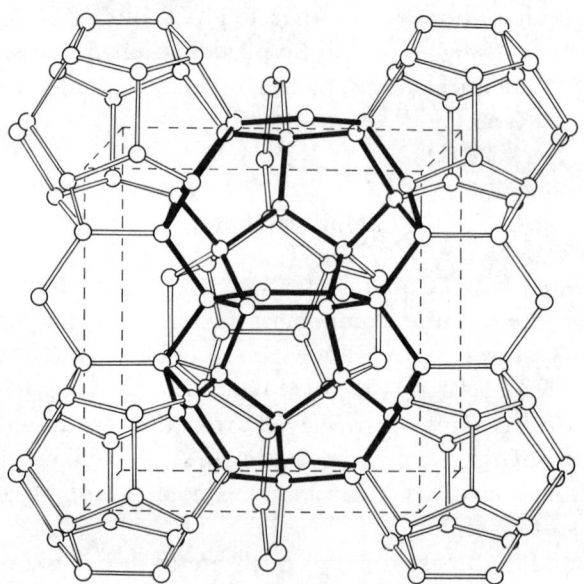

Figure 8. Host framework for Structure I clathrate hydrates showing oxygen positions in "ideal" systems.

The structure I hydrate (Fig. 8) has a cubic unit cell containing 46 water molecules, the positions of the hydrogen atoms being disordered. (McMullan and Jeffrey, 1965). Each unit cell has two almost spherical cages of radius = 3.91×10^{-8} cm, formed by pentagonal docahedra, and six larger holes, tetrakaidecahedra, of radius = 4.33×10^{-8} cm. If both sets of cages are filled by guests M, then the ideal stoichiometry is $M . 5.75 H_2O$. Large guests, e.g. ethylene oxide, may only occupy the larger holes; the composition is $M . 6.89 H_2O$.

There are 146 water molecules in the unit cubic cell of a structure II hydrate. Each cell has 16 small, pentagonal dodecahedral cages of

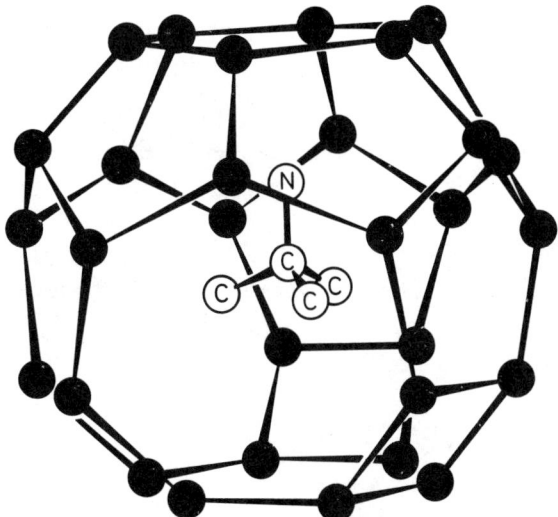

Figure 9. t-Butylamine hydrate; position of oxygen atoms only are indicated in the heptakaidecahedron (redrawn from Davidson, 1973).

radius ca. 3.9×10^{-8} cm and eight large cages, hexakaidecahedra, of radius 4.68×10^{-8} cm. When the larger holes are occupied by guests the composition is $M.17 H_2O$. Hydrates are also known where two guests are present, the small holes being occupied by small molecules and large holes by large molecules. In such cases, the small molecules act as a "help-gas", increasing the stability of the hydrate. For example, tetrahydrofuran forms a double hydrate with H_2S, an important help-gas, which occupies the small holes (Mak and McMullan, 1965). The proton arrangement in these hydrates is disordered (e.g., trimethylene oxide hydrate; Hawkins and Davidson, 1966).

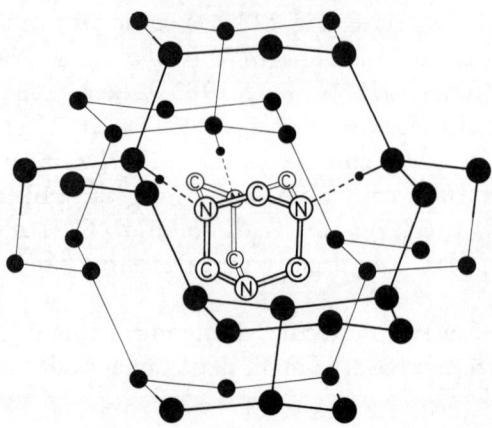

Figure 10. Hexamethylene tetramine hexahydrate (redrawn from Mak, 1965); oxygen atoms represented by ●.

In the absence of guest molecules, the water host framework is unstable. The stability stems from van der Waals interactions between guest and host and is dependent on the extent of occupancy of the cages by the guest.

In the t-butylamine hydrate (M . 9·75 H_2O), the guest molecules are held in heptakaidecahedral cages (Fig. 9), but guest and host are not linked by hydrogen bonds (McMullan *et al.*, 1967). This

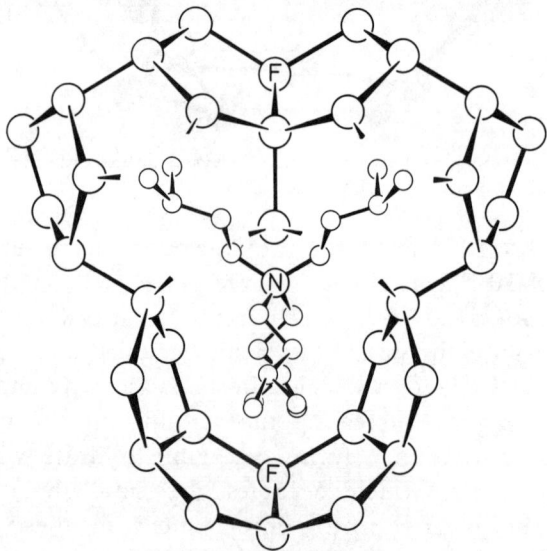

Figure 11. Tetra-iso-amylammonium fluoride hydrate (redrawn from Feil and Jeffrey, 1961).

behaviour is rare, and in most amine hydrates the amine is hydrogen-bonded to the host. For example, in the hexamethylenetetramine hydrate (M . 6 H_2O), the three nitrogen atoms from each guest are linked to the host by hydrogen bonds (Mak, 1965), but the rest of the molecule sits in a cage formed by water molecules, the guest hanging "batlike" to the cage wall (Fig. 10). This ability of water molecules to encapsulate but not interact with the apolar moiety of a guest is clearly seen in the structures of alkylammonium salt hydrates.

When alkylammonium salts are recrystallized from water, they often form solids with very low melting points (Fowler et al., 1940). An example is (iso-Am)$_4$N$^+$.F$^-$ 40 H_2O whose structure (Feil and Jeffrey, 1961) shows that the amyl groups are held in cages of water molecules located tetrahedrally around the N$^+$-charge centre of the "onium ion" (Fig. 11). Similar structures are observed in tetra-n-butylammonium (McMullan et al., 1963; Bonamico et al., 1962) and tri-n-butylsulphonium salt hydrates (Beurskens and Jeffrey, 1964).

5. WATER

Properties of Water

In this short survey, we examine briefly some properties of water which set it apart from other liquids. The pressure will be assumed to be atmospheric, thereby excluding the fascinating subject of super-critical water (Tödheide, 1973; Franck, 1970).

"pVT"-properties

Most systems expand by about 10% on going from the solid to liquid states, the expansion resulting from a decrease in co-ordination number rather than from a change in nearest neighbour distances. In contrast, the transition, ice to water, is accompanied by a decrease in molar volume and an increase in density of about 10%. Moreover, on raising the temperature, the volume decreases passing through a minimum near 277 K, the temperature of maximum density (TMD); subsequent increase in temperature results in an increase in volume

(Kell, 1967). With increase in pressure, the TMD decreases so that when $p > 400$ bar, no TMD is observed (Kell, 1970; 1975; Garnjost, 1974). That the TMD of water is 277 K has no special significance; other physical properties of water show extrema at other temperatures (Kell, 1973). Thus the compressibility, κ_T, and isentropic compressibility, κ_S, have extrema at 319 and 337 K respectively. The heat capacity, C_p^o, of water is twice that of the solid at 273K, the large C_p^o for water being vital to the environment (Franks, 1968). This heat capacity is relatively insensitive to temperature, having a shallow minimum near 308 K.

Viscosity

The shear viscosity, η_s, of water decreases with increase in temperature (Korson et al., 1969). For most liquids, η_s increases with increase in pressure, but for water below 300 K it decreases (Bett and Cappi, 1965). Above 300 K, η_s increases with increase in pressure.

X-ray scattering (Narten and Levy, 1973)

At low temperatures, just above the melting point of ice, the distance between nearest-neighbour oxygen atoms, $2 \cdot 84 \times 10^{-8}$ cm at 277 K, is only slightly larger than that in ice. This distance increases to $2 \cdot 94 \times 10^{-8}$ cm at 473 K. The second and third nearest neighbours occur at distances centred on $4 \cdot 5$ and 7×10^{-8} cm respectively. The first co-ordination shell is complicated but the co-ordination number is estimated to be $4 \cdot 4$. This 10% increase in co-ordination number over that in ice is consistent with a filling-in of space in the lattice. The X-ray data (Narten et al., 1967) can be interpreted in terms of an open network of hydrogen bonded water molecules having sufficiently large cavities to accommodate non-hydrogen bonded water molecules. This model also takes account of an absence of long-range correlation ($>8 \times 10^{-8}$ cm), and random occupancy of vacancies and network positions.

Spectroscopic properties

These have attracted considerable interest (Walrafen, 1973) because, in principle, detailed information should be forthcoming concerning the structure of water. For example, many models require that there exists in water both free and hydrogen bonded—OH groups. Consequently spectroscopic studies might be expected to identify absorption bands characteristic of these two types of OH-groups (Luck, 1973). In the event that two types of OH groups are detected, this does not mean that there are free water molecules in water; a given water molecule may have one non-hydrogen bonded OH group but still be involved in up to 3 hydrogen bonds with other water molecules. However, the interpretation of the spectra of water has aroused considerable debate and differences of opinion.

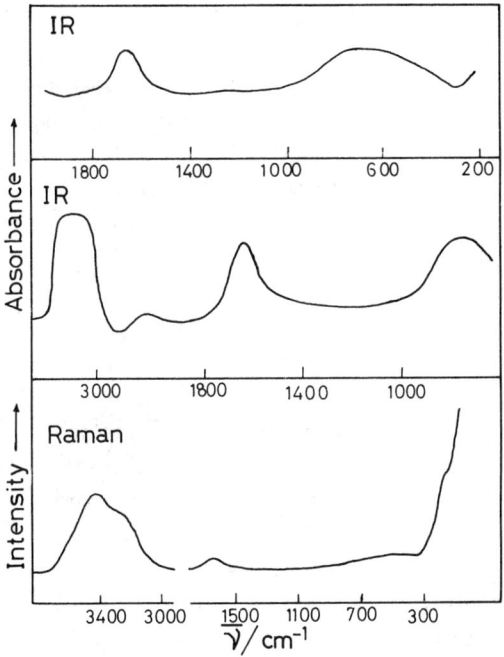

Figure 12. Raman and infrared absorption spectra of water at 298 K (redrawn from Eisenberg and Kauzmann, 1969).

The Raman and infra-red absorption spectra (Fig. 12) can be divided into two regions, high and low energy, corresponding to the intramolecular and intermolecular motions. A broad absorption band in the 3500 cm^{-1} region is assigned to the O—H stretching vibration, the shift of the band from the position in the gas phase being consistent with the influence of intermolecular interactions. This band is significantly broader than the corresponding band in methyl alcohol, indicating the complexity of water (Adams *et al.*, 1971). A band near 1600 cm^{-1} is assigned to the HOH bending motion and an "association band" in the 2000 cm^{-1} region is made up of overtones and combinations of the bending mode with librational modes. Intermolecular motions, libration and translation, produce bands in the 600, 200 and 60 cm^{-1} regions, these being observed in both Raman and neutron scattering experiments (Page, 1973). Walrafen (1973) assigns the 60 cm^{-1} and 175 cm^{-1} Raman bands to hydrogen bond bending and stretching motions respectively. The dependence of band intensities on temperature together with polarization effects can be thus explained in terms of the pentameric unit (Fig. 4). This arrangement describes that fraction of water molecules involved in hydrogen bonding, i.e. most water molecules at low temperatures.

The major controversy arises in the interpretation of spectra from the intramolecular region, particularly with reference to the search for evidence of free OH groups. It is, for example, tempting to conclude from the overtone infra-red spectra reported by Peron *et al.* (1971) that there is a contribution to the absorption from free OH, which grows as the temperature increases. Some years ago, Buijs and Choppin (1963) attempted to analyse the near infra-red spectra in terms of contributions to the absorbance from water molecules having different numbers of hydrogen bonds. However, such analyses are equivocal (Boettger *et al.*, 1967). Moreover, the breadth of the absorption bands in the fundamental OH stretch region may not stem from water molecules with different numbers of hydrogen bonds but from a distribution of absorptions from water molecules with distorted hydrogen bonds (Schiffer and Hornig, 1968). Vibrational spectra of aqueous systems can often be simplified by measuring the absorption spectra of isotopic mixtures. The infra-red spectra of the uncoupled OH vibration in D_2O show a broad absorption band, but this can be interpreted in terms either of a single peaked distribution of oscillators or of free and bonded OH groups (Falk and Ford, 1966; Ford and Falk, 1968). Also addition of trimethylamine, which forms a solid clathrate hydrate, produces no striking changes in the spectra of the liquid consistent with a structure forming action in which the proportion of free OH decreases (Falk, 1971). Consequently, many investigators conclude that water does not contain both free and hydrogen-bonded OH groups. However, a rather different conclusion was reached by Senior and Verrall (1969), who showed that superimposed spectra of uncoupled OD vibrations (i.e. HOD in H_2O) at several temperatures formed isosbestic points consistent with an equilibrium of the type

$(OD)_{free} \rightleftharpoons (OD)_{bonded}$ (Fig. 13). This interpretation has nevertheless been disputed on the grounds that the existence of isosbestic points cannot unequivocally be attributed to equilibria between different species (Falk and Wyss, 1969).

In the overtone region of the spectra of HOD in D_2O, the intensity of a band at 7062 cm^{-1} (ca. twice the fundamental stretching frequency for OH), increases as the temperature is raised, superimposed spectra showing isosbestic points (Worley and Klotz, 1966). This band was assigned to free OH groups (see also Luck and

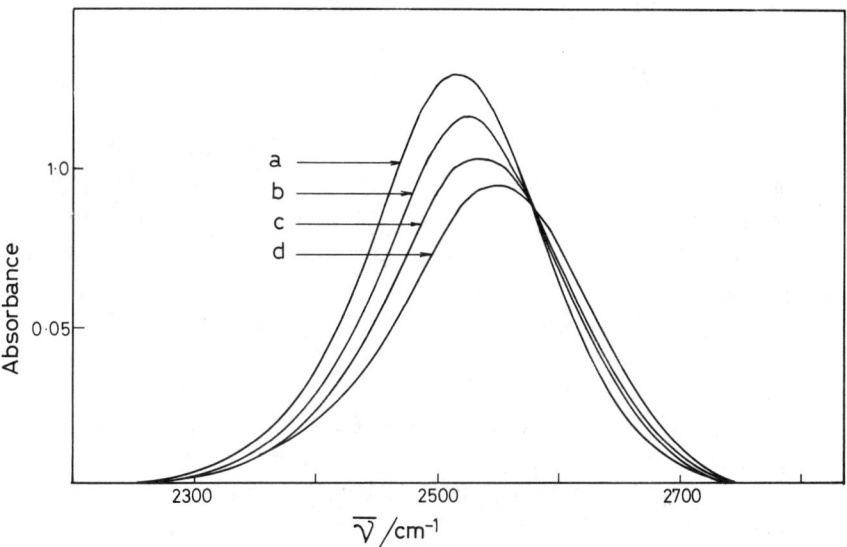

Figure 13. Infrared absorption spectra of dilute solutions of HOD in H_2O at four temperatures; (a) 302 K, (b) 320 K, (c) 340 K, and, (d) 360 K, with isosbestic point at ca. 2575 cm^{-1} (Senior and Verrall, 1969).

Ditter, 1967; 1969). Addition of t-butyl alcohol results in a fall in intensity of the 7062 cm^{-1} band, consistent with a structure forming action by this solute and providing further confirmation (Blandamer and Symons, 1968).

Raman spectra of uncoupled OH and OD stretch bands show a pronounced asymmetry on the high energy side (Walrafen, 1973). Analysis of the band shape reveals contributions from two bands assigned by Walrafen to non-hydrogen bonded and hydrogen bonded oscillators. Addition of sodium perchlorate, a water structure breaker, results in an increase in intensity of the high energy component, consistent with an increase in proportion of free OH.

However, this particular piece of evidence is not convincing in view of the dependence on anion of the position of the new band which is assigned to solvent molecules hydrating the anion (Adams *et al.*, 1971). Nevertheless the asymmetry of the band observed in the absence of salt appears consistent with the presence of free OH groups and this conclusion receives support from stimulated Raman spectra arising from a two photon interaction between incident laser and Stokes quanta. Walrafen (1972) has shown how the spectra change as the proportion of H_2O to D_2O is altered, and he assigns the bands in the OH stretch region, in some systems, to free OH and, in others, bonded OH (Coles *et al.*, 1970).

Dynamic Properties

The dielectric relaxation of water (Hasted, 1973) can be characterized by a relaxation time $\tau_D = 9\cdot3 \times 10^{-12}$ s at 293 K with activation energy 20 kJ mol^{-1}. The spread of relaxation times is remarkably small for such a complicated liquid. The data are interpreted in terms of rotation by water molecules having two hydrogen bonds, the spread of relaxation times showing that symmetrically hydrogen bonded and asymmetrically hydrogen bonded water molecules have slightly different relaxation times.

The self-diffusion coefficient for water has been measured using various techniques including tracer and nmr methods; $D = 2\cdot5 \times 10^{-5}$ cm s^{-1} at 298 K. The values of D and τ_D can be combined to yield a jump distance, $3\cdot7 \times 10^{-8}$ cm, which is the distance between equilibrium positions for water molecules in the liquid (Wang, 1965).

Quasi-elastic neutron scattering data can be interpreted (Page, 1973) in terms of a jump-diffusion model, according to which a water molecule vibrates about an equilibrium site for a time t before jumping to a new site in a much shorter time. Franks, Ravenhill, Egelstaff and Page (1970) calculate that at 298 K, $t = 4\cdot7 \times 10^{-12}$ s, which means that each molecule executes some 40 oscillations before jumping to the next position. The structural relaxation time, $\tau_P = 4\cdot3 \times 10^{-12}$ s, obtained from ultrasonic absorption studies (Davis and Jarzynski, 1967; 1973), is shorter than τ_D, a trend observed for H-bonded liquids. Nmr relaxation experiments (Glasel, 1973) yield a rotational reorientation time of $2\cdot5 \times 10^{-12}$ s (298 K).

This information can also be brought together using three structural representations (Eisenberg and Kauzmann, 1969). Consider a collection of water molecules in the liquid. At a given instant in time, their organization represents the instantaneous or I-structure. If the positions of the molecules are averaged over a time shorter than approximately 10^{-13} s, the organization is represented by the vibrationally averaged or V-structure. However if this time is longer than 10^{-11} s, we have to consider the diffusionally averaged or D-structure. Unfortunately we do not have a camera which can take molecular photographs with these various exposure times. Consequently the way is clear for considerable debate over what these I, D and V structures might be.

Computer experiments

The starting point for recent theoretical studies of water is an expression for the pair potential for two water molecules which recognizes the stereochemical features of the hydrogen bond (Stillinger and Ben-Naim, 1969) and the non-additivity of the interaction (Stillinger, 1970). A Bjerrum four-point-charge model has been used to mimic the characteristics of hydrogen bonding (Ben-Naim and Stillinger, 1971; Ben-Naim, 1973) in a molecular dynamic study of a system comprising 216 water molecules (Rahman and Stillinger, 1971). Stereoscopic pictures of the system show that this "computer-water" consists of a strained random hydrogen-bonded network of molecules, but the structure does not resemble any solid forms of water. There appears to be no evidence for a separation into two types of water molecules, network and interstitial, although free OH groups are present. However, there are indications that the Bjerrum model used here over stresses the directional nature of hydrogen bonding (Stillinger and Rahman, 1972), but, even after modification, the molecular dynamic calculations still show no similarity between the water structure and that found in the solid state (Rahman and Stillinger, 1973; Stillinger and Rahman, 1974). Rather the liquid seems to involve polygons of hydrogen-bonded water molecules (Lentz et al., 1974) with no marked preference for even or odd order polygons and with some polygons of 12 or more sides (Rahman and Stillinger, 1973). A related study of sound propagation in water (Rahman and Stillinger, 1974) has indicated that the density fluctuations in water have novel features.

The Bjerrum model for a water molecule has also been used in conjunction with a cell model (Angell, 1971) for water based on the ice-VIII structure (Weres and Rice, 1972). However, it is clear that the major problem in these studies is the formulation of the potential function and various solutions to the problem have been examined (Ben-Naim, 1972, 1972b, 1973, 1973b). For example, a Hartree-Fock potential for water-water interactions (Kistenmacher et al., 1974) has been used in conjunction with the Monte-Carlo method to simulate clusters of water molecules (Abraham, 1974). A new class of models has been examined by Lemberg and Stillinger (1975) in which the monomeric water molecule is represented by three effective point changes, i.e. a central force model. Molecular dynamic studies based on this model are apparently quite promising.

Structure for Water

Not unexpectedly, the widespread and continuing interest in water has prompted many molecular models for water (Perram and Levine, 1974; Ives and Lemon, 1968; Frank, 1973). We will not review all the attempts at formulating a structure for water. This brief account will reflect the opinions of the author and recognize that in chemists' terms we are interested in providing a basis for a discussion of the properties of aqueous solutions. The overiding difficulty is that an account of water structure also reflects the general problems involved in describing the liquid state (Powles, 1974).

The X-ray scattering, spectroscopic and thermodynamic properties confirm that in water a large proportion of the molecules are hydrogen bonded together in an open, low-density arrangement. Note that a strong interaction, hydrogen bonding, produces low local density (Ben-Naim, 1972c). Regions of water having this structure (part of the V-structure) might be formed as a result of the co-operative nature of hydrogen bonding. By the same token, local thermal fluctuations may result in a breakdown of this organization. Thus, water might comprise clusters of open low-density hydrogen bonded water molecules, $(H_2O)_b$, and dense non-hydrogen bonded molecules $(H_2O)_d$ (Frank and Wen, 1957). In these terms, water can be described in terms of equilibrium (16), where $\Delta V < 0$ and $\Delta H > 0$.

$$(H_2O)_b \rightleftharpoons (H_2O)_d \qquad (16)$$

The V-structure of water then corresponds to clusters of water molecules separated by non-hydrogen bonded water molecules. With increase in temperature, so the proportion of $(H_2O)_d$ increases (cf.

Raman spectra). The lifetime of the clusters is of the order 10^{-11} to 10^{-12} s, so that by superimposing many representations of the V-structure we obtain the D-structure. The co-operative hydrogen bonding may be one reason for the differences between relaxation times obtained by different techniques (Frank, 1973) so comparison between experiment and theory is not straightforward. However, X-ray scattering data show that water does not exist as distinct patches of dense and bulky water. Consequently considerable interest has been shown in interstitial models for water (Samoilov, 1965).

In the Frank–Samoilov model for water, $(H_2O)_d$ describes molecules which are interstitial guests in the $(H_2O)_b$ system, the actual structure of which is not specified. It may be based on, for example, pentagonal dodecahedra (p. 226) or on ice-Ih (p. 224), the latter idea being currently favoured especially for water at low temperatures. Calculations indicate that 20% of the water exists in the interstitial form.

There are several advantages, particularly in the context of aqueous solutions, in representing water using eqn (16). Thus, to a first approximation, a solute which increases $(H_2O)_b$ at the expense of $(H_2O)_d$ is a structure former; a structure breaker has the opposite effect. The large heat capacity for water can be attributed to the need to "melt" part of $(H_2O)_b$. In these terms, the partial molar heat capacity of solutes in water often indicates their effect on water structure.

The above account has provided sufficient background for analysis of the properties of aqueous solutions. The analysis has been restricted to bulk water; the properties of water near interfaces, including biological surfaces, is very interesting but outside the scope of this review. It should be noted, however, that the properties of "vicinal" water differ from those of bulk water, these differences being important in biological systems (Drost-Hansen, 1972; 1973). Thermal anomalies in the properties of water also seem explicable in terms of interfacial phenomena (Drost-Hansen, 1968).

6. SOLUTES IN WATER

Hydration

The chemistry of solutes in water is considered in general terms in this section. First we examine those properties which determine the standard state properties of solutes in solution, e.g. partial molar volume, $V^{\ominus}$, and partial molar heat capacity, $C_p^{\ominus}$. These are determined by the intrinsic properties of the solute and by the solute–water interactions. We can imagine, therefore, a solute molecule in water which is in effect infinitely far from any other solute molecule. In the next section we consider real solutions and the impact of solute–solute interactions.

The hydration characteristics of a solute are determined by a number of factors which can be summarized as follows (Franks, 1973a): (i) the electrical properties, e.g. charge, dipole moment, (ii) the number of hydrogen bonding sites on the solute, (iii) the size of any apolar residue, e.g. the size of the alkyl group R in a monohydric alcohol, ROH, (iv) the degree of unsaturation or aromaticity of the hydrocarbon part of the molecule, and (v) the relative positions of polar groups in the solute, together with their freedom to rotate and their conformation.

In each case, one or more of these factors determine the hydration characteristics of a particular solute. For example, to a first approximation, ion–water interactions can be discussed in terms which stress the importance of the sign and magnitude of the charge on the ion. However, the properties of alkylammonium ions in water are controlled to a considerable extent by the apolar alkyl groups. Indeed, a balance is struck between the demands of water–water interactions and water–solute interactions. Consequently a situation arises where a solute molecule can modify water–water interactions at some distance from the solute, the extent of this modification defining a co-sphere of solvent (Gurney, 1953). Co-spheres of ions probably comprise two parts, an inner and an outer co-sphere. The different forms of hydration produce different types of co-spheres. Friedman and Krishnan (1973b) have identified two main classes of co-sphere types (Table 2). In state I (hydration of the first kind) the structure of water is controlled by strong solute-solvent interactions, e.g. charge–dipole interactions in ionic hydration. In state II (hydration of the second kind) the water is perturbed by the proximity of the solute molecule.

The stereochemistry of a solute is also important, as is shown by the differences between the partial molar volumes and compressibilities of stereoisomers of tartaric acid and its potassium salts in water (Mathieson and Conway, 1975).

The summary of hydration types in Table 2 can be used as a basis for the description of the hydration of more complicated solutes. For example, with reference to the ion $C_4H_9N^+H_3$, state II_{a1}

describes the water around the alkyl group and state I_{hb}, that around the $-N^+H_3$ group.

An experimental basis has been suggested by Hepler (1969) for the classification of solutes; for structure formers, $(\partial^2 V_2^\ominus/\partial T^2) > 0$ and for structure breakers <0. More recently, Ben-Naim (1975) has demonstrated how the difference between solubilities of a solute in D_2O and H_2O can also be used in a similar fashion. Ben-Naim (1972a, 1973a) has also shown from theoretical arguments how it is possible for a solute to stimulate H-bond formation, i.e. structure-formation, between water molecules. However, the effect of a structure-former on water is not straightforward, the induced structure being apparently different from that of pure water at a lower temperature (Hertz, 1970).

TABLE 2

Water in Solute Co-spheres

STATE I. Hydration of the FIRST KIND
  State I_c: inner co-spheres of small cations
  State I_a: inner co-sphere of small anions
  State I_{hb}: inner co-spheres of hydroxyl groups and $R_3\overset{+}{N}H$ ions

STATE II. Hydration of the SECOND KIND
  State II_{rg}: co-sphere of rare gas atom
  State II_{al}: co-sphere of an alkyl group
  State II_{ar}: co-sphere of an aromatic group
  State II_{sb}: outer co-sphere of small ions.

The classification given in Table 2 (cf. Mathieson and Conway, 1947a,b) points to the importance of the organization around a solute of solvent molecules beyond the nearest neighbour solvation shell. Some indication of the interactions between ions and nearest neighbour water molecules can be obtained from mass spectroscopic studies (Kebarle, 1974).

Real solutions

The concept of a solvent co-sphere surrounding a solute particle leads to the consideration of what happens in real solutions where these co-spheres may overlap. Consider two solute particles, i and j, in solution which approach so that their co-spheres overlap (Fig. 14;

Friedman and Krishnan, 1973b). As a result of overlap, some of the solvent co-sphere is displaced and if, for example, the effect of solute j on the solvent dominates the process of overlap, then the overlap can be represented as in Fig. 14(a), the change in the solvent being summarized by the reaction in Fig. 14(b). This mutually destructive overlap can be characterized by the free energy change, A_{ij}, for the "solvent" reaction, where A_{ij} is related to the thermodynamic energy, U_{ij}, and entropy, S_{ij} by the expression $A_{ij} = U_{ij} - T \cdot S_{ij}$.

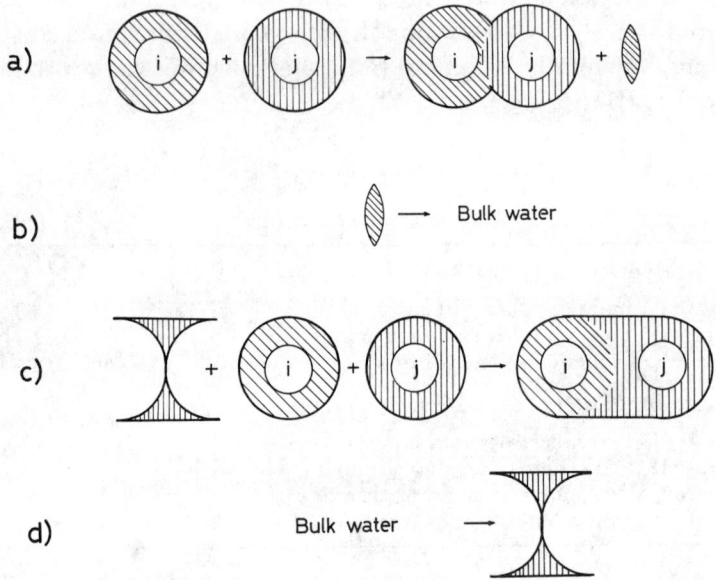

Figure 14. Diagrammatic representation of the process of co-sphere overlap as two solute species, i and j, come together; (a) mutually destructive co-sphere interaction where the co-sphere of solute j dominates the process leading to the solvent reaction shown in (b); in process (c) the interaction is mutually constructive, additional solvent being incorporated into the co-sphere as shown in reaction (d) assuming solute j dominates (Friedman and Krishnan, 1973).

The converse of the process constructive interaction, is represented in Fig. 14(c); the solvent reaction is in Fig. 14(d). Considerable efforts have been made to obtain thermodynamic parameters which quantify this overlap process.

Quantitative theories of salt solutions are well established. In dilute salt solutions, ion–ion interactions can be described to a first approximation, in terms of charge–charge interactions. These long-range interactions produce a stabilizing influence, the chemical potential of the salt in solution being less than in the corresponding ideal solution. For very dilute solutions, the dependence of the properties of a salt solution on concentration can be described in terms of the Debye-Hückel limiting law (DHLL). However, this limiting law often fails for salts in water because of the neglect of other important factors associated with the structures in the solvent co-spheres (Gurney, 1953; Frank, 1963; Frank and Robinson, 1940). The limiting law can be replaced by more sophisticated equations which take account of ion-size, as in the Debye-Hückel equation, or of specific ion–ion interactions as in the Guggenheim–Guntelberg equation. It is, however, convenient to use the DHLL as a reference and examine the properties of real solutions in terms of the extent to which they deviate from those predicted by the DHLL. This approach not only applies to the analysis of chemical potentials and related quantities (e.g. equilibrium constants), but also to other properties such as partial molar volumes and heat capacities.

With increase in salt concentration the approximations involved in the Debye-Hückel theory become less acceptable. Indeed it is noteworthy that before this theory was published a quasi-lattice theory of salt solutions had been proposed and rejected (Ghosh, 1918). However, as the concentration of salt increases so log $\gamma_\pm$, $\gamma_\pm$ being the mean ionic activity coefficient, appears as a linear function of $c^{1/3}$ (the requirement of a quasi-lattice theory) rather than $c^{1/2}$, the DHLL prediction (Robinson and Stokes, 1959). Consequently, a quasi-lattice theory of salt solutions has attracted continuing interest (Lietzke *et al.*, 1968; Desnoyers and Conway, 1964; Frank and Thompson, 1959; Bahe, 1972; Bennetto, 1973) and has recently received some experimental support (Neilson *et al.*, 1975).

The non-ideality of the solvent in an aqueous 1:1 salt solution can be expressed in terms of a practical osmotic coefficient, ϕ (17),

$$\mu_1 = \mu_1^\circ - 2\phi RT m_2 M_1/10^3 \tag{17}$$

where $\phi \to 1$ as $m_2 \to 0$, M_1 is the relative molecular mass of the solvent and μ_1 is the chemical potential of pure solvent at the same T and p. Of course, the chemical potentials of salt, μ_2, [eqn (14)] and solvent, μ_1, are not independent, their interrelation being described by the Gibbs-Duhem equation. A convenient way of examining the properties of a salt solution, "in toto", is through thermodynamic excess functions (Friedman, 1960). Thus eqns (14) and (17) can be combined (18) to express the difference per mole of salt between the

$$G^E = 2 RT(1 - \phi + \ln \gamma_\pm) \tag{18}$$

Gibbs functions for a real salt solution and the corresponding ideal salt solution containing m_2 moles of salt in a kilogram of solvent. G^E can be calculated for many aqueous salt solutions from published values of ϕ and $\gamma_\pm$. In the same way, the corresponding excess enthalpy H^E can be defined and this equals the apparent partial molar enthalpy. Thus the properties of salt solutions can be examined in plots of G^E, H^E, and $T \cdot S^E$ against m_2, where S^E is the

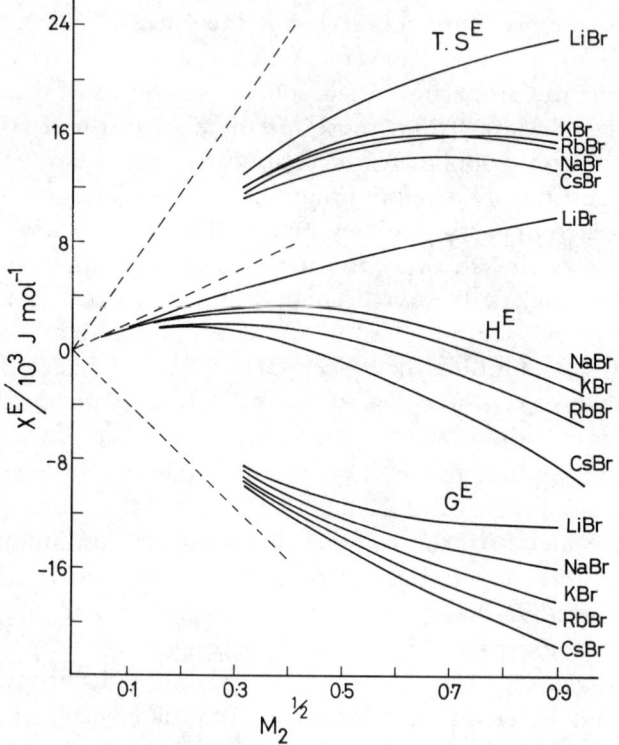

Figure 15. Excess thermodynamic functions for various 1:1 salts in water at 298 K; m_2 = molality of salt and the dotted lines indicate the behaviour predicted by the Debye-Hückel limiting law (Fortier et al., 1974).

related excess entropy. This form of analysis is noted here because it may prove useful in the analysis of salt effects on kinetics of reactions. These excess quantities have the merit that they reflect the non-ideality of both salt and solvent. Thus quite marked deviations are observed from the behaviour predicted by the DHLL for 1:1 salts in water at 298 K (Fig. 15; Fortier et al., 1974a). It is also noteworthy that H^E and S^E are more sensitive to the change from solution in H_2O to solution in D_2O than is G^E (Wu and Friedman,

1966), and that the pattern of the dependence on concentration is more complex for H^E and S^E than for G^E (Fig. 16). When the value of an excess function is not known at a particular salt concentration, it can be obtained by interpolation. Wu and Friedman have suggested a general equation for the dependence of X^E on m_2 for a salt solution and this is satisfactory even for aqueous solutions of alkylammonium salts (Blandamer and Waddington, 1970).

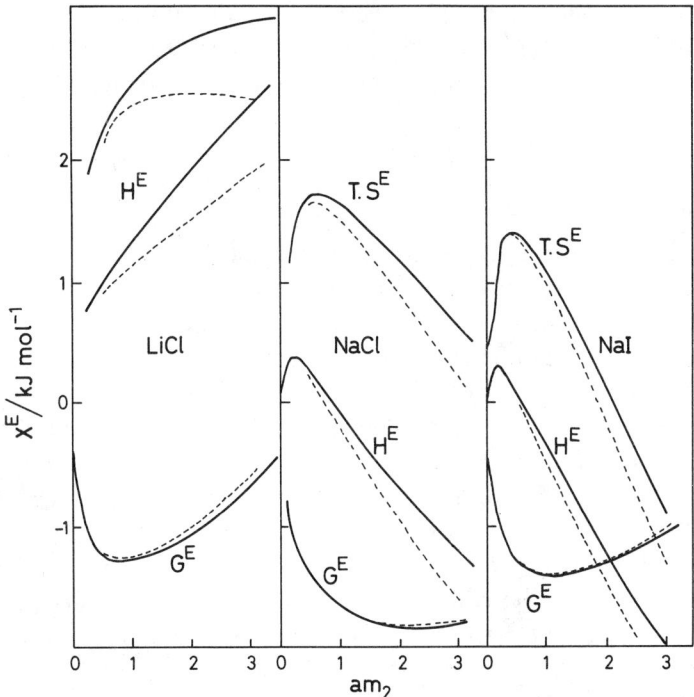

Figure 16. Excess functions for three 1:1 salts in water and deuterium oxide as a function of the aquamolality of the salt solution, am_2, at 298 K (Wu and Friedman, 1966).

Considerable information concerning structural effects on aqueous salt solutions has been provided by studies of the properties of mixed solutions (Anderson and Wood, 1973). In a mixed salt solution prepared by mixing $Y_A m$ moles of a salt MX (molality m) with $Y_B m$ moles of a salt NX (molality m) to yield m moles of mixture in 1 kg of solvent, if W is the weight of solvent, the excess Gibbs function of mixing $\Delta_m G^E$ is given by (19) where G^E is the excess function for

$$\Delta_m G^E/W = G^E/W - [Y_A G^E(MX)/W] - [Y_B G^E(NX)/W] \quad (19)$$

the mixture, G^E (MX) and G^E (NX) being the excess functions for the two salt solutions. To a first approximation, the dependence of $\Delta_m G^E$ on the product of the mole fractions, $Y_A Y_B$ is given by eqn (20) where I is the ionic strength. This equation defines an inter-

$$\Delta_m G^E/W = RTI^2 Y_A Y_B g_0 \qquad (20)$$

action parameter g_0 applying to Gibbs functions. In a similar fashion, enthalpy data yield h_0 and volume data v_0. The quantity g_0 can be related to a parameter b_{AB}^0 which describes interactions between pairs of like charged ions, e.g. MM, NN, and MN, and to another parameter which describes interactions involving triplets, e.g. MMX, NNX and MNX.

The analysis of the properties of mixed salt solutions and single salt solutions have shown how the non-ideality of real solutions can be understood in terms of overlap of the solvent co-spheres. In other words solute–solute interactions are transmitted by and through intervening solvent molecules. An example of the importance of such long-range effects in aqueous solutions is shown in the volumetric properties of monohydric alcohols in water. The partial molar volume of t-butyl alcohol, V_2, differs from the standard state value, $V_2^{\ominus}$ ($= V_2$ in the limit $x_2 \to 0$) even in very dilute solution, when $x_2 \simeq 10^{-3}$, i.e., when the solute molecules are on average 60×10^{-8} cm apart (Franks and Smith, 1968). In other words, the two alcohol molecules in water at this distance apart still "feel" each other's presence through the intervening water.

In other cases, the solute particles in water may come into close proximity (cf. ion pair formation in salt solutions). In aqueous solutions, there is an important phenomenon called hydrophobic association (p. 254).

The ideas outlined so far have been qualitative. It would be valuable to place the concept of co-sphere interaction on a quantitative footing, and this would seem to be the direction of future studies in the field of aqueous solutions.

The starting point is an expression for the intermolecular potential energy, U_{ij}, for two solute particles, i and j, distance r apart in solution. From this expression it is theoretically possible to calculate the thermodynamic properties of the solution. The quantitative link is provided by the radial distribution function, $g(r)$, which provides information concerning the distribution of particles in solution.

In a gas $g(r)$ is related to the pair distribution function, the probability of finding two molecules simultaneously at given distances from some reference point. In an ideal gas, $g(r)$ is unity and so the total correlation function, $h(r) = g(r) - 1$, is non-zero for real systems. This and related correlation functions can be linked to the thermodynamic properties of the system and the pair potential functions (Rowlinson, 1969). The equations are complicated but various standard approximations are made; one such is called the hypernetted chain approximation (HNC) to the results of which we shall refer later. Extension of these statistical theories to describe the properties of solutions was made possible by the work of McMillan and Mayer (1945). The HNC technique can be applied to salt solutions (Allnatt, 1964; Friedman, 1971). For example, Rasaiah (1970) obtained good agreement between observed and calculated thermodynamic properties for 1 : 1 salt solutions over the range $0 \leq m_2 < 1\cdot0$ mol dm^{-3} (well beyond the DHLL range) using a charged sphere model for ions. The possibility that a link can be obtained between a quantitative expression for U_{ij} and both $g(r)$ and the thermodynamic properties using the HNC approximation heralds a new era in this classical subject (Rasaiah, 1973), and the way is clear for an examination of more sophisticated expressions for U_{ij}.

Ramanthan and Friedman (1971) proposed the four term expression (21) for U_{ij}. The first term, $COUL_{ij}$, describes the Coulombic

$$U_{ij}(r) = COUL_{ij} + COR_{ij} + CAV_{ij} + GUR_{ij} \qquad (21)$$

interaction between ions i and j in a medium of permittivity ϵ_r. The term COR_{ij} describes the repulsive core potential between two hydrated ions having radii r_i and r_j. As two oppositely charged ions come together, $COUL_{ij}$ decreases but COR_{ij} increases rapidly when r is small. The quantity CAV_{ij} describes the electrical interaction energy resulting from (ion-cavity)-(ion-cavity) repulsion. The most interesting term is GUR_{ij}, the Gurney potential energy, which describes the effect of co-sphere overlap on $U_{ij}(r)$. Quantitatively, GUR_{ij} is related to the radii of the ions together with their co-spheres (Fig. 14) through the quantity A_{ij} (see above). A positive A_{ij} means that GUR_{ij} is also positive, increasing as r decreases so that the ions are repelled as their co-spheres overlap. The dependence of the separate contributions to $U_{ij}(r)$ on the ratio $r/(r_i + r_j)$ for NaCl in water is shown in Fig. 17 for $A_{ij} = 418$ J mol^{-1}. Thus A_{ij} is an adjustable parameter in the expression for U_{ij} but is an important quantity because it reflects the impact of solvent organization in the co-sphere on the solute properties. The computer calculations to obtain a best fit between calculated and observed thermodynamic properties are

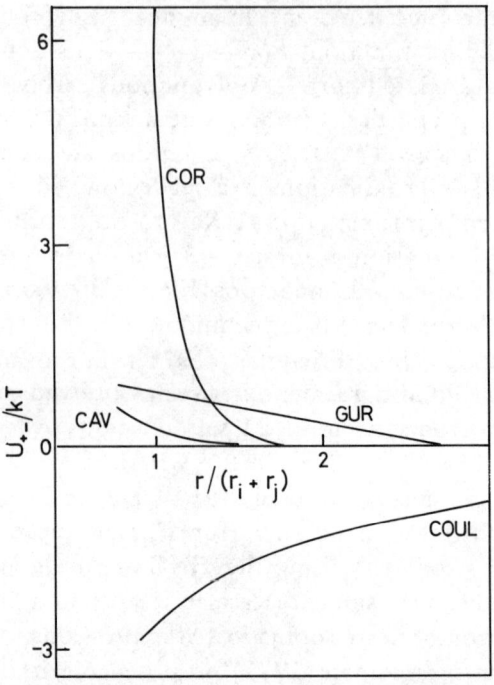

Figure 17. Contributions to the ion–ion pair potential as a function of distance apart, r, of two ions i and j, radii r_i and r_j; the contributions correspond to the terms in equation (21) for sodium chloride in water with $A_{ij} \simeq 400\,\text{J mol}^{-1}$.

formidable even with the HNC approximation. However, from the analysis, one can obtain a set of self-consistent A_{ij} parameters, i.e. A_{++}, A_{+-} and A_{--}, together with related entropy and volume quantities (Friedman and Krishnan, 1973a). We will describe some of these results in later sections of the review.

Nevertheless, we must guard against too ready an acceptance of "static" structural models for aqueous solutions. Just as in the case of water itself, a complete picture of these solutions requires information concerning the dynamic parameters describing the motions of solute and solvent molecules (Goldammer and Hertz, 1970; Hertz, 1964; 1970). Hertz in particular has emphasized the importance of dynamic models (Franks, 1973b).

Solvent Co-spheres and Kinetics of Reaction

The identification of a region of solvent molecules around a solute which has different properties from that of the bulk solvent has important repercussions in the analysis of kinetic data. The key point is that the solvent in the co-sphere (Fig. 14) is in thermodynamic equilibrium with the bulk solvent, and this applies to initial and transition states. Ives and Marsden (1965) explored the consequences of this equilibrium in an analysis of the thermodynamics of acid dissociation in water (Hepler and O'Hara, 1963; Hepler, 1971). They noted that $\Delta H^\ominus$ and $\Delta S^\ominus$ are more sensitive to the identity of the acid than is $\Delta G^\ominus$ (Ives and Marsden, 1970). These thermodynamic quantities can be divided into a chemical part, ΔX_r for the reaction part, and a hydrational component, ΔX_h for the changes in the solvent co-spheres. Arising from the equilibrium, co-sphere $\rightleftharpoons$ bulk water, $\Delta G_h = 0$, whence it follows that $\Delta H_h = T \cdot \Delta S_h$. In other words, for a set of allied equilibria (or reactions), ΔH quantities will be linear functions of ΔS quantities. Therefore, the source of *compensating* behaviour is attributed to the changes in hydration characteristics, i.e. changes in the solvent co-spheres.

These ideas have been developed further, particularly for reactions in aqueous solutions. Analysis of the chemistry of proteins in aqueous solutions led Lumry and Biltonen (1969) to ask whether in this particular subject, the role of water had been given enough importance. Indeed they concluded that a "protein is a machine preprogrammed in evolution to link chemical processes to liquid water". Part of their case was based on the discovery that the slope of $\Delta H/\Delta S$ plots, the compensation temperature T_c, is often close to 283 K. This point was probed in greater detail by Lumry and Rajender (1970) who examined thermodynamic parameters for many processes in aqueous solution. In numerous cases, T_c is close to 285 K and this behaviour was called Vaslow–Doherty compensation following these authors' study of enzyme-inhibitor complexes (Vaslow and Doherty, 1953; Doherty and Vaslow, 1952). Lumry and Rajender proposed that for processes in aqueous solution which were dominated by changes in solvent co-spheres, $250 < T_c < 320$ K. That such a wide range of phenomena (Eagland, 1973) exhibit such compensation led to the statement of Lumry's Law, viz., $T_c = 285$ K for processes in aqueous solution (Ramanthan et al., 1972). This law has been used to analyse the various steps in the α-chymotrypsin-catalysed hydrolysis of N-acetyl-L-tryptophan ethyl ester (Lumry and Rajender, 1971), and it seems to explain certain features in the solvent and temperature dependence of the quenching of anthracene fluorescence (Brooks et al., 1974). Not all phenomena in aqueous solution exhibit this Vaslow–Doherty compensation, and there are cases where T_c is very different from that required by Lumry's Law, indicating that the compensation is not always solvent based (Lumry, 1974).

Despite this enthusiasm for compensation phenomena, serious criticisms have been made by Exner and co-workers concerning the validity of this approach (Exner, 1973a; Wold and Exner, 1973). In particular, attention has been drawn to the need for a detailed statistical analysis of the kinetic data. Most of the examples in the literature do not bear out claims for a compensation temperature (Exner, 1972; 1973b). More recently, Good et al. (1975; Good and Ingham, 1975a) have re-examined entropy-enthalpy relations and pointed out the need to take into consideration possible reasons for a dependence between various calculated functions. Thus, in the context of aqueous mixtures, abrupt changes in the dependence of $\Delta H^{\ddagger}$ on $\Delta S^{\ddagger}$ as x_2 is increased are so large as to warrant a chemical explanation. In this sense compensation plots of $\Delta H^{\ddagger}$ against $\Delta S^{\ddagger}$ can be informative. It is recalled however that the error contour about a point on such plots is very eccentric (Leffler and Grunwald, 1963).

7. APOLAR SOLUTES IN WATER

Structure and Properties

Apolar solutes, e.g. rare gases, are almost insoluble in water; $\Delta G^{\ominus}$ (solute; gas $\rightarrow$ solution) is large and positive. The low solubility is a consequence of a large negative value for the corresponding entropy change, $\Delta S^{\ominus}$, which more than compensates for a negative value for $\Delta H^{\ominus}$, i.e. exothermic solvation (Table 3). The magnitude of $\Delta S^{\ominus}$ is much larger than expected from the behaviour of apolar solutes in non-aqueous solvents. It is explained (Frank and Evans, 1945) by an enhancement of water-water interactions around the solute, i.e. a structure forming action. (In general, entropy changes are very important for processes in aqueous solutions.) For this class of solutes, the term *apolar hydration* is used to describe the general situation in which $\Delta G^{\ominus} > 0$ and $|T\Delta S^{\ominus}| > |\Delta H^{\ominus}|$, and which is also characterized by a large positive value for the partial molar heat capacity, $C_p^{\ominus}$, for the solute. The quantity $\Delta C_p^{\ominus}$ at 298 K increases with increase in the size of the solute (Table 3), the most marked increase occurring between methane and butane. Thus for solutes smaller than methane, $\Delta C_p^\ominus \simeq 200$ J K^{-1} mol^{-1} and for solutes larger than butane, $\Delta C_p^\ominus \simeq 320\text{-}400$ J K^{-1} mol^{-1}, except for very large solutes such as phenanthrene where it is of the order 1 kJ mol^{-1} K^{-1}. Attempts have been made to estimate d$\Delta C_p^\ominus$/dT which appears to be generally negative (Wauchope and Haque, 1972).

The tendency for $\Delta C_p^\ominus$ to increase with the increase in the size of the solute indicates that the extent of structure forming action also increases. The positive value of $\Delta C_p^\ominus$ is attributed to the need to "melt" off this enhanced water structure as the temperature is raised. The two quantities, $\Delta H^\ominus$ and $\Delta S^\ominus$, are much more dependent on

TABLE 3

Thermodynamic Parameters for the Solution of Apolar Solutes in Water at 298 K[a, b]

Solute	$\Delta G^\ominus$/kJ mol^{-1}	$\Delta H^\ominus$/kJ mol^{-1}	$\Delta C_p^\ominus$/J mol^{-1} K^{-1}
Argon	26·25	−11·94	180
Methane	26·35	−13·52	272
Ethane	25·77	−16·94	288
Propane	26·04	−22·53	360
Butane	26·51	−25·57	447
Benzene	14·24	−31·89	301
H$_2$S	15·59	−17·47	146
D$_2$S(D$_2$O)	15·62	−17·78	170

[a] Gas phase: ideal gas, $p = 101325$ N m^{-2} (1 atm.)
Solution phase: hypothetical ideal solution where $x_2 = 1$.
[b] Data from Wauchope and Haque, 1972; Clarke and Glew, 1971.

the nature of the solute than is $\Delta G^\ominus$. Perfect compensation would require the solubilities of all apolar solutes in water to be the same. This is not so but it is noteworthy that, for pentane, hexane, heptane and octane, $\Delta G^\ominus$ (gas → solution) at 298 K is between 44 and 46 kJ mol^{-1}, $\Delta H^\ominus$ and $\Delta S^\ominus$ showing larger and systematic variations (Nelson and de Ligny, 1968). Horvath (1972) has compiled a valuable guide to the solubility data for halogenated hydrocarbons in water.

The actual organization of water molecules around an apolar solute and the mechanism of structure enhancement are not clear cut. Nemethy and Scheraga (1962b) suggested that a solute molecule stabilizes water molecules having four hydrogen bonds by means of dispersion interactions between the solute and water. Regions of enhanced water structure are envisaged as being adjacent to a solute molecule rather than surrounding it completely.

The structure enhancement can also be discussed as a perturbation of equilibrium (16) between $(H_2O)_b$ and $(H_2O)_d$, apolar hydration altering the equilibrium to favour $(H_2O)_b$. However, the analysis is not straightforward because the solute is partitioned between the two states; in $(H_2O)_b$ it is contained in interstitial sites and it forms a regular solution in $(H_2O)_d$ (Frank and Franks, 1968; Mikhailov, 1968; Mikhailov and Ponomarova, 1968). Thus Frank and Franks argue that propane actually depletes $(H_2O)_b$ but that the fraction of

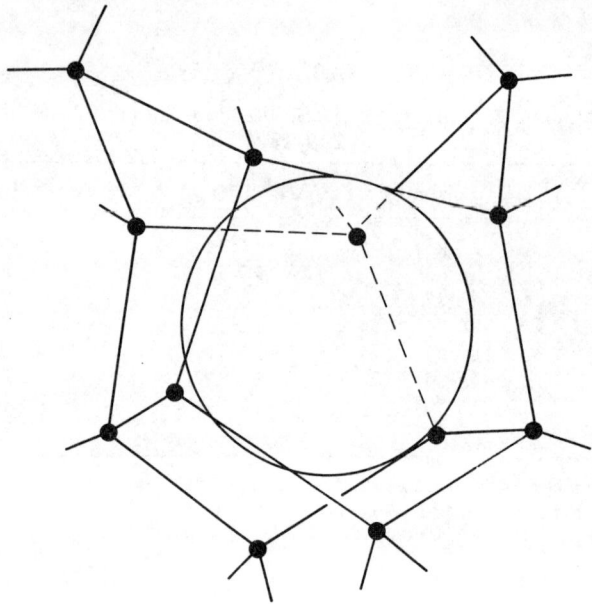

Figure 18. A cage of water molecules surrounding a spherical solute particle (redrawn from Stillinger, 1973).

solute which enters $(H_2O)_b$ is still sufficient for the large negative partial molar entropy of the solute in $(H_2O)_b$ to be the dominant factor in the overall entropy of solution.

The most conceptually attractive model for these solutions is to consider that the organization of water resembles that in the clathrate hydrates (p. 225), the structure being based on pentagonal dodecahedra of hydrogen bonded water molecules (Glew and Moelwyn-Hughes, 1953). This model receives some support from the observation that there is an optimum molecular radius of $4-5 \times 10^{-8}$ cm for solubility of apolar solutes in water (Franks and Reid, 1973).

This is also the optimum size for guests in clathrate hydrates, confirming the close link between methane solutions and methane hydrate discussed by Glew (1962). Indeed the mechanism of solid hydrate clathrate formation may involve the formation of this structure in the aqueous solution (Ewing and Ionescu, 1974).

In a recent application of scaled particle theory (Pierrotti, 1965), Stillinger (1973) has shown how a network of hydrogen bonded water molecules can surround an apolar solute, the largest complete cavity being formed by 60 water molecules, 12 pentagons and 2 hexagons. The bonds resemble the patterns found on a soccer ball (Fig. 18). The stabilizing effect of apolar solutes on water structure has also been examined using CNDO based calculations (Hojer and Keller, 1974). By way of contrast, hydrocarbon groups or small compact molecules are accommodated in methanol solutions between the chains of hydrogen-bonded solvent molecules without extensive breaking or reorganization of the solvent (Krishnan and Friedman, 1971).

Monofunctional Solutes

The molecular models for aqueous solutions of apolar solutes can be extended to simple monofunctional solutes such as monohydric alcohols with large apolar groups. The properties of such solutes are more amenable to investigation because the hydrophilic groups (e.g. OH in ROH) enhance the otherwise low solubility. The apolar residue plays an important part in determining the solute properties by enhancing water–water interactions around itself, i.e. apolar hydration. Thus for the monohydric alcohols (Alexander and Hill, 1969; Arnett, *et al.*, 1969), $\Delta G^{\ominus}$ (pure liquid → aqueous solution) is positive although $\Delta H^{\ominus}$ is negative, $\Delta S^{\ominus}$ being also negative and $\Delta C_p^{\ominus}$, positive (Table 4). Both $\Delta H^{\ominus}$ and $\Delta S^{\ominus}$ become more negative with increase in the size of the R-group, and $\Delta C_p^{\ominus}$ becomes more positive. The contribution from the apolar hydration of large alkyl groups to the enthalpy quantities exceeds that from intercomponent hydrogen bonding (Krishnan and Friedman, 1969). The enthaply of transfer of a methylene group from the gas phase to water is estimated to be $-3 \cdot 64$ kJ mol^{-1} (Krishnan and Friedman, 1973).

The thermodynamic properties for a wide range of solutes (Table 4; Franks and Reid, 1973) show a trend which again points to the importance of the apolar hydration. For example, within a given homologous series, the partial molar heat capacities show a positive increment per CH_2 group in R, independent of the nature of the functional group (Konicek and Wadso, 1971), except for the CH_2 group adjacent to the functional group (Nichols and Wadso, 1975). Consequently the observation that $\Delta C_p^\ominus$ for ethyl acetate is large and positive (384 J mol^{-1} K^{-1}) is interpreted in terms of a structure forming action (Stern and Herman, 1968). Cyclization, e.g. diethylamine to pyrrolidine, is accompanied by a fall in $C_{p2}^\ominus$ indicating that open chain molecules, having a larger hydrophobic surface area, enhance water structure more than the cyclic molecules. A marked fall in $C_{p2}^\ominus$ is observed when a second ether oxygen is introduced into a cyclic ether, e.g. on going from THF to 1,3-dioxolan, which indicates a further loss in structure-forming ability. Protonation of a

TABLE 4

Thermodynamic Properties of Alcohols and Related Compounds in Water at 298 K[a]

Solute	$\Delta G^\ominus$/kJ mol^{-1}	$\Delta H^\ominus$/kJ mol^{-1}	$\Delta C_p^\ominus$ /J mol^{-1} K^{-1}
MeOH	1·05	−7·25	75
EtOH	3·27	−10·1	142
i-PrOH	5·10	−12·98	206
t-BuOH	6·13	−17·30	264
THF	7·17	−14·9	160
1,4 Dioxan	4·26	−9·6	42
(Et)$_2$NH	2·80	−32·4	129

[a] Pure liquid → hypothetical ideal solution where $x_2 = 1$; taken from compilation given by Franks and Reid, 1973.

nitrogen atom in a solute does not in general alter the structure-forming ability of the hydrocarbon residue (Cabani et al., 1973), although considerable disruption in the solvent co-sphere occurs when diazabicyclo[2,2,2]octane (DABCO) is protonated (Wen et al., 1974). It is noteworthy that the calculated dipole moment of THF in water is larger than in benzene solution or in the gas phase (De Nooijer et al., 1971).

Similar conclusions concerning the hydration properties of these solutes follow from the analysis of partial molar volume, $V_2^\ominus$ (Friedman and Scheraga, 1965; Alexander, 1959; Franks and Smith, 1968; Nakanishi et al., 1967; Franks et al., 1970a) and compressibility data (Franks et al., 1972; Nakajima et al., 1975). Thus $V_2^\ominus$ (i.e. V_2 as $x_2 \rightarrow 0$) is generally less than the molar volume of the pure component, V_2^o, and V_2^E (= $V_2^\ominus - V_2^o$) becomes more negative with increase in size of the apolar group, again reflecting an increase in structure forming action. However, V_2^E is particularly sensitive to the number of hydrophilic groups (cf. other thermodynamic properties; Cabani et al., 1971) and the shape of the molecule. Thus $V_2^\ominus - V_2^\circ$ per methylene group is smaller in cyclic ethers than in alcohols or hydrocarbons (Cabani et al., 1972). In particular Franks et al. (1970a), modifying ethylene oxide by adding $-CH_2-$, or $-O-$ groups to the ring or a side chain $-CH_2OH$, have shown how subtly the solute hydration properties change. Although the hydration properties are determined by a range of factors (p. 238), we can envisage that, for this class of solutes, the hydrophilic group is hydrogen-bonded into the water structure, and that around the hydrophobic moiety the organization of water resembles the apolar hydration around hydrocarbons. The latter may, as noted in the previous section, approach the structure of water in the clathrate hydrates. In other words, for aqueous solutions this class of solutes should be regarded as bifunctional rather than monofunctional (Laiken and Nemethy, 1970).

While the clathrate model is attractive, it is not correct to assume that the water is organized in some long-lived structure; the observation that the self-diffusion coefficient for co-sphere water is larger than that for the solute rules this out. However, the rotational correlation time is shorter for ethanol and t-butyl alcohol in water (in the clathrate cage?) than in the pure liquid (Goldammer and Hertz, 1970; Goldammer and Zeidler, 1969). Nmr experiments show that in water the solvent dipole moments point away from the apolar groups (Hertz and Radle, 1973).

The solution properties of fluoro alcohols show some similarities with the hydrocarbon alcohols, although there is evidence for some $-F\cdots H_2O$ interaction which leads to a more compact arrangement of water molecules around the fluoroalkyl groups (Rochester and Symonds, 1973, 1974).

In real solutions, the properties of these solutes can be examined in terms of what happens as the solvent co-spheres approach and overlap. This aspect will be discussed in a later section except for a particular aspect of solute-solute interaction called hydrophobic association.

Hydrophobic Association

If apolar hydration is characterized by the conditions that $\Delta G^{\ominus} > 0$, $T\Delta S^{\ominus} < 0$ and $\Delta H^{\ominus} < 0$, then a process which minimizes exposure of apolar groups to water should be a thermodynamically favoured process. Then if two apolar groups of either the same or different molecules come together in water, ΔS for this process will be positive because some of the structured water is released into the bulk solvent. Such association is called hydrophobic, hydrophobic bonding or hydrophobic interaction (Kauzmann, 1959). The term "bond" is probably inappropriate because the association is due to entropy rather than to enthalpy effects, a consequence of the disruption of the clathrate structure around the apolar solute (Jolicoeur and Friedman, 1974). Despite the general acceptance of the concept of hydrophobic association, there are different approaches to the problem of understanding this phenomenon.

An important paper by Kozak et al. (1968) showed how the activity coefficients of water, f_1, in dilute aqueous solutions of non-electrolytes could be represented as a power series (22) in x_2.

$$\ln f_1 = Bx_2^2 + Cx_2^3 + \ldots \qquad (22)$$

The coefficient B for aliphatic alcohols, amines, ketones and acids is positive, increasing with increase in the size of the alkyl group. This quantity characterizes solute-solute pairwise hydrophobic association, which, because dB/dT is positive, becomes stronger with increase in the temperature (Nemethy and Scheraga, 1962c). In contrast, dB/dT is negative for urea (see p. 260). On going from water to D_2O, hydrophobic association is stronger for non-polar side chains of proteins but weaker for hydrocarbons (Krescheck et al., 1965) except in the case of benzene (Ben-Naim et al., 1973).

Ben-Naim (1972b, c) has examined hydrophobic association using statistical mechanical theories of the liquid state, e.g. the Percus-Yevick equations. He has also examined quantitative aspects of solvophobic interactions between solutes using solubility data for ethane and methane. The changes in thermodynamic parameters can be calculated when two methane molecules approach to a separation of, $1 \cdot 533 \times 10^{-8}$ cm, the C–C distance in ethane, and the solvophobic quantities $\delta_{SI}\mu_2^{\ominus}$, $\delta_{SI}H_2^{\ominus}$ and $\delta_{SI}S_2^{\ominus}$ can be calculated. In water (solvophobic = hydrophobic) $\delta_{SI}\mu_2^{\ominus}$ is more negative than in other solvents and decreases as the temperature rises; both $\delta_{SI}H_2^{\ominus}$

and $\delta_{S_1}S_2^{\ominus}$ are positive (Ben-Naim and Yaacobi, 1974). It is suggested that all added solutes (i.e. a third component) enhance hydrophobic interactions. This conclusion is at odds with other statements in the literature where, for example, urea is said to weaken such interactions. However Ben-Naim and Yaacobi suggest that changes in the solubilities of solutes as new solutes, salts, or co-solvents are added do not necessarily indicate how the strength of the hydrophobic association changes.

Friedman and Krishnan (1973c) using the HNC method (p. 245) calculate that A_{ij} for the hydrophobic interaction between the alkyl groups of alcohols ROH in water (R = Me to t-Bu) is around -418 J mol^{-1}. In general, for overlap between solvent co-spheres of methylene groups, $A_{XX} \simeq -360$ J mol^{-1}, which characterizes hydrophobic association.

Calorimetric studies have probed enthalpy changes associated with hydrophobic association. As noted above, we anticipate that hydrophobic association is endothermic. Indeed Cassel and Wood (1974) found that for pair interactions between t-butyl alcohol molecules, $h_{XX} = +656$ J mol^{-1}, the corresponding triplet coefficient, h_{XXX}, being $+372$ J mol^{-1} at $c_2 = 1$ mol dm^{-3}. The triplet term for t-BuOH-Bu$_4$N$^+$Cl$^-$ is notably large, 1·8 kJ mol^{-1}. Franks, Pedley and Reid (1975) find that an average h_{XX} for methylene pair interactions is ca. $+380$ J mol^{-1} but that there is evidence for marked co-operative effects in hydrophobic association, h_{XXX} being larger the larger the apolar molecules. One can contrast these endothermic values with the exothermic values, $h_{XX} < 0$, for solutes such as urea where pair interaction involves H-bonding or dipolar interactions (Cassel and Wood, 1974).

A rather different and attractive model of hydrophobic association has been proposed by Franks (1973b). In contrast to the Nemethy-Scheraga-Kauzmann model reviewed above in which the apolar parts of associated solute molecules are in contact, Franks suggests that one or more layers of water molecules still separate the solute molecules. In other words, hydrophobic association is viewed as long-range rather than short-range. Thus one thinks of clathrate hydrate structures and the interaction between guests in adjacent guest sites. Franks has discussed the evidence for this model and we mention one such piece of evidence which is persuasive. We recall that for monohydric alcohols, V_2^E ($= V_2^{\ominus} - V_2^{\circ}) < 0$. Then for the reverse process, hydrophobic association, $\Delta V > 0$. However, for dilute solutions of monohydric alcohols in water, as we shall see later, V_2^E decreases as x_2 increases. In other words, solute–solute interactions in dilute solutions operate in the opposite way to that predicted from the generally accepted model for hydrophobic association. We shall return to this point.

Kinetics of Reactions involving Apolar Solutes in Water

We have now established sufficient background to consider briefly the kinetics of reactions in water where apolar solutes are involved. For example, if the hydration characteristics of t-butyl alcohol in water are controlled to a marked extent by the hydration of the apolar t-butyl group, then it is likely that the same state of affairs exists for, say t-butyl chloride and other alkyl halides and related compounds in water. In other words, the hydration properties can be characterized by the general statement that, in the solvent co-sphere, water–water interactions $\gg$ water–solute interactions, but that in the activation process water–solute interaction will increase. Since for apolar solutes, $C_{p3}^{\ominus} > 0$, and assuming that in the transition state, $C_{p3}^{\ddagger} \simeq 0$, then a tentative prediction is that $\Delta C_p^{\ddagger} < 0$ and $\simeq -C_{p3}^{\ominus}$. Indeed, this behaviour is observed for the hydrolysis of methyl halides, e.g., for MeCl, $\Delta C_p^{\ddagger} = -205$ J mol^{-1} K^{-1} (Robertson, 1967) and $C_{p3}^{\ominus} = +180$ J mol^{-1} K^{-1} (Glew and Moelwyn-Hughes, 1953). This close agreement and the general observation that, for hydrolysis of this class of solutes, $\Delta C_p^{\ddagger} < 0$ prompted the proposal that activation in hydrolysis required the breakdown of water–water interaction in the neighbourhood of the solute.

For $S_N 1$ reactions, the large negative $\Delta C_p^{\ddagger}$-values suggest that the extent of solvent reorganization on going from the initial state to the transition state is extensive [see Table 5, e.g. t-butyl chloride (Moelwyn-Hughes *et al.*, 1965) and 2,2-dibromopropane (Queen and Robertson, 1966)]. It is proposed (Robertson, 1967) that the developing carbonium ion–water interactions in the $S_N 1$ transition state balance the adjacent water–water interactions. Beyond this level of charge development, cation–solvent interactions are sufficiently exothermic for the reaction to run downhill. In other words, a "trimethylcarbonium chloride" ion pair model for the transition state in the t-butyl chloride hydrolysis is not favoured by Robertson.

For $S_N 2$ reactions, the extent of solvent reorganization is not so marked as indicated by smaller values for $|\Delta C_p^{\ddagger}|$ (Table 5). In the transition state for this mechanism the developing anion–water

TABLE 5

Thermodynamic Parameters for Hydrolysis Reactions[a] in Aqueous Solution

Compound	T/K	$\Delta H^{\ddagger}$/kJ mol^{-1}	$\Delta S^{\ddagger}$/J mol^{-1} K^{-1}	$\Delta C_p^{\ddagger}$/J mol^{-1} K^{-1}
CH$_3$Cl	348	100·4	−51·5	−205
t-Butyl Cl	283	99·6	60·2	−347
2,2-dibromo-propane	313	107·6	33·5	−338
t-pentyl Cl	283	91·9	42·6	−385
Et Br	343	97·8	−37·2	−205
n-Pr Br	358	92·5	−56·9	−179

[a] Data from Robertson, 1967.

interactions balance adjacent water–water interactions. Again, beyond this stage, exothermic hydration of the anion allows the reaction to run downhill. Consequently the general hypothesis is that $|\Delta C_p^{\ddagger}(S_N 1)| > |\Delta C_p^{\ddagger}(S_N 2)|$. These arguments mark a break with more classical viewpoints of the course of reaction (p. 205) but would appear to be more consistent with the current models for aqueous solutions. Exceptions to the patterns are noteworthy. Introduction of a hydrophilic centre into the apolar solute leads to the expected fall in $|\Delta C_p^{\ddagger}|$ (Robertson, 1967), while complications arise where there is anchimeric assistance (Blandamer et al., 1969).

In the above description of the hydrolysis reactions we made a sharp distinction between $\Delta C_p^{\ddagger}$ values for $S_N 1$ and $S_N 2$ reactions. However, as more data become available, it is becoming clear that there is a gradation in $\Delta C_p^{\ddagger}$ as the reaction moves towards a limiting ion-pair $S_N 1$ mechanism (Koshy et al., 1973) so that the magnitude of $\Delta C_p^{\ddagger}$ can provide some indication of the extent of solvent reorganization in the activation process (Ko and Robertson, 1972).

TABLE 6

Thermodynamic Transfer Functions[a] for Hydrocarbons from Water to an Aqueous Solution of Urea, 7 mol dm^{-3}, at 298 K

Solute	$\Delta G^{\ominus}$/kJ mol^{-1}	$\Delta H^{\ominus}$/kJ mol^{-1}	$\Delta S^{\ominus}$/J mol^{-1} K^{-1}
Propane	−0·54	7·1	26·3
Butane	−1·04	7·9	30·1
Neopentane	−1·21	7·1	27·6

[a] Standard states—hypothetical solutions containing unit mole fraction of hydrocarbon; data from Wetlaufer et al., 1964.

In addition, $\Delta C_p^{\ddagger}$ is also a function of the leaving group, being more negative for nitrates than for halides, which indicates that $\Delta C_p^{\ddagger}$ is also dependent on the water-structure surrounding the transition state (Koshy and Robertson, 1974). However, we must be wary of reactions which proceed by two reaction pathways (p. 212) because $\Delta C_p^{\ddagger}$ can then be misleading (see, for example, hydrolysis of chloroformates; Queen, 1967).

In this discussion we have concentrated attention on $\Delta C_p^{\ddagger}$ because this quantity seems to probe the solvent contribution to the activation process. Indeed, an important aspect of these studies is the dependence of $\Delta C_p^{\ddagger}$ on composition in mixed solvents (see p. 318). However, the question naturally arises as to the significance of $\Delta H^{\ddagger}$ and $\Delta S^{\ddagger}$ in Table 5. A considerable contribution to $\Delta S^{\ddagger}$ from the solvent reorganization can reasonably be expected, but $\Delta S^{\ddagger}$ also contains other contributions and analysis is not straightforward. This is clearly seen when we turn to $\Delta H^{\ddagger}$. Although $|\Delta C_p^{\ddagger}|$ may be of similar size and magnitude to $C_{p3}^{\ominus}$ for the initial state, this does not hold for $H_3^{\ominus}$ and $\Delta H^{\ddagger}$, the latter being significantly larger. Some small part (cf. $\Delta H_2^{\ominus}$ for butane $\simeq -25$ kJ mol^{-1}) of the measured $\Delta H^{\ddagger}$ (99 kJ mol^{-1}) for the t-butyl chloride must be due to reorganization in the co-sphere, the remainder stemming from the reaction part.

The exciting suggestion that $\Delta H^{\ddagger}$ for the hydrolysis of benzyl chloride is negative at temperatures below 277 K (Hills and Viana, 1971) turned out to be incorrect (Albery and Curran, 1972). The more recent proposal that in the hydrolysis of t-butyl chloride there is an induction period which depends on the hydrogen ion concentration (Adams et al., 1973) is puzzling. The possibility that these observations arise from practical problems associated with the low solubility of t-butyl chloride in water warrants consideration.

When chemical reaction involves more than one apolar solute, short-range hydrophobic association may be important in determining the rate and products of reaction. For example, hydrophobic association between apolar reagent and apolar substrate can enhance the rate of reaction in water (Cayley and Margerum, 1974). Similarly, hydrophobic interactions between enzyme and substrate are important in enzyme catalysed interactions (Jencks, 1969). The role of hydrophobic association in these reactions has been studied using model systems. For example, aminolysis of 8-quinolyl octanoate by octylamine or decylamine is faster than predicted on the basis of the behaviour of lower amines, the rate enhancement being possibly due to hydrophobic association between ester and nucleophile (Maugh and Bruice, 1971). Similarly, the apparent second-order rate constant for the reaction between p-nitrophenyl decanoate and N-decylimidazole (a "long-long" reaction) is 29 times faster than that between acetate and N-ethyl derivative (a "short-short" reaction). Moreover the acceleration is mainly an entropy based effect, a trend expected if hydrophobic interaction is important (Blyth and Knowles, 1971). The rate enhancement of the imidazole-catalysed hydrolysis of alkanoate esters increases with increase in the length of the ester alkyl chain, and while the trend is not so dramatic as in the previous example, it agrees with that predicted on the basis of increasing importance of short range hydrophobic interactions (Guthrie and Ueda, 1974).

8. HYDROPHILIC SOLUTES

Rather less information is available concerning hydrophilic solutes than for apolar solutes (Section 7), although considerable effort has been made in recent years to overcome this deficiency. The chemistry of aqueous solutions of these solutes is equally fascinating, the importance of the different factors which affect hydration being readily apparent (p. 239). In molecules having two or more hydrophilic groups, the stereochemistry of the solute plays an important part in their hydration.

Small Molecules

In this class, we include such molecules as H_2O_2 and dimethyl sulphoxide (DMSO). The thermodynamic properties of these solutes (Franks and Reid, 1973) show clearly that their solution properties are dominated by strong solute-solvent hydrogen-bonding interactions; for example, $\Delta C_p^{\ominus}$ (liquid → solution) for H_2O_2 is negative (ca. -37 J K^{-1} mol^{-1}). In DMSO solutions, strong S=O$\cdots$H$_2$O interactions disrupt the long-range order in water and DMSO is therefore classified as a structure breaker. Similarly the introduction of a second oxygen atom into H_2O to form H_2O_2 produces a molecule whose structure is not compatible with the organization of hydrogen-bonded water molecules.

The introduction of a hydrophilic group into a solute dramatically changes its hydration properties consistent with the destruction of apolar hydration. Thus $\Delta C_p^{\ominus}$ (pure component → solution) drops markedly on going from propane to 1-propanol, 1,3-propanediol, and glycerol (Franks et al., 1973). A glycerol molecule can adopt a conformation in which the distance between the first and third O-atoms is about $4 \cdot 85 \times 10^{-8}$ cm (p. 262). In practice there is no clear distinction between apolar and hydrophilic solutes, but rather a gradual change proceeding from hydrocarbon through ethers, ketones, alcohols, amides, nitriles to H_2O_2 and urea.

Urea

This solute deserves special mention in view of its extensive use in protein chemistry as a denaturant. The thermodynamic properties (Stokes, 1967; Egan and Luff, 1966) show that urea exerts a structure breaking action on water (Franks and Reid, 1973). This behaviour can be accounted for using the equilibrium given in eqn (16) (Frank and Franks, 1968). The shape and size of urea means that it is debarred from $(H_2O)_b$ but dissolves in $(H_2O)_d$. In mixing with $(H_2O)_d$, the chemical potential of the latter is lowered and so equilibrium (16) shifts to favour $(H_2O)_d$ at the expense of $(H_2O)_b$. Thus urea acts as a "statistical water structure breaker" (Frank and Franks, 1968). Certainly nmr spectra of aqueous solutions show that urea breaks the long range structure in water, addition of 6 mol dm^{-3} shifting the water resonance by an amount equivalent to an increase in temperature from 293 to 313 K (Finer et al., 1972). Raman (Walrafen, 1966) and near-infrared spectroscopic studies (Barone et al., 1970) together with enthalpy interaction terms (Cassel and Wood, 1974), heat capacity data (Philip et al., 1974) and the effect of urea on thermodynamic properties of ions in water (Desrosiers et al., 1974; Kundu and Mazumdar, 1975) and on partial molar volumes of solutes (Hargreaves and Kresheck, 1969) appear to confirm that urea breaks water structure.

The structure-breaking influence of urea accounts for a remarkable set of solubility data for hydrocarbons in aqueous solutions (Wetlaufer et al., 1964). Added urea "salts-in" propane so that transfer of one mole of propane from water to an aqueous solution of urea (7 mol dm^{-3}) is thermodynamically favourable at 298 K (Table 6). However, the transfer is endothermic, increase in solubility in the urea solutions arising from the large positive entropy change. By breaking water structure and favouring $(H_2O)_d$ urea destroys that character of water which produces the unfavourable entropy change when a gas dissolves in water. A similar trend is observed when urea is replaced by guanidinium chloride. It is noteworthy, however, that at 278 K, urea salts out propane and ethane (Wen and Hung, 1970). When alkylammonium bromides are added to urea in water, both urea and the salt are salted-in, the effect increasing with increase in cation size. While this trend probably reflects the effect of these solutes on water structure, the possibility that some direct salt-urea interaction is involved cannot be ruled out (Wen and Chen, 1969). A similar interaction may account for the complex dependence of vapour pressure on urea concentration in salt-urea-water systems (Sarnowski and Baranowski, 1965). A great deal remains to be learnt about urea solutions. It is noteworthy, for example, that the pair correlation length for aqueous urea systems calculated from the HNC theory (p. 245) does not appear to agree with that predicted by the Frank and Franks (1968) model (Friedman and Krishnan, 1973a).

In contrast to urea itself, its N-methylated derivatives enhance water-water interactions, i.e. lower the structural temperature; hexamethylene tetramine produces similar marked effects (Barone et al., 1968). Glycine and β-alanine appear to be structure breakers (Devine and Lowe, 1971) according to their effect on the viscosity of water (Herskovitz and Kelly, 1973). The viscosities and diffusion properties of urea solutions show striking changes as the concentration increases (MacDonald and Guerrera, 1970).

Large Molecules

The importance of matching the positions of hydrophilic groups in a solute to the structure of water was noted by Warner (1965). The oxygen spacings in the hydrophilic groups of many water-soluble compounds often correspond to the second-nearest neighbour distance (4.8×10^{-8} cm) in ice (Berendsen, 1967). Compatibility between solute and water structures means that the solute can be built into, say, the $(H_2O)_b$ system, i.e. as part of the host framework rather than as a guest in a clathrate hydrate (Fig. 19). For example, nmr and dielectric studies (Tait et al., 1973) show how glucose is surrounded by $(H_2O)_b$ and glucose can be thought as a short-range structure former. Quite small changes in the structures of these solutes can produce quite dramatic changes in solution properties; for example, mucic and saccharic acid differ only in the configuration of one —OH group but the former is insoluble and the latter soluble (Franks 1973a). Indeed, the conformation of sugars in aqueous solution and thus their properties (Allen *et al.*, 1974) are

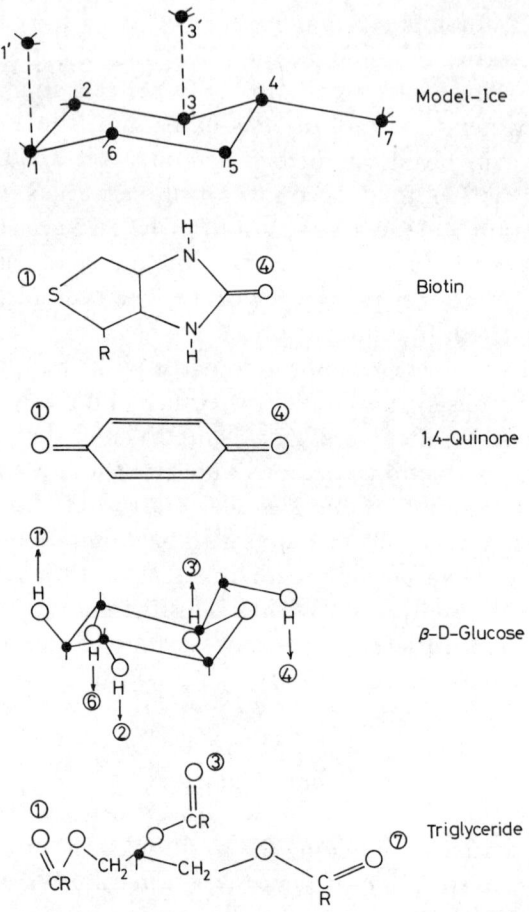

Figure 19. Correlations between oxygen spacings in ice-lh lattice and those in various organic molecules; the numbers indicate interaction points with the ice-lattice (Franks, 1968).

determined to a considerable extent by their ability to maximize solute-water interaction (Franks *et al.*, 1973). The intensity of solvent-solute interactions is shown by the similarity between the partial molar compressibilities in water of these solutes and those of salts such as sodium chloride (Franks *et al.*, 1972).

The importance of the stereochemistry of solutes is shown in the solubility data for bipyridyls and biphenyl (Bennetto and Letcher, 1972). It is also confirmed by the discovery that the transfer of sodium chloride to dilute solutions of mannitol is endothermic whereas the transfer to solutions of sorbitol is exothermic (Stern and O'Connor, 1972).

The importance of the $4 \cdot 8 \times 10^{-8}$ cm spacing between hydroxyl oxygen atoms is clearly demonstrated in the aqueous chemistry of polysaccharides (Suggett, 1973). In some instances, the reactivity of a mismatched OH group can differ from one which is matched to this spacing. However, the lack of extensive and specific hydration shells around macromolecules makes the significance of this repeating distance in biopolymers mysterious (Berendsen, 1973).

9. IONIC SOLUTIONS

Structure and Properties of Aqueous Salt Solutions

In this brief review of salt solutions, we start by examining ionic hydration. We then examine from the standpoint of ion-ion interactions the properties of real solutions and how these are affected by ionic hydration (Blandamer, 1970).

Ionic hydration

For alkali metal and halide ions, the enthalpies of hydration, $\Delta H^{\ominus}$ (ion; gas $\rightarrow$ solution), are negative, behaviour which can be readily understood in terms of intense ion-solvent dipole interactions. The entropies of hydration, $\Delta S^{\ominus}$, are also negative but, suprisingly, not very different in magnitude from those for apolar solutes. Although it might be anticipated that electrostriction of solvent molecules around these ions should lead to markedly negative values for $\Delta S^{\ominus}$, the combined entropy of hydration for K^+ and Cl^- ions is less negative than that for two (isoelectronic) atoms of the apolar solute argon. Frank and Evans (1945) reasoned that there was around an ion in solution a region in which the water had "less structure" than water itself at the same T and p. The corresponding model for ionic hydration (Frank and Wen, 1957) contains three zones (Fig. 20). In zone A (states I_a and I_c in Table 2), water molecules hydrate the ion in an electrostricted layer of solvent molecules, the number of such molecules being the primary hydration number. NMR data show that in the hydrated fluoride ion, the H-atom is collinear with F^- and the O atom of water (Hertz and Radle, 1973). Zone C contains water whose structure is essentially unperturbed from water itself at the same T and p, save that it is subjected to the electric field from the ion. The key proposal concerns zone B (state II_{sb} in Table 2) which separates zones A and C. This is a fault zone in which the water structure is broken and it arises because water is such a highly ordered liquid (Millero, 1969; 1971). Zone B increases with increase in size of the ions. Thus the larger the ion, the more water structure

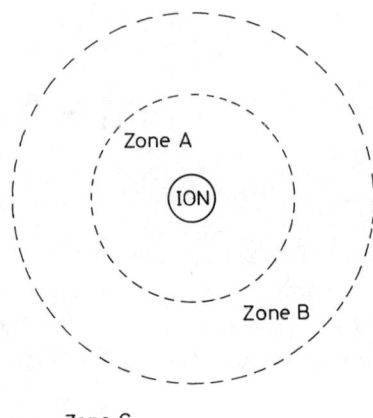

Figure 20. Frank–Wen model for ionic hydration in aqueous solutions.

is broken (Safford and Leung, 1973); ions such as Cl^- Br^- and I^- are therefore called "electrostrictive structure breakers". This mechanism for structure breaking is therefore quite different from the statistical structure breaking action of urea (p. 260). Indeed, there is evidence for this distinction based on the different effects which sodium chloride and urea have on the partial molar heat capacities of alkylammonium salts in water (Ahluwalia and Chawla, 1973; Chawla and Ahluwalia, 1972). For small ions, e.g. Li^+ and F^-, zone B is absent, the hydrated ion matching into and enhancing water-water interaction. These ions are called "electrostrictive structure formers" because their mechanism of structure forming is quite different from that for apolar solutes (Section 7). Complex ions, e.g. cyano complexes, are structure breakers (Mathieson and Curthoys, 1975).

Recently attempts have been made to set up computer models of salt solutions (Kistenmacher et al., 1974; Fromm et al., 1975; Watts et al., 1974). A molecular dynamic study has been reported for a system comprising 198 H_2O + $9Li^+Cl^-$ (Heinzinger and Vogel, 1974). The results show good agreement with experiment. For example, a water molecule hydrating Li^+ rotates faster than water molecules in bulk (Hertz et al., 1971).

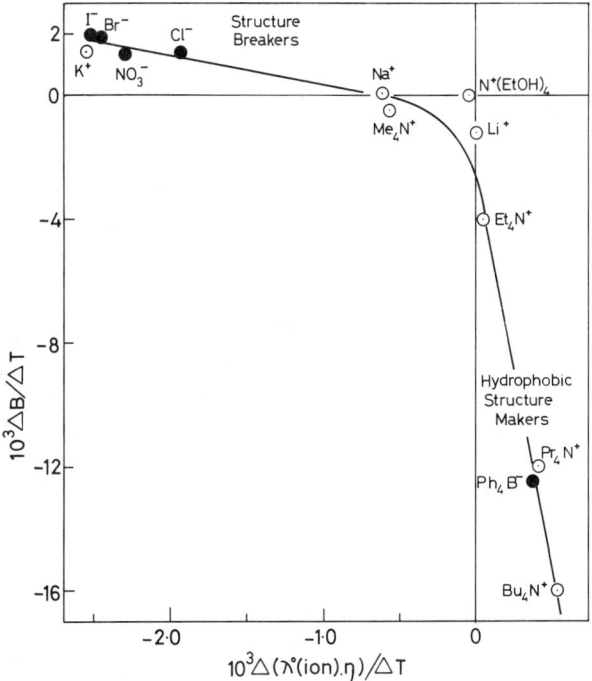

Figure 21. Comparison of the temperature coefficient of the B-viscosity coefficient and that of the Walden product for various ions in water (Kay, 1968).

The simple model (Fig. 20) can be criticized because it cannot readily be quantified. However, it does account for a wide range of properties, such as the tendency for the partial molar heat capacity and the viscosity B-coefficient to become more negative with increase in ion size (Fortier et al., 1974a; McDowell and Vincent, 1974; Kay, 1968; 1973). Kay has collated conductance and viscosity data and shown how these lead to a classification of ionic properties (Fig. 21). The effects of added salts on the self-diffusion of ions is consistent with the Frank–Wen structural model (Hertz et al., 1974). It is noteworthy that in D_2O, which is argued to be more structured than H_2O at the same T and p, the structure breaking effect of these ions is larger (Krishnan and Friedman, 1970; Greyson and Snell, 1969; Fortier et al., 1974b). The often substantial entropy change on transferring an ion from water to another solvent reflects the importance of the B-zone in water (Cox et al., 1974).

There is an important group of ions, however, for which the above model is unsatisfactory. For example, the trends outlined above suggest that the viscosity B-coefficients and partial molar heat capacities of alkylammonium ions, R_4N^+, would become more negative as R was changed through the series, methyl, ethyl, propyl, butyl. In fact the reverse trend is observed, the heat capacities being comparable to those observed for apolar solutes and the B-coefficients increasing along this series (Desnoyers and Perron, 1972). The hydration of these ions is controlled by the apolar alkyl groups (Wen, 1973) and as such these ions are hydrophobic structure formers, the extent of structure forming increasing with increase in the size of the alkyl group (Fig. 21). For example, $C_p^\ominus$ ($Bu_4N^+Br^-$) is large and positive in water. By way of contrast, C_p for this ion is negative in N-methylformamide, presumably because of the disruption of the hydrogen bonded non-aqueous solvent (De Visser and Somsen, 1973). Indeed $\Delta G^\ominus$ (ion; gas $\rightarrow$ water) for R_4N^+ becomes less negative through the series, R = Me to Pr, because $\Delta S^\ominus$ also becomes more negative, a trend accounted for in terms of increased structure forming but not accounted for in terms of ion size and solvent permittivity (Johnson and Martin, 1973). The first member of the series, Me_4N^+, is usually classed as a structure breaker (cf. Cs^+), since $\Delta C_p^\ominus$ (salt $\rightarrow$ solution) for $Me_4N^+Br^-$ and $Me_4N^+I^-$ is negative. In contrast these values are positive when R is larger than Me (Sarma et al., 1969). The next member, Et_4N^+, appears to have no marked effects on water structure (Kay, 1968) but when R is larger than Et, the ionic properties are dominated by the hydrophobic alkyl groups. Viscosity data indicate that n-propyl is the smallest alkyl group in R_4N^+ which can enhance water–water interactions (Lowe and Rubienski, 1974). The nmr spectra of water in dilute alkylammonium salt solutions at low temperatures have confirmed the structure forming action of these ions (Davies et al., 1972). Hsieh et al. (1974) have suggested a general rule, namely that the temperature dependence of the slope of the plot of solvent shift, δ_s, against solute molarity, $\delta(d\delta_s/dm)/\delta T$, is positive for structure formers (e.g. $Bu_4N^+Br^-$ and DABCO) and negative for structure breakers (e.g. $Me_4N^+Br^-$ and triethylene diamine). Proton magnetic relaxation times indicate that around the Et_4N^+ ions in water, the water protons preferentially point into the bulk water (Hertz and Wen, 1974).

The properties of other large organic ions (Lindenbaum, 1970) including alkyl sulphates (Endo, 1972; Tamki et al., 1972) and asymmetric alkylammonium ions, e.g. pyridinium derivatives (Lowe and Rendall, 1971) in water can be explained along similar lines. It is noteworthy that Bu_4N^+ is a more efficient structure former than Ph_4X^+ and Ph_4X^- ions (Jolicoeur et al., 1972). However, 1,1'-dimethyl-4,4'-dipyridinium ion is a structure breaker, the rigidity of the solute ion hindering adoption into the water-structure (Perron and Desnoyers, 1972). Both structure making and structure breaking appear to be more extensive in D_2O than in H_2O (Conway and Laliberte, 1968; Levine and Wood, 1973).

The importance of the apolar alkyl group is clearly indicated by the change in properties which occur when the terminal methyl groups in $(C_3H_7)_4N^+$ are replaced by OH groups to form $(HOCH_2\text{-}CH_2)_4N^+$. The latter shows no exceptional properties because now the ion can hydrogen bond to the solvent (Kay, 1968). However, there still remain many problems to be resolved. For example, no satisfactory explanation has been offered for the various patterns shown by the temperature dependence of the molar heat capacities (Sarma and Ahluwalia, 1973; Sunder et al., 1974).

Real solutions

The degree of success resulting from the extensive application of the Debye–Hückel theory to the analysis of the properties of salt solutions has generally meant that the importance of other factors affecting ion–ion interactions have been ignored. Indeed, it would be expected that the characteristics of ionic hydration would also be reflected in the non-ideality of a given salt solution. The presence of some underlying pattern to ionic activity coefficients, $\gamma_\pm$, not readily accounted for by the DHLL has been known for many years. Such patterns are apparent when values of $\gamma_\pm$ for salt solutions at fixed molality and temperature are examined as a function of the anion (Fig. 22). For example, we note that $\gamma_\pm(Pr_4N^+F^-) > \gamma_\pm(Cs^+F^-)$ but $\gamma_\pm(Cs^+I^-) > \gamma_\pm(Pr_4N^+I^-)$. These and related trends can be understood in terms of the effect on $\gamma_\pm$ of overlapping solvent co-spheres (Desnoyers et al., 1969). As a general rule, two solute ions will attract each other if their structural influences or tendencies to orient water molecules are compatible, but they will repel each other if their influences are incompatible. Attraction will lower and repulsion raise the activity coefficients. For example, overlap between co-spheres of Bu_4N^+ ions in aqueous solution (hydrophobic hydration) is attractive and this is the dominant influence in $Bu_4N^+I^-$ solutions, cation-cation interactions leading to a low value

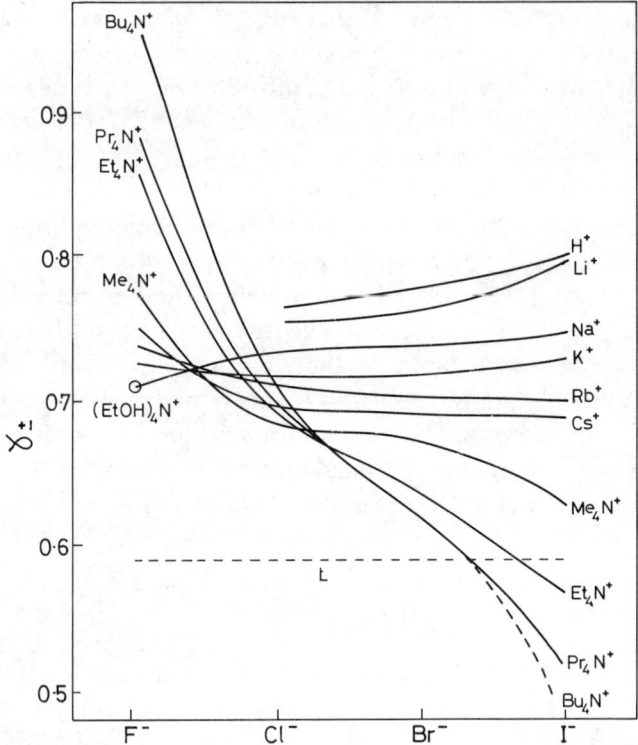

Figure 22. Mean ionic activity coefficients, $\gamma_\pm$, for various salts in water, $m_2 = 0.2$ at 298 K; the dotted line indicates the value required by the Debye-Hückel limiting law (Desnoyers and Jolicoeur, 1969).

of $\gamma_\pm$, i.e. mutual salting-in (Desnoyers, 1969). In contrast, overlap between Bu_4N^+ co-spheres (hydrophobic hydration) and F^- co-spheres (hydrophilic hydration) leads to repulsion, the cation-anion interaction raising $\gamma_\pm$, i.e. salting out. As a general rule, interactions between solvent co-spheres of alkylammonium ions in their salt solutions determine the properties of these solutions, (Wen, 1973).

The success of the DHLL is often limited to the analysis of chemical potentials and related quantities, e.g. the effect of ionic strength on solubility products. The failure of the DHLL to account for the dependence of enthalpy, entropy and volume properties on salt concentration is often striking, compensation between H- and S-quantities minimizing the changes in Gibbs functions. Indeed the dependence of these derived parameters on concentration highlights the importance of co-sphere interactions, the latter being the source of the compensation. We have indicated how excess functions for salt solutions (Fig. 15 and 16) can be used to characterize the properties of aqueous solutions. For alkylammonium salt solutions bewildering patterns are observed (Lindenbaum, 1966), bearing out the complexity of these systems. Even more complex patterns emerge for aqueous solutions of tetra-n-butylammonium salts of carboxylic acids, e.g. butyric acid. There is a loss of entropy as the solution is diluted because hydration spheres are formed around the alkyl groups as these large ions separate (Lindenbaum, 1971). This can be likened to the pulling apart of a hydrophobic bond.

As the concentration of salt increases, the properties of the salt solutions often change more or less abruptly. We have commented (p. 241) on the change from a $c_2^{1/2}$ to a $c_2^{1/3}$ dependence for activity coefficients. Volume and enthalpy data for 1 : 1 salt solutions show transition points near characteristic concentrations (e.g. 0·75 mol dm^{-3} for NaCl), but the precise nature of these phenomena is not understood (Vaslow, 1969, 1971).

Extrema in the properties of alkylammonium salt solutions are often observed. For example, the relative apparent molar volume of Bu$_4$N$^+$Br$^-$ has a minimum near m_2 = 1·0 mol kg^{-1} (Wen and Saito, 1964). Up to this concentration, the system approaches a structure resembling the solid clathrate hydrate (p. 229) involving long range hydrophobic interaction, which can be disrupted by added N-methylacetamide (Falcone and Wood, 1974). At higher concentrations of Bu$_4$N$^+$Br$^-$, there is insufficient water to maintain the pseudoclathrate structure. Beyond a concentration of 1 m, the ultrasonic absorption increases rapidly (Blandamer and Waddington, 1970), as does the solubility of benzene in these solutions (Wirth and Lo Surdo, 1968).

Quite striking extrema are observed in the dependence of apparent partial molar heat capacities on salt concentration for alkylammonium salts and related compounds, e.g. Bu$_4$N$^+$ octanoate, and these can be understood in terms of long-range hydrophobic interactions (Leduc and Desnoyers, 1973). However, the properties of n-alkylamine hydrobromides have indicated that there are still unresolved problems concerning the hydrophobic interactions between long alkyl chains (Leduc et al., 1974; Jolicoeur and Philip, 1975, Jolicoeur and Boileau, 1974).

Analysis of ion–ion interactions in aqueous salt solutions using the pair potential summarized in eqn (21) in conjunction with the HNC technique (Friedman, 1971) has proved interesting. The value of this approach is indicated by the good agreement between observed and theoretical MacMillan–Mayer osmotic coefficients for lithium chloride in water at 298 K as shown in Fig. 23 (Friedman and Krishnan 1973a). For alkali metal halides in water, A_{ij} parameters are

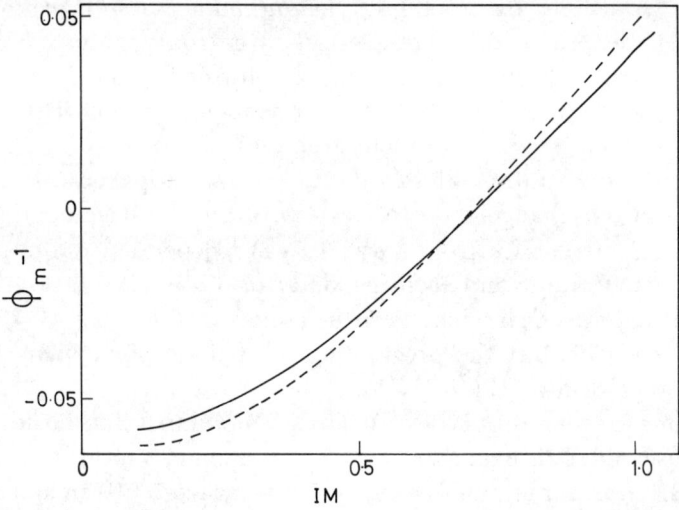

Figure 23. Dependence of MacMillan–Mayer osmotic coefficient on concentration for lithium chloride aqueous solutions at 298 K; dotted line indicates calculated and full line indicates experimental dependence on IM, the ionic strength defined in terms of MacMillan–Mayer variables (Friedman and Krishnan, 1973a).

small and generally negative. Solvation in aqueous solution results in solutes attracting each other to a larger extent than would be expected if the solvent were an idealized structureless medium. Some Gurney A_{ij} parameters, calculated assuming that $A_{--} = 0$, are summarized in Fig. 24 (Ramanthan et al., 1972).

The quantity A_{++} for Li^+ is positive but decreases with increasing ion size. Thus for small ions, water molecules tightly bound to the ion are released as the co-spheres overlap (cf. electrostrictive structure formers). The more negative values for Na^+, K^+, Rb^+ and Cs^+ show also that a water molecule in the co-sphere, i.e. in the B-zone, is in a higher energy state than one in bulk water. However, one must be wary of this simple explanation because it does not account for the negative values of A_{++} for R_4N^+ ions. A more important factor is the decrease in the surface area of the solute cavity as the co-spheres overlap, but the values of A_{++} for Et_4N^+, Pr_4N^+ and Bu_4N^+ are the same, indicating that the co-spheres of these ions are similar. The sensitivity of A_{+-} to the anion where the cation is R_4N^+ indicates that the stability of the co-sphere around the ions increases from iodide to fluoride. The pair correlation functions show impor-

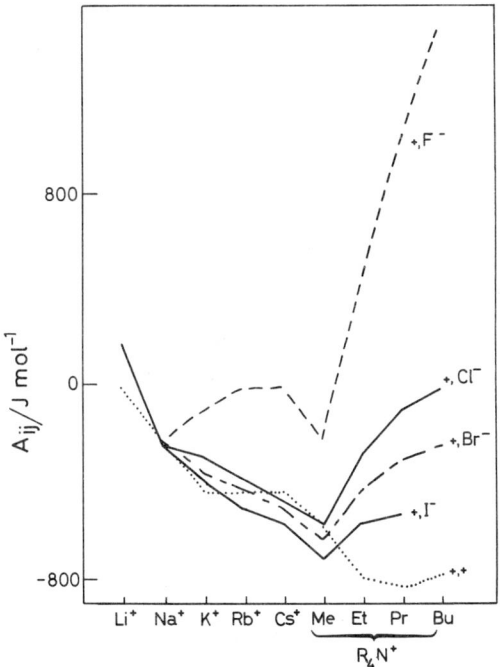

Figure 24. Gurney parameters, A_{ij}, for aqueous salt solutions at 298 K; calculated assuming $A_{--} = 0$ (Ramanthan, Krishnan and Friedman, 1972).

tant patterns. One example is shown in Fig. 25 for $Et_4N^+Cl^-$ solutions. The peak in the cation-cation pair correlation function indicates the presence of cation-cation ion pairs. The g_{++} maxima for Me_4N^+ and Bu_4N^+ occur at 9 and 12 x 10^{-8} cm respectively, these distances being more than twice the radius of the ions. This behaviour is in line with the Frank Model for hydrophobic interactions discussed on p. 255. The peak in g_{+-} is assigned (Ramanthan et al., 1972) to water-structure enforced ion pairs (Diamond, 1963). Wen (1973) suggests that +− ion pairs are more important than ++ ion pairs when $c_2 < 0.2$ mol dm^{-3}. The +− interaction increases

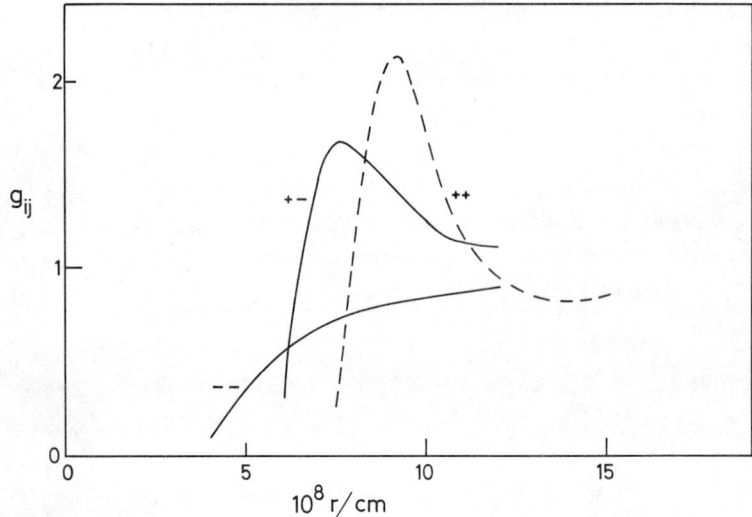

Figure 25. Computed pair-correlation functions for tetraethylammonium chloride in water at 298 K where $c_2 = 0.4$ mol dm^{-3}.

with increase in the size of the cation, which is interesting when it is recalled that association of Bjerrum coulombic ion-pairs tend to increase with decrease in ion size.

Apolar Solutes in Aqueous Salt Solutions

Chemical reactions studied in aqueous salt solutions often involve one or more neutral solutes. The question then arises as to the effect that the salts have on these solutes and how in turn the rate of reaction is affected. The effects of added salts on the solubilities of apolar solutes have been extensively studied (Conway and Novak, 1975). The literature up to 1952 was reviewed by Long and McDevit (1952) and more recent data for aqueous solutions documented by Masterson et al. (1971). The information can often be summarized using eqn (23).

$$\log (S_3/S_3') = K_s c_2 \qquad (23)$$

Here S_3 is the solubility of the non-polar solute in the solvent and S_3', the solubility in a salt solution of concentration c_2. K_s is the Setschenow coefficient. A positive value for K_s means that the solute is "salted-out" by added salt.

Simple apolar solutes, e.g. H_2 and O_2, are salted out by inorganic salts and the effects of a given salt can be separated into relative contributions of anion and cation. The effectiveness of salting out varies considerably with salt but the order is often the same for different solutes. The problem of understanding solubility patterns is closely linked with that of understanding ion-solvent interactions. Thus large organic ions often "salt-in" an apolar solute, a phenomenon termed hydrotropism. The situation becomes more complex when the solute contains both hydrophobic and hydrophilic centres. For example, γ-butyrolactone in water at 298 K is salted out by NaCl but salted in by NaI. The importance of the solvent in such solubility studies is shown by the discovery that, while α-naphthoic acid in water is salted out by KI and salted in by $Me_4N^+Cl^-$, the order is reversed in 50% (w/w) aqueous dioxan (Grunwald and Butler, 1960).

Interest in the solubility of non-electrolytes in salt solutions has increased following the growth of interest in aqueous alkylammonium salt solutions. There has also been a considerable effort to obtain related enthalpy and entropy data as well as Gibbs functions, the results often being tabulated as transfer functions, δ_m, describing the change on going from a solution in water to a solution in an aqueous salt solution.

For argon in solutions of alkali metal halides or ammonium chloride, $\delta_m \mu_3^\ominus > 0$ and the transfer is endothermic, $\delta_m H_3^\ominus > 0$. This behaviour reflects a breakdown of water structure by the salts, shifting equilibrium (16) to favour $(H_2O)_d$, thereby decreasing the amount of bulk water, $(H_2O)_b$, which can accommodate apolar solutes in the cavities (Ben-Naim and Egel-Thal, 1965). The salting out of hydrocarbons by sodium chloride is also endothermic, e.g. $\delta_m H^\ominus$ (butane) = 2·5 kJ mol^{-1} when $c_2 = 1$ mol kg^{-1}. The heat capacity change depends on the size of the solute, being positive for hydrogen but, for example, negative for butane and argon (Clever and Holland, 1968).

Rather more complex patterns emerge for the transfer of nitromethane between water and salt solutions at $c_2 = 1$ mol dm^{-3} (Stern and Hermann, 1967). Thus, $\delta_m \mu_3^\ominus$ is positive for LiCl solutions (salting out) but decreases with increase in cation size, and is negative for CsCl (salted in). For the enthalpies, $\delta_m H_3^\ominus$ is positive for LiCl, again decreasing on going to CsCl, and is exothermic in the latter case. However, for perchlorates (e.g. sodium), $\delta_m \mu_3^\ominus$ is negative but the transfer is endothermic. A similar complex pattern occurs where the solute is acetic acid. For example, for transfer from water to sodium chloride solution (1 mol dm^{-3}) $\delta_m \mu_3^{\ominus} > 0$ but the transfer is now exothermic (Stern et al., 1967). The important effect of introducing an apolar alkyl group into a solute is indicated by data for the enthalpy of transfer of amides. Thus for formamide, $\delta_m H_3^{\ominus} < 0$, becoming more exothermic as the size of the alkali metal or halide ion increases. However, with increasing alkyl substitution the transfer becomes less exothermic and eventually endothermic, e.g. N-methyl propionamide in KBr solutions (Stimson and Schrier, 1974).

The patterns which emerge from these studies are complex. They become even more so when we turn our attention to salt solutions containing alkylammonium ions. A systematic study of the solubilities of hydrocarbons in these solutions has been reported by Wen and Hung (1970). Large alkylammonium salts, e.g. $Pr_4N^+Br^-$ and $Bu_4N^+Br^-$, generally enhance the solubility, while $NH_4^+Br^-$ and $(HOCH_2CH_2)_4N^+Br^-$ lower it. However, the pattern depends significantly on temperature. For example, at 298 K, $\delta_m \mu_3^{\ominus}$ is more negative in solutions of $Bu_4N^+Br^-$ than in solutions of $Et_4N^+Br^-$ (0·1 mol dm^{-3}) but the order is reversed at 278 K. Also $\delta_m \mu_3^{\ominus}$ for a hydrocarbon becomes more temperature-dependent the larger the hydrocarbon. The transfer is generally endothermic and the entropy of transfer is clearly very important. For example, in the case of butane in 0·1 mol dm^{-3} $Bu_4N^+Br^-$ solution at 298 K, $\delta_m \mu_3^{\ominus} = -0·163$ kJ mol^{-1}, $\delta_m H_3^{\ominus} = +2·18$ kJ mol^{-1} and $T\delta_m S_3^{\ominus} = +2·34$ kJ mol^{-1}, the latter being the quantity which determines the sign of $\delta_m \mu_3^{\ominus}$.

Wen and Hung explain their observations in terms of two factors; (i) indirect long-range hydrophobic interaction between alkylammonium ion and hydrocarbon and (ii) direct contact between these two components with short-range hydrophobic association, the latter being important at high temperatures. The fact that the $(HOCH_2CH_2)_4N^+$ ions behave like Me_4N^+ rather than Pr_4N^+ is in keeping with the ideas discussed on p. 267).

The proposal that these solubility trends result from the effects of cations on water-structure has been criticised (Feillolay and Lucas, 1972). Instead, dispersion forces between salt and solute are used to explain why added $Bu_4N^+Br^-$ raises the solubility of helium and methane (Lucas and de Trobriand, 1971).

A detailed study has been reported of the thermodynamic quantities for transfer of alcohols from water to alkylammonium salt solutions (Aveyard and Heselden, 1974). Both $\delta_m S_3^\ominus$ and $\delta_m H_3^\ominus$ for the alcohols are positive, increasing with increase in the size of the alkyl group of either the alcohol or the cation (Table 7). However $\delta_m \mu_3^\ominus$ becomes more negative with increase in the size of the alkyl group, the transfer being therefore favourable because the entropy term is dominant. In this system, the salting-in arises from hydrophobic interactions.

Quite dramatic changes in solubility are observed for benzene in water when alkylammonium salts are added. The solubility increases with increase in the chain length of the alkyl group in R_4N^+, ammonium bromide salting-out benzene (Desnoyers et al., 1965). Actually the solubility of benzene increases rapidly when the salt concentration exceeds a characteristic value, e.g. 1 mol dm^{-3} for $Bu_4N^+Br^-$ (Wirth and Lo Surdo, 1968), indicating that some change

TABLE 7

Thermodynamic Transfer Quantities[a] for Alcohols, from Water to Aqueous Salt Solution (1 mol Kg^{-1}) at 293 K

Salt	Solute	$\delta_m\mu_3^\ominus$/kJ mol^{-1}	$\delta_m H_3^\ominus$/kJ mol^{-1}	$\delta_m S_3^\ominus$/J mol^{-1} K^{-1}
$Et_4N^+Br^-$	BuOH	−0·48	1·8	7·8
$Pr_4N^+Br^-$	BuOH	−0·62	4·4	17·1
$Bu_4N^+Br^-$	BuOH	−1·01	10·8	40·3
$Bu_4N^+Br^-$	PrOH	−0·62	6·9	25·6

[a] Mole fraction scale: data from Aveyard and Heselden, 1974.

occurs in the organization of the salt solution at this point (p. 269). This change is more readily apparent in the enthalpy function, $\delta_m H_3^\ominus$ for benzene (Arnett et al., 1970). With increase in salt concentration, $\delta_m H_3^\ominus$ is positive (but $\delta_m \mu_3^\ominus$ is negative) and reaches a maximum when the salt concentration is 1·0 mol dm^{-3} before decreasing. This maximum is difficult to understand without reference to the structure of the aqueous solution, the role of hydrophobic interactions and the fact that this change occurs near the composition of the solid clathrate hydrate. A distinction is nevertheless drawn between this behaviour and the increase in solubility of benzene when perchloric acid is added. In this case, a specific benzene-perchloric acid complex is involved (Rehfeld, 1973).

Numerous attempts have been made to account for the solubilities of gases in salt solutions using theoretical models, e.g. scaled particle theory (Masterson et al., 1971). According to HNC calculations

(Krishnan and Friedman, 1974), the main contribution to the salting out stems from the COR term [eqn (21)], i.e. an excluded volume effect. For alkane — R_4N^+ co-sphere interactions, hydrophobic association is important.

Ions in Aqueous Salt Solutions

The analysing of kinetic data for reactions of non-polar solutes in salt solutions require the solubilities of apolar solutes in salt solutions as background information. By the same token, analysis of the effects of added salts on the rates of chemical reactions involving ions requires consideration of the effects which these salts have on the properties of ions in water. It is clear from the complexity of the properties of single salt solutions that understanding the properties of mixed salt solutions in water presents a formidable problem. We have indicated in Section 6 how the properties of such solutions can be quantified using mixing functions, e.g. g_0, h_0, v_0. In a review of these systems, two generalizations are given (Anderson and Wood, 1973). First, enthalpies of interaction between cations are almost independent of anions; and second, the dependence of RTg_0, RTh_0 and v_0, on concentration is often small. The latter approximation is valid for alkali metal halide systems but is not correct for alkylammonium salt solutions. Anderson and Wood tabulate g_0, h_0, and v_0 parameters for a wide range of systems. Analysis of these mixing parameters is a subject of active research (Cassel and Wood, 1974; Friedman and Krishnan, 1974; Pitzer and Kim, 1974; Falcone et al., 1973; Friedman et al., 1973), and it is difficult to identify any clear cut patterns. However, several points are noteworthy. For example, at unit ionic strength, RTg_0 is small for KBr/NaBr solutions, is slightly more negative for $Et_4N^+Br^-/Pr_4N^+Br^-$, but large and positive for $NaBr/Pr_4N^+Br^-$. Further, h_0 for alkylammonium salt solutions is large and negative, e.g. for $Et_4N^+Cl^-/Pr_4N^+Cl^-$, much more so than g_0. This indicates that $R_4N^+-R_4N^+$ interactions are predominantly entropy controlled, i.e. structural factors are dominant.

Kinetics of Reactions in Salt Solutions

Reactions involving ions

The contrast between the hydration of apolar and polar solutes is indicated by the small value of $\Delta C_p^\ddagger$, -30 J mol^{-1} K^{-1}, for the hydrolysis of t-butyldimethylsulphonium ion (Robertson, 1967).

Thus, for this small ion, activation is not accompanied by significant reorganization of water in the solvent co-sphere as in the case of apolar solutes (p. 256). The volume of activation for the hydrolysis of t-BuEtMeS$^+$ is positive, 6 cm^3 mol^{-1} at 313 K (Brower and Wu, 1970) from which it has been concluded that the reaction involes breaking of a C–S bond. However, the complexity of volume properties of solutes in water indicates that $\Delta V^{\ddagger}$ values should be interpreted with caution. It would not be surprising if $\Delta V^{\ddagger}$ were controlled to a considerable extent by changes in the solvent co-sphere.

An exciting development in this subject has been the application of the Frank–Wen model for ionic hydration to solvent exchange reactions (Caldin and Bennetto, 1973). In these reactions, a solvent molecule must pass from bulk solvent through the disordered layer of zone B, a process which resembles evaporation. Consequently activation enthalpies for these reactions are directly related to the solvation characteristics of the ion.

Salt Effects on Kinetics

The effect of added salts on the rate constant of a given ionic reaction has been studied for many years. The Brønsted-Bjerrum treatment of these salt effects has been particularly successful, the rate constant being related to the ionic strength of the solution. The observed trends can be quantitatively accounted for using the DHLL or a related expression for the activity coefficients of reactants and transition state. This subject has been reviewed in detail (Perlmutter-Hayman, 1971). The ionic-strength principle appears satisfactory when the reaction involves ions of opposite charge but less so when it involves ions of the same charge.

Nevertheless one might have expected that the kinetic data would reflect clearly the hydration properties of the added salt. This is not usually the case because charge–charge interactions are dominant. Moreover, structural effects often compensate and their influence on rate constants is minimised (p. 247). If, however, the reactants are neutral solutes, it might be anticipated that the kinetic parameters would be markedly sensitive to the particular salt added to the solution (p. 272). This prediction is borne out in practice, striking differences often being observed between the effects of added alkali metal cations and alkylammonium ions as the following examples show.

The first-order rate constant for the base-catalysed decomposition of diacetone alcohol increases rapidly when $Pr_4N^+OH^-$ is added, less rapidly when $Me_4N^+OH^-$ is added, and decreases on addition of KOH. The conclusion that the variation in rate constant stems from the effects of the added cation is confirmed by observing that, at a fixed concentration of KOH, added $Et_4N^+I^-$ increases and added KI decreases the rate constant (Halberstadt and Prue, 1952). More recently, Hibbert and Long (1972) observed that the rate constant for the water-catalysed detritiation of t-butylmalononitrile at 298 K

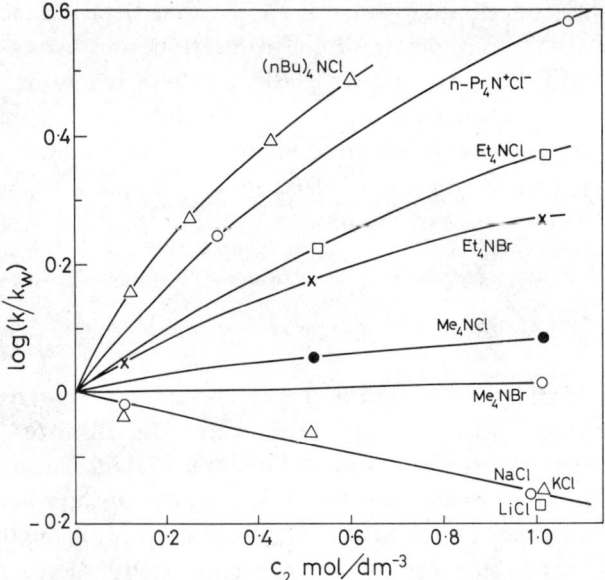

Figure 26. Relative rates of detritiation of t-butylmalononitrile in aqueous salt solutions at 298 K as a function of added salt concentration; k and k_W are rate constants for reaction in salt solution and in water respectively, all solutions containing 0·001 mol dm^{-3} HCl (Hibbert and Long, 1972).

is very sensitive to the nature and the concentration of added salt (Fig. 26). Thus through a series of chlorides, the rate constant increases as the hydrophobic character of the cation increases. A similar behaviour is reported for salt effects on the base-catalysed hydrolysis of p-$CH_3C_6H_4SO_2CH_2OClO_3$ (Menninga and Engberts, 1973). The rate enhancement of structure-forming ions such as n-Bu_4N^+ was explained in terms of a stabilization of the transition state, the enhanced water structure facilitating proton transfer from the substrate to water and co-operative interactions within the water structure enhancing the acceptor ability of water. This is a satisfactory mechanism, but some factors still remain unexplained. It is not apparent why here and in other examples (cf. Fig. 26), plots of relative rates against salt concentration become more curved as the cation becomes more hydrophobic. Clearly it would be valuable to have information concerning the effect of added salts on both initial and transition states. In this respect, the study by Clarke and Taft (1962) of the effect of added salts on the rate of hydrolysis of t-butyl chloride is noteworthy. By measuring the solubility of t-butyl chloride in salt solutions Setschenow parameters [eqn (23)] were obtained for initial and transition states. Clarke and Taft report that a classical analysis of salt effects based on a calculation of activity coefficients for solute dipoles using an extension of the Debye-Hückel theory (Bateman et al., 1940) was inadequate, the salts having quite specific effects. Addition of salts to the reacting system

TABLE 8

Salt Effects on the Hydrolysis of Acetic Anhydride[a]

Salt	Concentration/mol dm^{-3}	$\delta_m \mu^{\ominus}(Ac_2O)$	$\delta_m \mu^{\cdot}(H_2O)$	$\delta_m \mu^{\ddagger}$
LiCl	5	$-2{\cdot}74$	$-0{\cdot}57$	$-0{\cdot}42$
NaCl	5	$-2{\cdot}49$	$-0{\cdot}32$	$+0{\cdot}78$
Me$_4$N$^+$Cl$^-$	5	$-3{\cdot}48$	$+0{\cdot}57$	$+0{\cdot}80$
Bu$_4$N$^+$Cl$^-$	1$\cdot$86	$+2{\cdot}51$	$+1{\cdot}55$	$+7{\cdot}51$

[a] Thermodynamic quantities expressed in kJ mol^{-1}; data from Oakenfull (1971).

results in a salting-out of the initial state but, at least for the 1 : 1 salts, there is a cancellation of the specific effects of the salts between initial and transition states. An attempt was made to separate the effects of salts into ionic contributions. In these terms, the alkylammonium ions do not follow the pattern established by the inorganic cations, although it is unfortunate that the study does not extend to alkylammonium ions, R$_4$N$^+$, with R larger than Et. Nevertheless, alkali metal salts enhance the rate constant for hydrolysis more markedly than do alkylammonium salts.

In contrast, addition of 1 : 1 salts lowers the rate constant for the hydrolysis of acetic anhydride in water (Bunton et al., 1962; Oakenfull, 1971). Combination of kinetic and solubility data leads to a clearer understanding of these trends. If the reaction is bimolecular, osmotic data for salt solutions can be used to calculate the change in the chemical potential of the water. Thus it turns out (Table 8) that the salting in of the initial state, acetic anhydride, is the dominant factor in LiCl and NaCl solutions (Oakenfull, 1971). However, the dramatic fall in rate constant when $Bu_4N^+Cl^-$ is added stems from a marked destabilization of the transition state. Clearly, the dependence of the rate constant on the nature of the added salt can be a complex function of effects on both initial and transition states, and it is difficult to generalize. For example, addition of $Bu_4N^+Br^-$ increases the rate of hydrolysis of trimethylamine sulphur trioxide while the opposite trend is observed when KBr is added (Blandamer et al., 1975a). Clearly a great deal more needs to be learnt about salt effects on rates of reactions in water.

10. AQUEOUS MIXTURES

A challenging subject in physical organic chemistry is the interpretation of changes in activation parameters for a given reaction as an organic co-solvent is added to the reaction in water. In order to understand these systems, we start by examining some thermodynamic properties of aqueous mixtures. These provide a convenient basis for classification of diverse systems.

Thermodynamic Properties

In a mixture formed from two liquids, components 1 and 2, the chemical potentials μ_1 and μ_2 can be expressed in terms of the chemical potentials of the pure components at the same temperature and pressure, μ_1^* and μ_2^*, the mole fractions x_1 and x_2 and the rational activity coefficients, f_1 and f_2 [eqns (24) and (25)]. As x_1

$$\mu_1 = \mu_1^* + RT \ln x_1 f_1 \tag{24}$$

$$\mu_2 = \mu_2^* + RT \ln x_2 f_2 \tag{25}$$

$\to 1$ $f_1 \to 1$, and $f_2 \to 1$ as $x_2 \to 1$. In the corresponding ideal mixture, $f_1 = f_2 = 1$. Consequently, for a real mixture with mole fractions x_1 and x_2 ($= 1 - x_1$), an excess molar Gibbs function can be defined by (26) which expresses the extent to which the real

$$G^E = x_1 RT \ln f_1 + x_2 RT \ln f_2 \tag{26}$$

mixture differs from the corresponding ideal mixture. G^E can be calculated, for example, from the dependence of the vapour pressure of the components and of the mixture on composition (Barker, 1953). These determinations require patient and careful work. Moreover the calculations are not straightforward, particularly for aqueous solutions where G^E can be miscalculated if it is assumed that the vapour behaves as an ideal gas (Treiner et al., 1973).

For an ideal mixture, the enthalpy of mixing is zero and so a measured molar enthalpy of mixing is the excess value, H^E. The literature concerning H^E-values is more extensive than for G^E-values because calorimetric measurements are more readily made. The dependence of H^E on temperature yields the excess molar heat capacity, while combination of H^E and G^E values yields S^E, the molar excess entropy of mixing. The dependences of G^E, H^E and $T \cdot S^E$ on composition are conveniently summarized in the same diagram. The definition of an ideal mixture also requires that the molar volume is given by the sum, $x_1 V_1^* + x_2 V_2^*$, so that the molar volume of a real mixture can be expressed in terms of an excess molar volume V^E (Battino, 1971).

The dependence of any excess function, X^E on mole fraction can be fitted to an algebraic equation such as the Guggenheim–Scatchard eqn (27) (Guggenheim, 1937a; Scatchard, 1949). The data are fitted

$$X^E = x_2(1 - x_2) \sum_{i=1}^{n} A_i(1 - 2x_2)^{i-1} \qquad (27)$$

to (27) by a least-squares technique using n coefficients, A_1, A_2, $A_3 \cdots A_n$. More sophisticated equations can be used (Bale and Pelton, 1974). For example, for tetrahydrofuran + water mixtures, Glew and Watts (1973) found the plot of H^E against x_2 to be S-shaped. It proved advantageous to fit the exothermic and endothermic parts of the curve separately using (27).

The dependence of X^E on mole fraction can be analysed to obtain the dependence on mole fraction of the relative partial molar quantities, $X_1 - X_1^*$ and $X_2 - X_2^*$, which are calculated from the intercepts on the X_1- and X_2-axes of the tangent to the X^E-curve. Consequently the X^E-data must be precise if these partial molar values are to be meaningful. For example, one of the features of some aqueous mixtures is a minimum in the value of $V_2 - V_2^*$ at low values of x_2. This means that the V^E-curve has a point of inflexion at this mole fraction.

Before surveying aqueous mixtures, it is informative to examine briefly the thermodynamic excess functions for two particular non-aqueous mixtures, (a) acetone + chloroform (Fig. 27) and (b) methyl alcohol + carbon tetrachloride (Fig. 28).

The thermodynamic functions for the acetone + chloroform system reflect the importance of inter-component association. Here

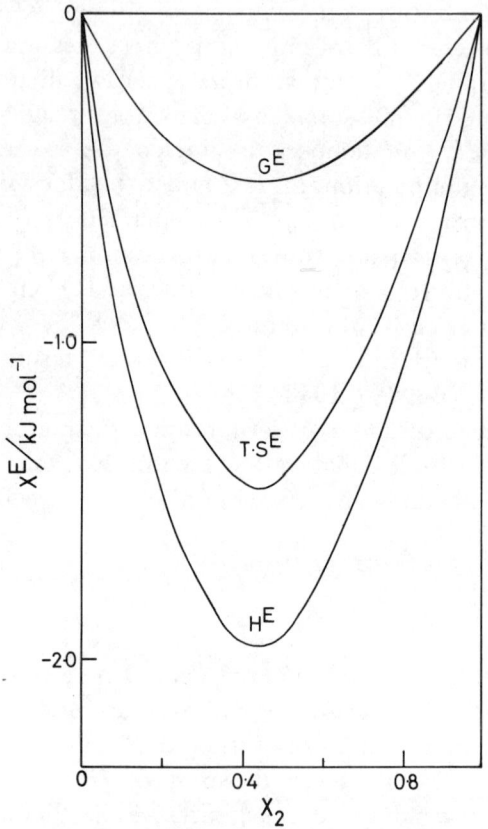

Figure 27. Thermodynamic excess functions for acetone + chloroform mixtures at 298 K; x_2 = mole fraction of chloroform (Franks and Ives, 1966).

hydrogen bonding, $(CH_3)_2CO\cdots HCCl_3$, determines the non-ideal properties, both G^E and H^E being negative which indicates favourable and exothermic mixing. In contrast, G^E is positive for the system $CH_3OH + CCl_4$, and this is attributed to the depolymerization by carbon tetrachloride of the hydrogen-bonded methyl alcohol. Thus these two systems provide examples of two extreme types of behaviour, association and disruption.

In both examples summarized in Figs. 27 and 28, the sign and magnitude of G^E is for the most part determined by H^E, i.e. $|H^E| > |T \cdot S^E|$. This behaviour is called "typically non-aqueous" (Franks, 1968a). However for many aqueous mixtures a rather different type of behaviour is observed in which G^E is positive because $T \cdot S^E$ is large and negative (although H^E can be negative) i.e. $|T \cdot S^E| > |H^E|$.

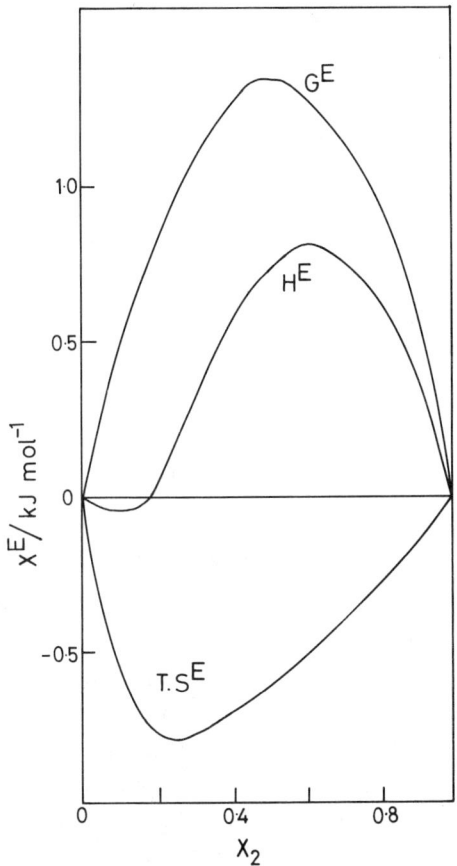

Figure 28. Thermodynamic excess functions for methyl alcohol + carbon tetrachloride at 298 K; x_2 = mole fraction of carbon tetrachloride (Franks and Ives, 1966).

The dependence of H^E and S^E on mole fraction can be complex although the overall effect on the G^E-curve is not. The importance of the entropy of mixing warrants the description of such mixtures as "typically aqueous" (Franks, 1968b). In the following account of the kinetics of reactions in these mixtures, we examine these systems under three headings: (i) "typically aqueous", TA; (ii) "typically

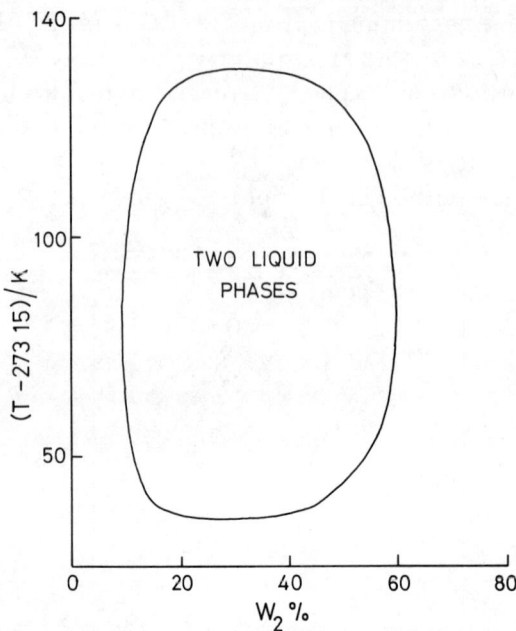

Figure 29. Phase diagram for 2-butoxyethanol + water mixtures; w_2 = weight per cent of 2-butoxyethanol (e.g. when $w_2 = 58$, $x_2 = 0 \cdot 17$) and LCST = 321·9 K and UCST = 403·3 K (Ellis, 1967).

non-aqueous" where G^E is negative, TNAN; and (iii) typically non-aqueous where G^E is positive, TNAP.

The importance of the excess entropy of mixing in aqueous mixtures explains why many of these systems show phase separation with a lower critical solution temperature (LCST). This phenomenon is rarer—though not unknown—in non-aqueous mixtures (for an example, see Wheeler, 1975). The conditions for phase separation at a critical temperature can be expressed in terms of the excess functions of mixing (Rowlinson, 1969; Copp and Everett, 1953).

For a binary mixture, the condition for a critical point is given by (28).

$$(\partial^2 G^E / \partial x_2^2)_{T, p} + RT/x_1 x_2 = 0 \qquad (28)$$

At an LCST, the additional conditions in (29) and (30) apply.

$$(\partial^2 H^E / \partial x_2^2) > 0;\ (\partial^2 S^E / \partial x_2^2) > 0 \qquad (29)$$

where

$$|T(\partial^2 S^E / \partial x_2^2)| > |(\partial^2 H^E / \partial x_2^2)| \qquad (30)$$

Thus, as noted above, the entropy of mixing is the key quantity for the system to show an LCST. At an upper critical solution temperature, UCST, an important condition is that expressed in (31).

$$\partial^2 H^E / \partial x_2^2 < 0 \qquad (31)$$

Thus the enthalpy of mixing is a key quantity for a system to show a UCST. For the other excess functions, e.g. V^E and C_p^E, there are no restrictions, but generally at a UCST, $C_p^E < 0$ and $V^E > 0$, while at a LCST, $C_p^E > 0$ and $V^E < 0$ (Rowlinson, 1969). If C_p^E is negative at an LCST and remains so as the temperature increases, then H^E and S^E may change in such a way that the conditions for a UCST are met. Such systems show a closed solubility loop. The mixture water + nicotine is a classic example of such a system. The behaviour of another example, the mixture water + 2-butoxyethanol, is shown in Fig. 29 (Ellis, 1967).

Solutes in Aqueous Mixtures

Analysis of solvent effects on activation parameters, $\delta_m \Delta X^{\ddagger}$, requires information concerning the behaviour of solutes in aqueous mixtures. Although some such information has been published in recent years, a great deal more is certainly desirable. The division between the different classes of solvent mixtures (p. 283) has been made, however, because the properties of solutes in these mixtures also reflect this subdivision. Although interpretation of the changes in solute properties with change in solvent composition is not straightforward, we can nevertheless predict some of the trends to be expected.

For apolar solutes in water, we have seen that $\Delta G^{\ominus} > 0$, $\Delta H^{\ominus} < 0$, $T\Delta S^{\ominus} < 0$ and $|T\Delta S^{\ominus}| > |\Delta H^{\ominus}|$ (p. 248). At the other end of the scale when $x_2 = 1$, we are concerned with the thermodynamic properties of an apolar solute in a non-aqueous solvent. Because the solubility is usually larger for such solutes in these solvents, $\Delta G_3^{\ominus}$ (gas → solvent) will be less. Thus the effect of going from a solution in water to a solution in a pure co-solvent should be as predicted in (32). Also $\Delta S_3^{\ominus}$ (gas → co-solvent) will not be as negative as in the

$$\delta_m \mu_3^{\ominus} = \Delta G_3^{\ominus}(\text{gas} \to \text{co-solvent}) - \Delta G_3^{\ominus}(\text{gas} \to \text{water}) < 0 \qquad (32)$$

aqueous solution, so that for transfer between the two solvents condition (33) holds. If we follow through this argument using for

$$\delta_m S_3^{\ominus} > 0 \qquad (33)$$

TABLE 9

Thermodynamic Transfer Quantities[a] for Three Apolar Solutes from Water to Organic Co-solvent[b]

Solute	Co-solvent	$\delta_m\mu_3^\ominus$	$\delta_m H_3^\ominus$	$T\delta_m S_3^\ominus$
Propane	ethyl alcohol	−21·51	22·40	43·91
Butane	ethyl alcohol[c]	−21·54	27·23	48·77
Methane	acetone[d]	−10·72	11·93	22·65

[a] Expressed in kJ mol^{-1}.
[b] Standard states—hypothetical unit mole fraction at 298 K.
[c] Calculated using solubility data reported by Kretschmer and Wiebbe, 1951.
[d] Calculated using data reported by Lannung and Gjaldbaek, 1960.

example the data given in Table 9, the diagram shown in Fig. 30 can be constructed showing the behaviour expected for an apolar solute in a solvent mixture consisting of water + organic co-solvent. The full lines join the two points for the solute properties in water and pure co-solvent, and indicate the underlying trend. The slopes of the lines

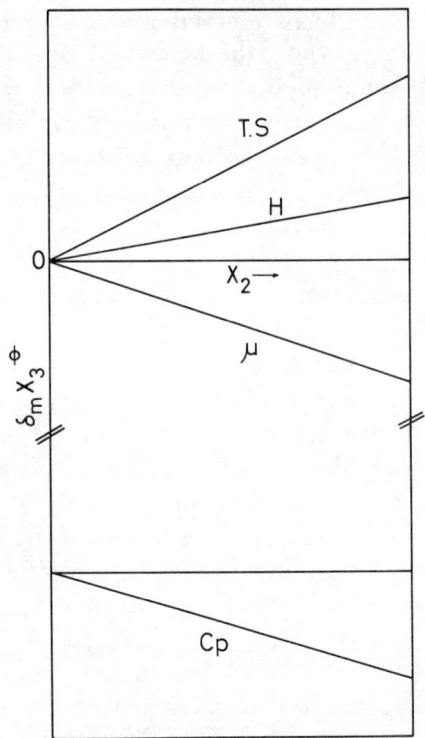

Figure 30. Predicted dependence of the thermodynamic properties of an apolar solute on mole fraction in a water + organic co-solvent mixture.

will depend upon the solute and co-solvent. The impact of this behaviour on the kinetic parameters can be illustrated as follows. Consider a reaction in which the reactant is apolar. Assume that the reaction is unimolecular and that the properties of the transition state are independent of solvent composition, i.e. $\delta_m X_3^{\ddagger} = 0$. Then, from eqn (9), it follows that the variation of $\delta_m \Delta X^{\ddagger}$ will be the mirror image of that shown in Fig. 30. Thus we anticipate that the rate constant will decrease, the change in $T\Delta S^{\ddagger}$ will be more dramatic than in $\Delta H^{\ddagger}$ and $|\Delta C_p^{\ddagger}|$ will decrease (Blandamer et al., 1974). Further because the solvation parameters are often more dependent on the size of the solute in water than in organic co-solvents, we anticipate that $\delta_R \Delta X^{\ddagger}$ will become less marked as x_2 increases.

The key questions are of course the extent to which the full lines in Fig. 30 approximate to the actual behaviour and how deviations from such lines might be accounted for. Intuitively, it seems reasonable that if the two solvent components interact (i.e. $G^E < 0$), then the solute will be less soluble in the mixtures than in the ideal mixture where $G^E = 0$. Alternatively, if $G^E > 0$, the solubility of the solute may be larger than if the mixture were ideal.

Some support for these qualitative ideas comes from an extra-thermodynamic analysis of solubilities in liquid mixtures. Krishchevsky (O'Connell 1971) showed that if the mixture is ideal then the Henry's law coefficient for a solute in the mixture, L_3 (1, 2), is related to the Henry's law coefficients for the solute in the two pure components, L_3 (1) and L_3 (2) by eqn (34)

$$L_3(1, 2) = x_1 L_3(1) + x_2 L(2) \tag{34}$$

$$\Delta G_3^{\ominus} \text{ (gas} \rightarrow \text{solvent)} = RT \ln (L_3/p^{\ominus}) \tag{35}$$

$p^{\ominus}$ being the standard pressure. Attempts have been made to obtain quantitative expressions for real liquid mixtures (Tiepel and Gubbins, 1972). If the dependence of G^E on mole fraction can be fitted using only one term in eqn (27) for the three systems 1-2, 1-3 and 2-3, then eqn (34) can be modified by adding a further term, $-G^E/RT$, on the right-hand side. The resulting expression predicts a pattern of solubilities similar to that discussed qualitatively above. Consequently, we have a reference point for a detailed analysis of the solubilities of apolar solutes in aqueous mixtures. As one example, we note that $\delta_m \mu_3^{\ominus}$ for adenine and thymine in binary aqueous mixtures containing methyl, ethyl and propyl alcohol is more negative when $x_2 \simeq 0.5$ than when $x_2 = 0$ or 1; for such systems $G^E > 0$, as we show below (Herskovits and Harrington, 1972).

The analysis of the behaviour of ions in aqueous mixtures, is not so straightforward. As noted on p. 218, to obtain $\delta_m X^{\ominus}$ for an ion, it is necessary to make an extrathermodynamic assumption in order to subdivide the transfer value for salt into the ionic components. Consequently, different conclusions may be reached depending on the method of analysis, especially when $\delta_m X^{\ominus}$ (ion) is small. For example, Abraham (1973), assuming $\delta_m \mu^{\ominus}$ ($Me_4 N^+$) = 0, predicts that $\delta_m \mu^{\ominus}$ ($Et_4 N^+$) is -6.1 kJ mol^{-1} at 298 K on the molar scale for transfer from water to methyl alcohol whereas Cox et al. (1974), assuming $\delta_m \mu^{\ominus}$ ($Ph_4 As^+$) = $\delta_m \mu^{\ominus}$ ($Ph_4 B^-$), predict that $\delta_m \mu^{\ominus}$ ($Et_4 N^+$) is $+0.84$ kJ mol^{-1}. Thus it is not clear whether this ion is stabilized or destabilized. Perhaps this implied criticism is harsh if the quantities are small, but even when there seems general agreement over the sign of $\delta_m \mu^{\ominus}$, (ion), the magnitudes differ considerably. Thus most authors agree that, for chloride ion, $\delta_m \mu^{\ominus} > 0$ on going from water to methanol. However Cox (1973) calculates a value of 12.5 kJ mol^{-1} at 298 K [molar scale assuming $\delta_m \mu^{\ominus}$ ($Ph_4 As^+$) = δ_m ($Ph_4 B^-$)], Abraham (1973) suggests a value of 20.3 kJ mol^{-1} [assuming $\delta_m \mu^{\ominus}$ ($Me_4 N^+$) = 0], while Andrews et al. (1968) calculate from the dependence of transfer quantities on ionic radii, a value of 35.9 kJ mol^{-1}.

Nevertheless various patterns seem to emerge. The transfer of a hydrophobic ion such as $Bu_4 N^+$ from water to a non-aqueous solvent is favourable as a result of a large positive value for $\delta_m S_3^{\ominus}$ although the transfer is endothermic (cf. Fig. 30). For the alkali metal cations, the transfer is generally unfavourable (Cox et al., 1974), i.e. $\delta_m \mu^{\ominus}$ (ion) > 0 for water to methyl alcohol, but favourable for transfer to dimethyl sulphoxide. For the anions, $\delta_m S^{\ominus}$ (ion) is generally negative but $\delta_m H^{\ominus}$ (ion)-values cover a wide range, being positive in some cases and negative in other. However, $\delta_m \mu^{\ominus}$, (ion) is positive for halide, perchlorate and azide ions.

These single ion parameters can prove extremly valuable in the analysis of kinetic data. For example, suppose the reaction involves nucleophilic attack by Cl^- on a substrate in water and the rate is to be compared with that in methyl alcohol. Then on the molar scale, if $\delta_m \mu^{\ominus}$ (Cl^-) = $+12.5$ kJ mol^{-1} (Cox, 1973), there is a contribution of -12.5 to $\delta_m \Delta G^{\ddagger}$, the predicted rate constant increasing. The change in rate constant on going from water to solutions in DMSO should be more dramatic because here $\delta_m \mu^{\ominus}$ (Cl^-) = $+38.6$ kJ mol^{-1}.

The information discussed above can, however, yield only the overall trend in transfer quantities, and it would be unrealistic to expect, for example, that in aqueous mixtures, $\delta_m \mu^{\ominus}$ (ion) is a linear function of mole fraction. Nevertheless it is noteworthy that the transfer quantities between pure solvents are not in agreement with trends expected from a simple Born model for ionic solvation.

Activation Parameters and Thermodynamic Properties of Mixtures

Ultimately it should be possible to relate the properties of a solute quantitatively to the properties of a given mixture and thus to explain the "roller coaster" patterns shown by kinetic data. As a first step, it should be possible to relate the dependence of $\delta_m \Delta X^{\ddagger}$ on x_2 with that of X^E on x_2. There is, however, an important stumbling block. In a three-component system, water + co-solvent + solute, the problem arises as to the extent to which the solute modifies the properties of the solvent mixture. It would be advantageous therefore to calculate a set of activation parameters for a given reaction in the solvent mixture in which the components have the same properties as in the reaction mixture. A method of calculating such a set of parameters, called "endostatic activation parameters", has been described by Grunwald and Effio (1974). A set of thermodynamic parameters for solutes in binary mixtures is defined under the condition that the ratio of the activities, α ($=a_1/a_2 = x_1 f_1/x_2 f_2$) is identical to that in the binary mixture alone; thus the endostatic chemical potential, $\mu_{\alpha 3} = (\partial G/\partial n_3)_{T,p,(n_1+n_2),\alpha}$. It can be shown therefore that the endostatic activation Gibbs function $\Delta G^{\ddagger}$ is related to the calculated $\Delta G^{\ddagger}$ by equation (36). The thermodynamic

$$\Delta G_{\alpha}^{\ddagger} = \Delta G^{\ddagger} + (\ln \alpha)(d\Delta G^{\ddagger}/dx_1)/(d \ln \alpha/dx_1) \qquad (36)$$

non-ideality of the mixture is linked into this equation by the two terms $\ln \alpha$ and $d \ln \alpha/dx_1$. The analysis combines two sets of information, kinetic and thermodynamic.

In practice $\Delta G^{\ddagger}$ is known for a given T, p and x_2 so that the other quantities based on the ratio α must be calculated from the excess functions for the mixture. Differentiation of eqn (26) with respect to x_1 yields, using the Gibbs–Duhem equation, $\ln (f_1/f_2)$, and hence α. A second differentiation yields $d \ln \alpha/dx_1$. If eqn (27) is used to fit the G^E data then these quantities can be calculated from the A_i-coefficients. The arithmetic is tedious but a computer program can be used to advantage here (Blandamer et al., 1975b). Because the analysis requires the calculation of the second differential, $\partial^2 G^E/\mathrm{d}x_1^2$, the thermodynamic data must be very precise if the numbers obtained are to be meaningful.

A similar analysis leads to expression (37) for the endostatic enthalpy of activation, $\Delta H_\alpha^\ddagger$. Here the quantities $H_1 - H_1^*$ and

$$\Delta H_\alpha^\ddagger = \Delta H^\ddagger - [(H_1 - H_1^*) - (H_2 - H_2^*)](\mathrm{d}\Delta G^\ddagger/\mathrm{d}x_1)/RT(\mathrm{d}\ln\alpha/\mathrm{d}x_1) \tag{37}$$

$H_2 - H_2^*$ are the relative partial molar enthalpies of the two co-solvents which can be calculated from the dependence of H^E on x_2 as described on p. 281. Grunwald and Effio (1974) list some of the important quantities for four solvent mixtures.

11. TYPICALLY AQUEOUS MIXTURES

Properties of the Mixtures

Thermodynamic excess functions

In these mixtures, the excess function G^E is positive while $T \cdot S^E$ is large and negative such that $|T \cdot S^E| > |H^E|$. Co-solvents forming such mixtures include monohydric alcohols, acetone, tetrahydrofuran and dioxan which are often used in kinetic studies.

The dependence of G^E, H^E and $T \cdot S^E$ on mole fraction show the complexity of these mixtures; e.g. ethyl alcohol + water in Fig. 31, (G^E-data from Linderstrøm-Lang and Vaslow, 1968; Dobson, 1925; H^E-data from Bertrand, Millero, Wu and Hepler, 1966; Lama and Lu, 1965; Boyne and Williamson, 1967; Larkin, 1975). At 298 K, the mixing is exothermic over the whole range and yet G^E is positive. The dependence of H^E on x_2 requires the relative partial molar enthalpy of ethyl alcohol to increase rapidly as x_2 increases to a slight endothermic maximum near $x_2 = 0\cdot3$. The dependence of H^E on x_2 for methyl alcohol (Murakami et al., 1974; Benjamin and Benson, 1963), and propan-2-ol (Lama and Lu, 1965) show similar patterns except that the mixing at 298 K becomes less exothermic in the water-rich mixtures and more endothermic in the alcohol-rich mixtures as the size of the R-group in ROH increases. Nevertheless, as x_2 increases from zero, H^E becomes more exothermic more rapidly as the size of the alkyl group increases. The dependence of H^E on x_2 is particularly sensitive to temperature (Fig. 32; Larkin, 1975). Thus, on raising the temperature from 298 to 383 K, the mixing becomes endothermic over the whole range, the curves at intermediate temperatures having complex shapes. With reference to kinetic studies, Fig. 32 shows how the nature of an aqueous mixture depends markedly on temperature so that one might anticipate a measurable value of $\delta_m \Delta C_p^{\ddagger}$ in the systems. Values of G^E have been reported for n-propyl alcohol + water mixtures by Dawe et al. (1973). The excess functions for t-butyl alcohol + water are shown in Fig. 33 Kenttamaa et al., 1959) together with the phase diagram for

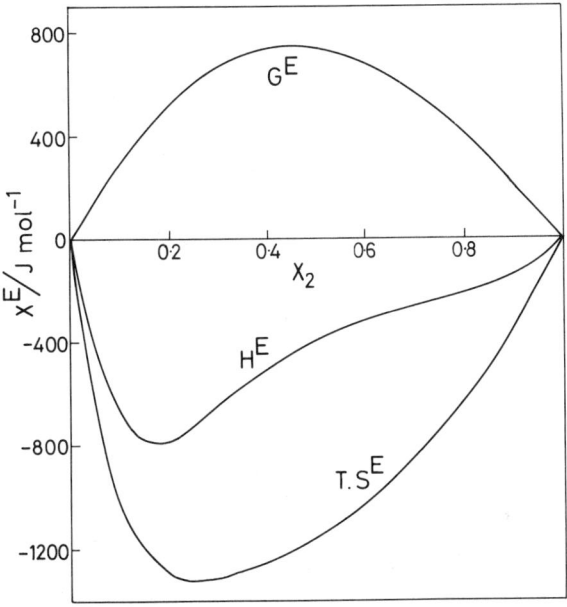

Figure 31. Excess thermodynamic functions of mixing for ethyl alcohol + water mixtures at 298·15 K.

this system (Rosso and Carbonnel, 1968). In this mixture the excess functions show more marked extrema than in the case of ethyl alcohol + water mixtures (cf. Fig. 31).

The thermodynamic excess functions for a mixture containing a cyclic ether, THF, are summarized in Fig. 34, calculated using the G^E-data given by Treiner et al. (1973) (see also Matous et al., 1972) and the H^E-data reported by Glew and Watts (1973). An interesting feature is the almost linear dependence of H^E on x_2 over the range $0·2 < x_2 < 0·8$ where the mixing changes from exothermic to endothermic; evidently as x_2 increases so water–water hydrogen bonds are replaced by water–THF hydrogen bonds. When D_2O

replaces H_2O, the mixing is more exothermic in D_2O-rich mixtures and more endothermic in THF-rich mixtures. When THF is replaced by ethylene oxide, the mixing is less exothermic in water mixtures (Glew and Watts, 1971). The excess functions for tetraethylene glycol diethyl ether + water are summarized in Fig. 35 (Nakayama, 1972). The thermodynamic excess functions for amine + water

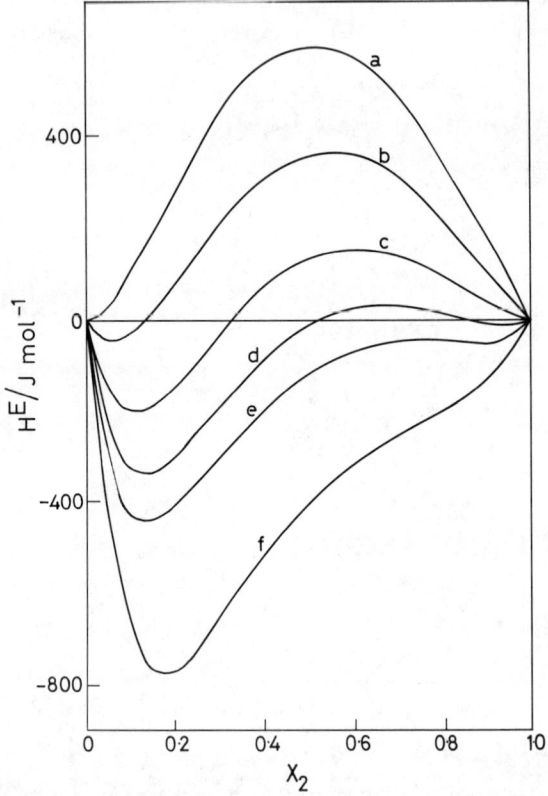

Figure 32. Excess enthalpies of mixing for ethyl alcohol + water mixtures at various temperatures; T/K = (a) 383·15, (b) 363·15, (c) 343·15, (d) 333·15, (e) 323·15 and (f) 298·15 (Larkin, 1975).

mixtures have been carefully studied by Copp (1955) and Copp and Everett (1953). The behaviour of, for example, methyldiethylamine + water clearly places this mixture in the TA class (Fig. 36). As a final example, we consider, acetone + water (Fig. 37, based on the diagram given by Wells, 1974). In these diagrams a striking feature is the S-shaped H^E curves, indicating how the interactions between components in the mixture change as the mole fraction is varied.

Similar complex dependences for H^E are observed in aqueous mixtures of 1,4-dioxan, 1,3-dioxan, trimethylene oxide, ethylene glycol dimethyl ether (Morcom and Smith, 1970; Nakayama and Shinoda, 1971), or 1,3-dioxolan (Blandamer et al., 1969b). However, not all TA mixtures have this S-feature, e.g. alkyl substituted amides + water (Assarson and Eirich, 1968), and t-butylamine + water at 313 K (Duttachoudhury and Mathur, 1974). The H^E dependences for TA systems are very complex. Even for ethanol-water mixtures at 298 K

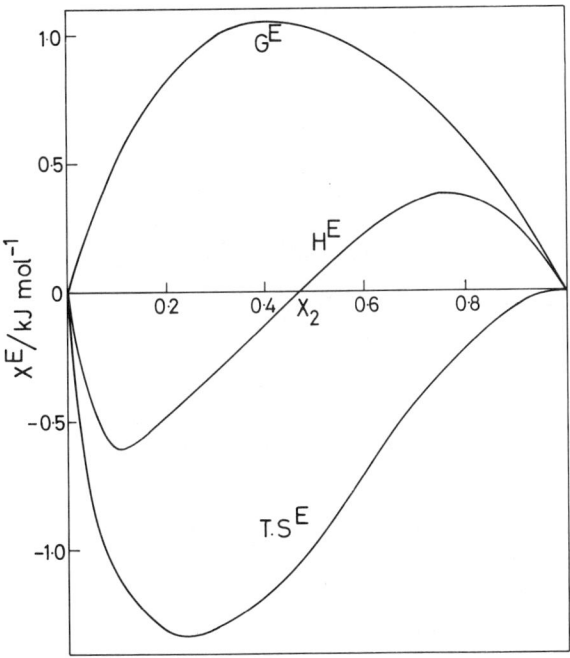

Figure 33. Excess thermodynamic functions of mixing for t-butyl alcohol + water mixtures at 299·15 K (Kenttamaa et al., 1959).

there are three points of inflexion in the curve as well as a minimum. Data published before such complexities were recognized (before about 1965) should be used with caution. Indeed the analysis of H^E and C_p^E data at low x_2 is providing a great deal of current interest. The broad outline of the C_p^E-curves is known for a few mixtures. Thus for t-butyl alcohol + water, C_p^E shows a sharp maximum near $x_2 = 0·1$. (Arnaud et al., 1972) (Fig. 38), but at low x_2 the dependence of C_p^E and, particularly, of the relative partial molar heat capacity of the co-solvent $(C_{p2} - C_{p2}^{\cdot})$ on x_2 may not be straightforward. The apparent partial molar heat capacity of t-butyl

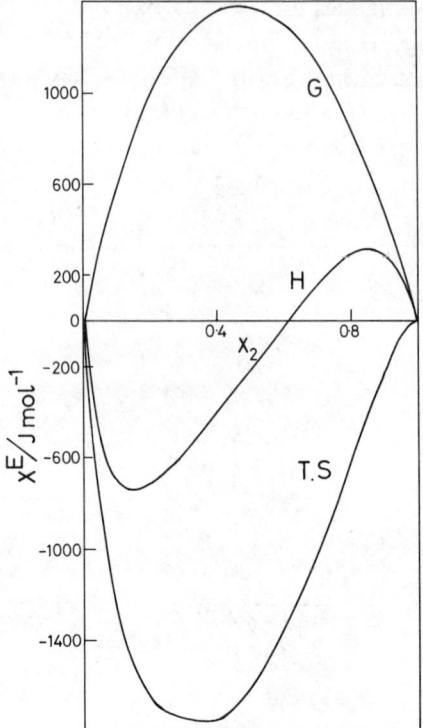

Figure 34. Excess thermodynamic functions of mixing for tetrahydrofuran + water mixtures at 298·15 K.

alcohol in aqueous mixtures has a maximum near $x_2 = 0\cdot04$, the composition where the partial molar volume has a minimum (see below; Avedikian et al., 1975).

Interpretation of this extensive information is not an easy task, nor are all the features completely understood. Nevertheless some patterns can be discerned. Thus the exothermic mixing in water-rich mixtures can be largely attributed to enhancement of water–water interactions by the added co-solvent, together with a contribution from intercomponent hydrogen bonding (cf. acetone + chloroform, Fig. 27). The endothermic mixing at high x_2 is attributed to disruption of H-bonds (cf. methyl alcohol + carbon tetrachloride, Fig. 28).

Partial miscibility

A most striking feature of TA mixtures is the dominant role which $T \cdot S^E$ plays in these systems. As noted on p. 284 this trend combined with a positive G^E can satisfy the conditions for an LCST. The mixture triethylamine + water (LCST = 291 K) is a classic example of this behaviour; at 288 K the mixing is strongly exothermic (Bertrand *et al.*, 1968). Even when the mixture is completely miscible at ambient pressure and temperatures, partial miscibility may set in when the conditions are slightly altered. Thus the mixture

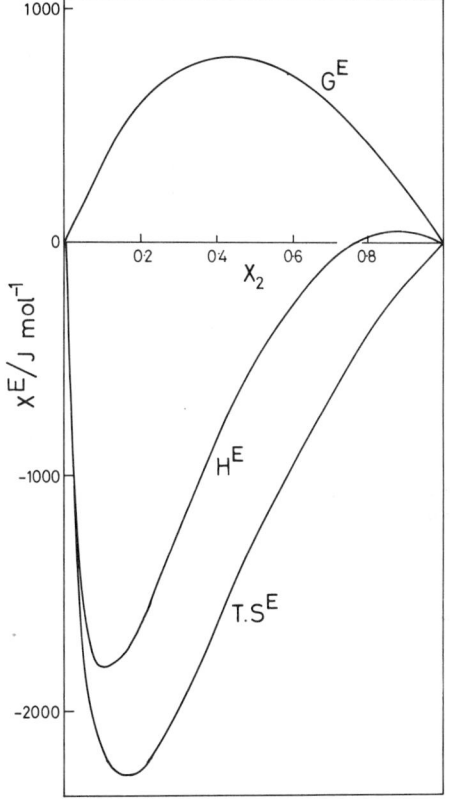

Figure 35. Excess thermodynamic functions of mixing for tetraethylene glycol diethyl ether + water mixtures at 298·15 K (Nakayama, 1972).

dioxan + water separates into two phases when small amounts of acid are added (Malcolm and Rowlinson, 1957). The mixture THF + water has an LCST at 345 K, just above the boiling point at ambient pressure (Matous *et al.*, 1970).

The miscibility properties of TA mixtures show interesting patterns as a function of pressure (Schneider, 1966; 1973). An increase in pressure can result in an increase in LCST (e.g. triethylamine + water), a maximum in LCST (e.g. 4-methylpiperidine +

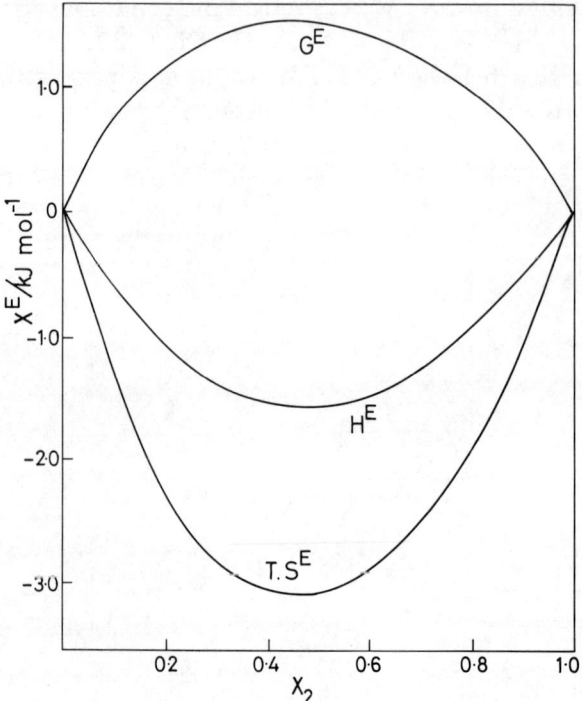

Figure 36. Excess thermodynamic functions for methyldiethylamine + water mixtures at 320 K (Copp, 1955).

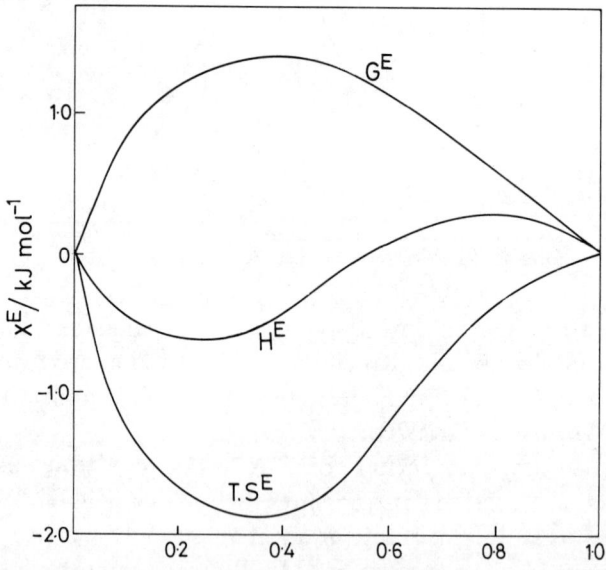

Figure 37. Excess thermodynamic functions for acetone + water mixtures at 298 K (Wells, 1974).

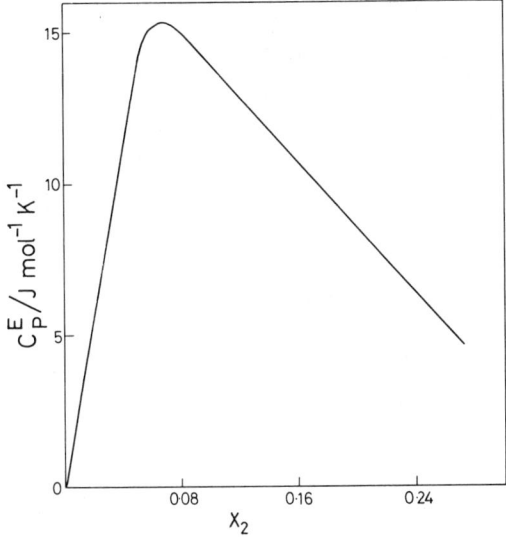

Figure 38. Excess heat capacities of mixing for t-butyl alcohol + water mixture (Arnaud et al., 1972).

water), or a decrease in LCST (e.g. 2,4,6-trimethylpiperidine + water). For systems which have both a UCST and an LCST a closed miscibility loop is found. In some systems (e.g. methyl ethyl ketone + water and 2-butoxyethanol + water), the loop shrinks in size as the pressure increases and ultimately the system becomes completely miscible. In other systems, the loop contracts but then expands as the pressure increases (e.g. 3-methylpiperidine + water).

The properties of a liquid mixture at or near a critical point (Stein and Allen, 1974) are complicated (Rowlinson, 1974) and will not be commented on further. Nevertheless, it seems likely that the kinetics of reactions in solvent mixtures near an LCST or a UCST may prove interesting in view of the report, admittedly not concerned with aqueous mixtures, that the rate of a Diels–Alder reaction increases by 30% within 0·01 K of the UCST for reaction in hexane + nitrobenzene mixtures (Wheeler, 1972). Measurement of the kinetics of reaction in such systems may prove difficult by spectrophotometric techniques because systems close to a critical point scatter light, but should be possible by electrical conductance measurements (Stein and Allen, 1973; Gammell and Angell, 1974).

Phase diagrams

Where explanations of the properties of the liquid mixtures have been couched in terms of either clathrate–hydrate structures or specific water–co-solvent associates, it is interesting to ascertain whether such species are found in the solid state. However, the factors involved in determining organization in a solid (e.g. molecular packing) are such that the failure to observe a solid clathrate hydrate may not rule out such structures as models for the liquid mixture. Where solid clathrates are detected, the observations provide useful support for such models. As noted on p. 225, many apolar solutes form solid clathrate hydrates. Analyses of phase diagrams have shown, for example, that clathrate hydrates are formed by cyclic ethers (Carbonnel and Rosso, 1972), trimethylene oxide (Rosso and Carbonnel, 1972) 1,3-dioxan (Rosso and Carbonnel, 1972) pyrrolidine (Carbonnel et al., 1973) and pyrroline (Rosso and Carbonnel, 1973).

Volumetric properties

At very high pressures the excess molar volume V^E can become positive (Haman and Smith, 1971). An example is water + 3-methylpyridine at $2 \cdot 5 \times 10^8$ N m^{-2} and 354 K when $x_2 < 0 \cdot 05$ (Engels and Schneider, 1972). However in general, the excess molar volume is negative, and, for TA mixtures, the plot of V^E against x_2 has a point of inflexion in water-rich mixtures. As noted on p. 281, this means that the partial molar volume of an organic co-solvent has a minimum in the water-rich mixtures. The value of $V_2^{\ominus} - V_0^*$ is negative (p. 253) and $V_2 - V_2^*$ decreases as x_2 increases. The latter feature is more characteristic of TA mixtures than the negative value for $V_2^{\ominus} - V_2^*$. (Fig. 39). Minima in partial molar volumes of the non-aqueous component are observed for aqueous solutions containing, for example, monohydric alcohols (Franks and Smith, 1968), cyclic ethers (Franks et al., 1970a; Cabani et al., 1971), amines (Kaulgud and Patil, 1974) and N-methylpropionamide (Hoover, 1969; see also Assarsson and Eirich, 1968b). The slope of the plot of the apparent partial molar volume against x_2 increases with increase in $V_2^{\ominus}$, thus prompting the remark that the solute–solute interactions (responsible for the dependence of V_2 on x_2) are not determined by the functional groups but by the hydrophobic parts of the molecule (Cabani, et al., 1974). It is noteworthy for example that V_2 for hexamethylene tetramine, which forms a clathrate hydrate, decreases with increase in solute concentration (Crescenzi et al., 1967).

Many properties of TA mixtures show extrema, e.g. the viscosity (Hayduk et al., 1973) and the activity of water in THF +

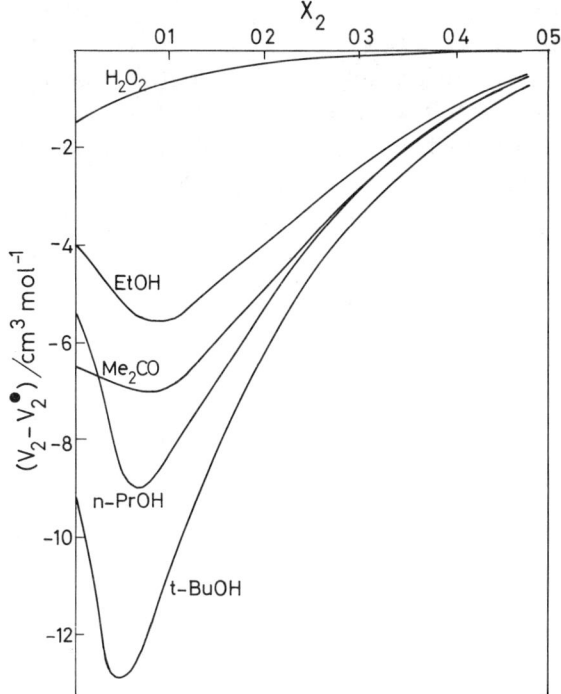

Figure 39. Dependence of relative partial molar volume, $V_2 - V_2^{\circ}$, for co-solvents in various aqueous mixtures as a function of co-solvent mole fraction at 298 K (Franks and Reid, 1973).

water mixtures (Pinder, 1973). These extrema often occur in the water-rich region, which has led to the suggestion that the properties of these TA mixtures can be understood by considering three mole fraction ranges $0 \leqslant x_2 < x_2^*$, $x_2^* < x_2 < x_2^{**}$, and $x_2^{**} < x_2 < 1\cdot 0$, defined by "signpost" mole fractions x_2^* and x_2^{**} (see Table 10). One example serves to justify this subdivision, based on the ultrasonic absorption properties of these mixtures. A plot of α_a/ν^2 (where α_a is

TABLE 10

Some "Signpost" Mole Fractions for TA Mixtures

Co-solvent	T/K	x_2^*	T/K	x_2^{**}
Ethyl alcohol	298	0·09	298	0·2
n-Propyl alcohol	298	0·05	298	0·1
t-Butyl alcohol	298	0·04	298	0·1
t-Amyl alcohol	298	0·02		
Acetone	298	0·06	273	0·38
THF			273	0·3

the amplitude absorption coefficient and ν is the frequency of the sound wave) against x_2 at fixed T, p and frequency ν (Fig. 40) shows three such regions (Blandamer, 1973). In the very water-rich mixtures, α_a/ν^2 is almost insensitive to x_2 although the velocity of sound rises sharply (Baumgartner and Atkinson, 1971). As more alcohol is added and $x_2 > x_2^*$, α_a/ν^2 rises sharply, reaching a maximum when $x_2 \simeq x_2^{**}$ and then decreasing as $x_2 > x_2^{**}$. Because the values of x_2^* and x_2^{**} depend to some extent on the property of the system being

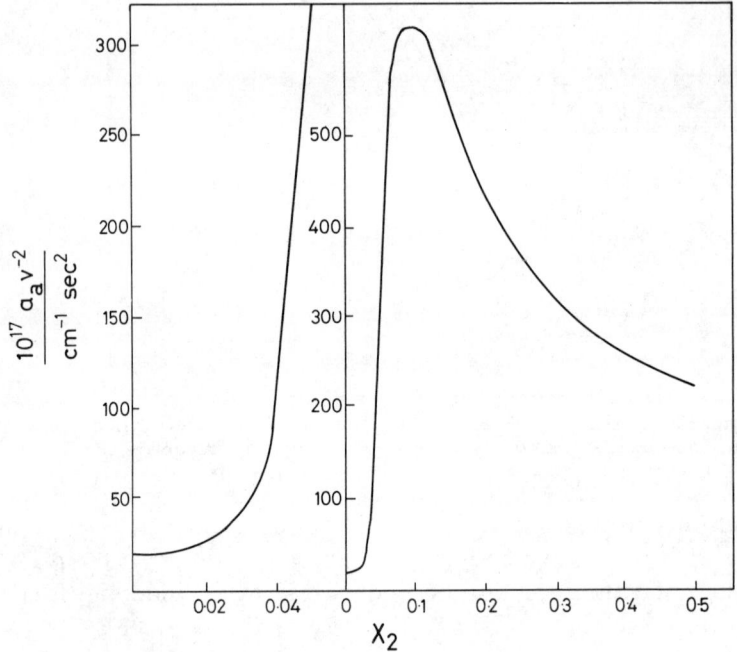

Figure 40. Dependence of α_a/ν^2 (where α_a is the amplitude absorption coefficient and ν the frequency of the sound wave) on mole fraction of t-butyl alcohol in aqueous mixtures at 70 MHz and 298 K.

measured and on the temperature, they cannot be regarded as absolute values. However, they do provide a basis for examining the properties of these systems and act as "signposts" along the mole fraction range.

Water-rich mixtures

In the solution standard state, apolar solutes are surrounded by a region of enhanced water-structure (p. 250). As x_2 increases, the solvent co-spheres overlap and water–water interactions in the overlap region are mutually enhanced (Fig. 14). In terms of the clathrate hydrate model for aqueous solutions, the mixture appears to develop a clathrate hydrate type of organized system. This reinforcement of water–water interactions is more marked the larger the hydrophobic alkyl groups in the organic co-solvent; for example, the slope $d(V_2 - V_2^{\bullet})/dx_2$ becomes more negative and the minimum in $V_2 - V_2^{\bullet}$ occurs at lower mole fractions of co-solvent (Fig. 39). It is here that we see some of the evidence for the Frank model for hydrophobic association (p. 255). This model also accounts for the extremum in apparent partial molar heat capacity for t-butyl alcohol in these mixtures (Avedikian et al., 1975). The HNC calculation (p. 245) has been extended to consider interactions between methylene groups can be characterized for these systems by $A_{ij} = -400$ J mol^{-1}, the co-sphere of a CH_2 group containing two water molecules (Friedman and Krishnan, 1973c).

The increased structural integrity of water in the region $0 < x_2 < x_2^*$ is indicated by numerous observations. A clear-cut example is the effect of these co-solvents on the temperature of maximum density (TMD) for water. For an ideal mixture, TMD should decrease as x_2 increases, and one can evaluate the difference between the actual and calculated change in the TMD, $\Delta\theta(\text{st})$ (Franks and Watson, 1967). In general, a TA co-solvent increases $\Delta\theta(\text{st})$ and in one case, t-butyl alcohol, the observed TMD increases. The effect of a co-solvent on water is also revealed by proton nmr measurements. At low x_2, there is a shift of the OH resonance to low field, consistent with more/stronger hydrogen bonding; the extent of this shift is a maximum for t-butyl alcohol when $x_2 \simeq x_2^*$ (Blandamer and Symons, 1968; Anderson and Symons, 1969). Similar behaviour is observed in aqueous mixtures containing 1,4-dioxan, acetone, THF (Glew et al., 1968; Glew and Watts, 1973) and triethylamine (Kingston and Symons, 1973). The proton shift has been analysed in terms of separate contributions from the hydrophobic and hydrophilic groups, the low-field shift being a maximum for alcohols when the molecule is spherical, with a diameter of ca. 5×10^{-8} cm (Wen and Hertz, 1972). Spectroscopic studies of solutes acting as probes of solution structure support the model for these systems discussed above (Blandamer et al., 1970), as does, for example, spin-relaxation measurement for dioxan + water mixtures (Lee and Jones, 1973).

Structure-broken region

With increase in co-solvent mole fraction, a point is reached, $x_2 > x_2^*$, where there is insufficient water to accommodate the co-solvent in a clathrate-like structure. Many properties show a marked change in this region, e.g. a minimum in $V_2 - V_2^{\bullet}$ (Fig. 39), an increase in α_a/ν^2 (Fig. 40), a maximum C_p^E (Fig. 38), an extremum in energy-volume coefficients (MacDonald and Hyne, 1971) and minimum diffusion coefficients (Pratt and Wakeham, 1975). At this mole fraction, interference between solvent co-spheres produces a structure-breaking action. This interference occurs at lower mole fractions, x_2, with increase in size of the apolar group. Although the organization in mixtures where $x_2 > x_2^*$ is not well understood, there is evidence for component clustering to form microscopic heterogeneities, i.e. regions which are either water-rich or co-solvent-rich. It seems that the water molecules endeavour to maintain local networks of hydrogen-bonded water-molecules by ejecting excess co-solvent. In the limit, these systems separate into two liquid phases at an LCST (see above), but even for completely miscible systems there is a strong tendency to immiscibility which is greatest when $x_2 \simeq x_2^{**}$. A variety of observations, e.g. a maximum in sound absorption (Fanning and Kruus, 1970), a maximum in X-ray scattering (Bale *et al.*, 1970) and in light scattering (Vuks and Shurupova, 1972) can be understood in terms of the model outlined above. Extrema in C_p^E, light scattering and sound absorption can be quantitatively understood in terms of correlations between fluctuations in local thermodynamic properties (Fixman, 1962; Blandamer *et al.*, 1970a).

The clathrate-hydrate/long range hydrophobic interaction model is not to be interpreted as implying some long lived organization which exists when $x_2 < x_2^*$ but which disappears when $x_2 > x_2^*$. Such an idea is ruled out by the dynamic parameters obtained from nmr relaxation experiments (Goldammer and Hertz, 1970). Nevertheless, the emphasis on water as a key component of the system is borne out by the dependence of the rotational correlation time on mole fraction. For example, when acetone is added to water, the correlation time increases markedly, τ(co-sphere water) $> \tau$(bulk water), reaches a maximum and then decreases. In contrast, τ for the solute molecule, as measured by the behaviour of the CH_3-group, remains almost unaffected. The results for t-butyl alcohol + water mixtures do not exclude the possibility of a relatively long-lived hydration cage at low x_2. Extrema in activation energies for the various relaxation times (e.g. in acetone + water) also indicate the presence of some microheterogeneity in the mixture.

As the mole fraction of the co-solvent increases and the mixture becomes rich in co-solvent, there is insufficient water to establish networks of hydrogen-bonded water molecules; the properties of the mixture now resemble those of mixtures of more conventional polar molecules.

Solutes in Typically Aqueous Mixtures

The complexity of the properties and structures of aqueous mixtures is not unexpectedly carried over to the properties of solutes in these TA mixtures. In reviewing the properties of some of these three component systems, we shall consider first apolar solutes and finally ionic solutes.

In this section of the review, it may be helpful to keep in mind the various mole fractions identified in Table 10.

Apolar solutes in TA mixtures

Yaacobi and Ben-Naim (1973) have reported the solubilities of methane and of ethane in water + ethyl alcohol mixtures. They used Ostwald coefficients and report the thermodynamic quantities using molar standard states. As expected, the overall trend is, in the case of ethane for example, for $\delta_m \mu_3^\ominus$ to decrease as x_2 increases (i.e. solubility increases), and for $\delta_m H_3^\ominus$ and $\delta_m S_3^\ominus$ to increase (Fig. 30). However, the detailed pattern is certainly more complex in the case shown in Fig. 41; $\delta_m \mu_3^\ominus$ decreases as x_2 increases until $x_2 \simeq 0.03$, then increases to a maximum when $x_2 \simeq 0.15$ before decreasing smoothly when $x_2 > 0.2$. Maxima in $\delta_m \mu_3^\ominus$ for apolar solutes at low mole fraction of co-solvent are observed for methane (Yaacobi and Ben-Naim, 1973), oxygen (Schukarev and Tolmacheva, 1968), and argon (Ben-Naim and Baer, 1964) in ethyl alcohol + water mixtures and for argon in methyl alcohol + water mixtures (Ben-Naim, 1967). The effect of added t-butyl alcohol on the solubility of argon in water is particularly striking (Cargill and Morrison, 1975). The minimum in $\delta_m \mu^\ominus$ for argon occurs at progressively lower mole fractions through the series CH_3OH, C_2H_5OH and $(CH_3)_3COH$, as does the maximum in $\delta_m \mu^\ominus$ [0.4 (CH_3OH), 0.22 (C_2H_5OH) and

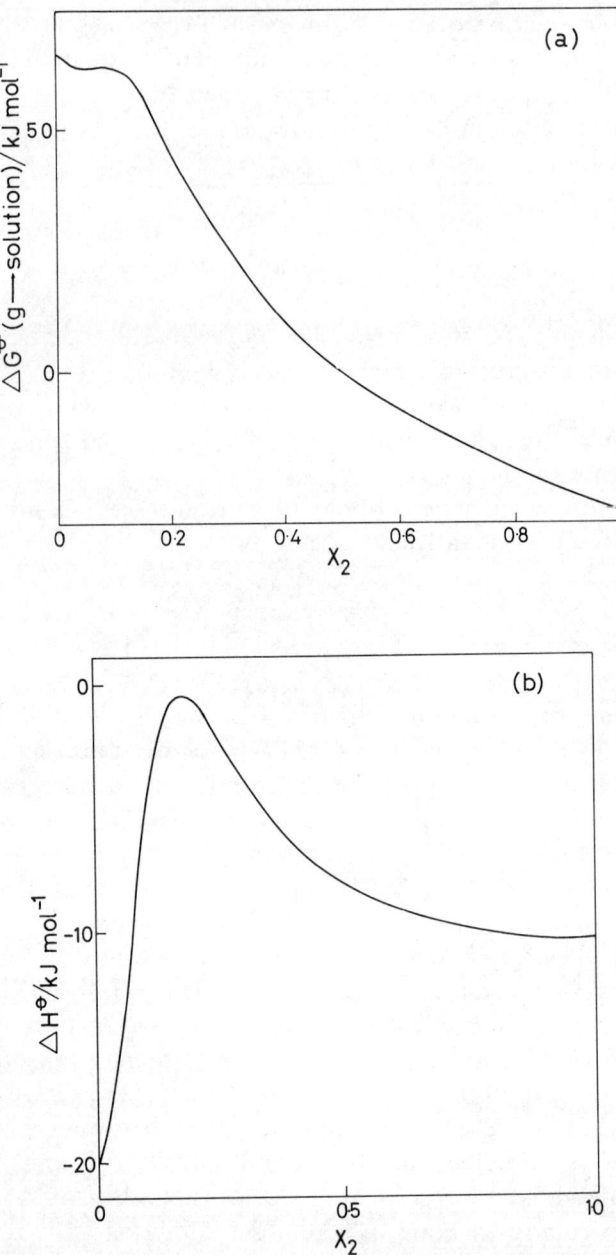

Figure 41. Thermodynamic parameters for solutions of ethane in ethyl alcohol + water mixtures; (a) $\Delta G^{\ominus}$ (gas phase → solution) at 283 K, (b) $\Delta H^{\ominus}$ at 298 K (Yaacobi and Ben Naim, 1973).

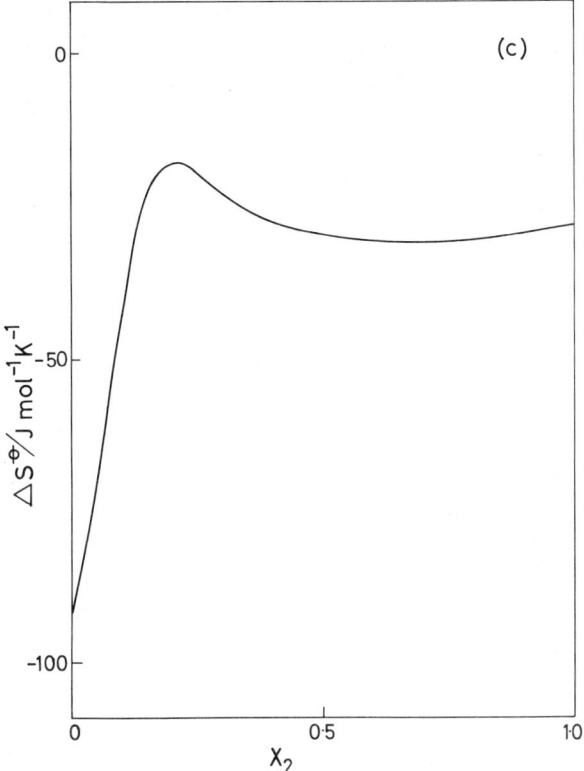

Figure 41(c). $\Delta S^{\ominus}$ for solutions of ethane in ethyl alcohol + water mixtures at 298 K (Yaacobi and Ben Naim, 1973).

0·11 ($(CH_3)_3COH$) at 293 K]. With increase in temperature, the extrema in $\delta_m \mu_3^{\ominus}$ curves at low mole fractions become almost lost (e.g. ethane in water + ethyl alcohol mixtures at 303 K).

The associated enthalpy and entropy quantities, $\delta_m H_3^{\ominus}$ and $\delta_m S_3^{\ominus}$, show more marked variations. Both $\delta_m H_3^{\ominus}$ and $T \cdot \delta_m S_3^{\ominus}$ increase rapidly as x_2 increases and, in the case of ethane in ethyl alcohol + water mixtures, there is an endothermic maximum near $x_2 = 0·2$ (Fig. 41). In water-rich mixtures the solubility data reflect the impact of water structure (Cargill and Morrison, 1975). A key observation is the tendency for the solubility of an apolar solute to decrease (i.e. $\delta_m \mu_3^{\ominus}$ rises to a maximum) as an organic co-solvent is slowly added. This salting-out of an apolar solute accounts for the enhancement of protein structure by low mole fractions of TA solvents, for example (Brandts and Hunt, 1967). A further clear example of this tendency is shown by the decrease in solubility of argon when dioxan is added, there being no initial increase in solubility at very low x_2 (cf. Fig. 41) (Ben-Naim and Moran, 1965).

The solubility data can be qualitatively understood as follows. In the very water-rich mixtures, both the apolar solute and the co-solvent enhance water–water interactions (Ben-Naim, 1965) so that

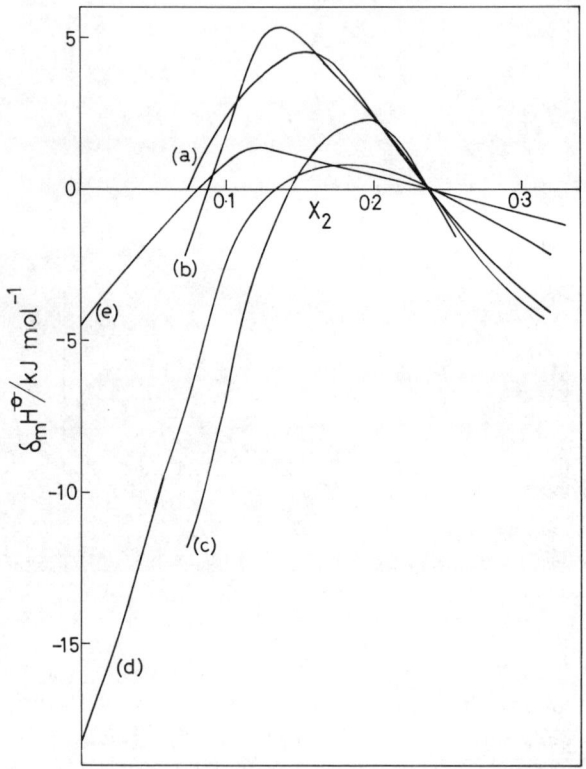

Figure 42. Relative partial molar enthalpies for solutes in ethyl alcohol + water mixtures at 298 K; reference solvent is a mixture where $x_2 = 0.235$: (a) t-BuCl, (b) i-PrBr, (c) CCl$_4$, (d) ethyl acetate and (e) KCl (Arnett et al., 1965).

the apolar solute may be accommodated in the "clathrate structure" defined by the co-solvent, the latter acting in the manner of a "help-gas" (p. 227). Thus the solubility of the apolar solute increases. As more co-solvent is added the water-structure breaks down and the solute cannot be accommodated in the above fashion; the solubility therefore decreases ($\delta_m \mu_3^{\ominus}$ increases). The solubility continues to decrease until the alcohol mole fraction is sufficient for the alcohol to control the solubility (Fig. 30).

Transfer of a wide range of solutes, including salts, from water to ethyl alcohol + water mixtures has an endothermic maximum near $x_2 = 0.1$ (Fig. 42; Arnett, 1967; Arnett and McKelvey, 1965; Arnett et al 1965). A less marked extremum is observed for 1-methyl-2-pyrrolidone in methyl alcohol + water mixtures where a hint of a maximum in $\delta_m H_3^{\ominus}$ occurs near $x_2 = 0.3$ at 298 K (Murakami et al 1974). In ethyl alcohol + water mixtures, the endothermic maximum is more intense for neutral solutes than for salts, and becomes more intense the larger the apolar solute. Arnett attributes the behaviour to enhancement of water-water interaction by the co-solvent. The fact that endothermic extrema are produced by both structure-formers and structure-breakers prompted the following conclusions. If the solute is a structure-former, it is more effective at structure-forming in water than in the mixture because water is less structured to begin with. If the solute is a structure breaker, then it is more effective in the mixture where there is more structure to break. Thus, if apolar solutes enter cavities prepared by the co-solvent, the contribution to the enthalpy of solution from changes in organization of water decreases as x_2 increases, $\delta_m H_3^{\ominus}$ being therefore positive.

Endothermic maxima are also observed in plots of $\delta_m H_3^{\ominus}$ for carboxylic acids in t-butyl alcohol + water mixtures at $x_2 \simeq 0.05$. The importance of the hydrophobic side chain in determining the height of these maxima is confirmed by the observation that these become more intense as the size of the alkyl group increases. Indeed the function $\delta_R \delta_m H^{\ominus}$, calculated for replacement of H- by CH$_3$-, has an endothermic maximum near $x_2 = 0.05$ (Avedikian et al., 1973).

Salts in TA mixtures

When a salt is added to an aqueous mixture, the vapour pressures of the mixture and the components change; for example, salts raise the vapour pressure of acetone in acetone + water mixtures (Sada et al., 1972).

Meranda and Furter (1974) classify the patterns observed and note that often the addition of salt to water-rich mixtures will result in a decrease in the mole fraction of co-solvent in the vapour, whereas for alcohol-rich mixtures there is an increase. This cross-over behaviour is related to the structural changes occurring in the liquid as x_2 is varied. In some cases the system may become partially miscible when a salt is added, but it is difficult to generalize. Thus sodium perchlorate enhances the miscibility of glycerol triacetate + water while sodium chloride lowers it (Raridon and Kraus, 1965). However, in the present context, the changes in the properties of salts in water with addition of co-solvent are of greater interest.

For most salts, the standard state chemical potential of the salt changes gradually with increase in the mole fraction of co-solvent (Bates, 1968). Thus at 298 K, plots of $\delta_m \mu^\ominus$ (salt) against x_2 for

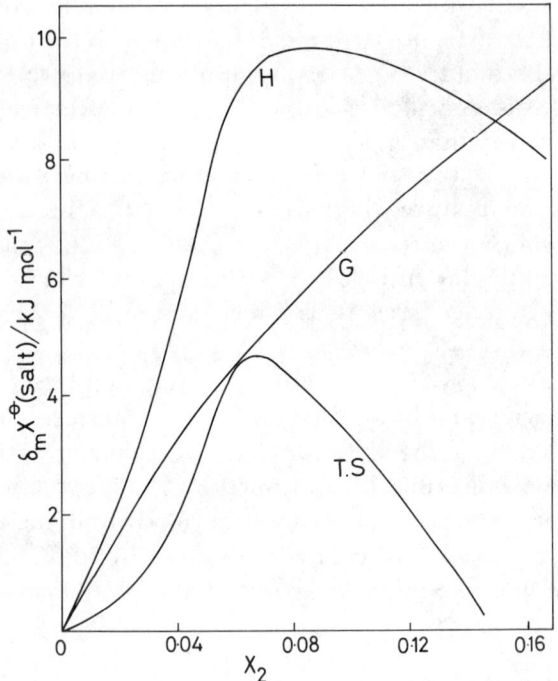

Figure 43. Thermodynamic transfer properties for sodium chloride on going from a solution in water to t-butyl alcohol + water mixtures at 298 K (Pointud et al., 1974).

Li^+, Na^+, K^+, Rb^+, Cs^+, Ag^+ and H^+ chlorides are almost linear, $\delta_m \mu^\ominus$ increasing with increase in $w_2\%$ in t-butyl alcohol + water mixtures. The slope is smallest for HCl and largest for NaCl solutions (Pointud et al., 1974b). However, the dependence of enthalpy and entropy functions is much more complex. There are extrema in $T\delta_m S_3^\ominus$ and $\delta_m H_3^\ominus$ near $x_2 = 0.06$ and 0.08 respectively (Fig. 43) for Na^+, K^+, Rb^+ and Cs^+ chlorides in this mixture (Pointud et al., 1974b). A similar pattern emerges for HCl in other TA mixtures (Roy and Bothwell, 1971). Thus in water mixtures containing methyl alcohol, ethyl alcohol and n-propyl alcohol, the extrema in say $T\delta_m S^\ominus$ for HCl become more marked and at lower mole fractions through this series. Interestingly, for particular mixtures, it is possible to choose a value of x_2 where transfer of HCl from water to this mixture is accompanied by zero change in the entropy of the solute, e.g. when $x(C_2H_5OH) = 0.16$ at 298 K (Stern and Hansen, 1971).

In certain cases, $\delta_m \mu^\ominus$ does show an extremum value. For example, $\delta_m \mu^\ominus$ (HI) has a maximum near $x_2 = 0.15$ in methyl alcohol + water mixtures followed by a minimum near $x_2 = 0.3$ where $\delta_m \mu^\ominus$ (HI) = 0 (Feakins, 1967) while $\delta_m \mu^\ominus$ (HCl) in ethyl alcohol mixtures has a point of inflexion which is less marked in aqueous mixtures of isopropyl alcohol and t-butyl alcohol (Morel and Morin, 1970). Similar points of inflexion are apparent in plots of $\delta_m \mu^\ominus$ against volume fraction of methyl alcohol for NaCl, NaBr, and NaI, becoming more marked in that order (Feakins and Voice, 1973). Transfer data for alkali-metal chlorides from water to three dioxan + water mixtures show that $\delta_m \mu^\ominus$ (MCl) increases with increase in cation size and $w_2\%$ (Feakins et al., 1975). A similar trend is observed for $\delta_m \mu^\ominus$ (NaCl) when THF is added (Treiner, 1973).

Analysis of these patterns of behaviour into single ion contributions suggests that, broadly, anions are destabilized and cations are stabilized when TA co-solvents are added to aqueous solutions. This behaviour is shown by plots of $\delta_m \mu^\ominus$ (ion) for acetone + water mixtures as given by Bax et al. (1972) and for methyl alcohol + water mixtures by a number of workers (de Ligny and Alfenaar, 1965; Feakins et al., 1967; Andrews et al., 1968; Feakins and Voice, 1972). Rochester (1972) predicts that $\delta_m \mu^\ominus$ (H$^+$) is negative (molality scale) on going to methyl alcohol + water mixtures, with a negative extremum near $x_2 = 0.8$. The H$^+$ ion is also stabilized when t-butyl alcohol is added to aqueous solutions (Bose et al., 1975). Rochester also concludes that $\delta_m \mu^\ominus$ (OH$^-$) is positive in these mixtures, an observation which is clearly important in the analysis of the rates of alkaline hydrolysis. Wells (1973) has calculated that, as x_2 increases in methyl alcohol + water mixtures $\delta_m \mu^\ominus$ (Na$^+$) is first positive, changing to negative when $x_2 > 0.35$. Similarly $\delta_m \mu^\ominus$ (K$^+$) changes to negative when $x_2 > 0.50$, but $\delta_m \mu$ (Li$^+$) and $\delta_m \mu^\ominus$ (H$^+$) are negative over the range investigated, $0 \leqslant x_2 < 0.4$ (Fig. 44). In summary, $\delta_m \mu^\ominus$ for anions decreases with increase in ion size although the pattern for the cations is not so clear cut.

A similar general pattern emerges in other TA mixtures, e.g. chloride ions are destabilized on addition of acetone (Wells, 1974) or dioxan (Feakins *et al.*, 1975; Bennetto *et al.*, 1968); hydroxide ions are similarly affected (Feakins and Turner, 1965; Villermaux *et al.*, 1972; Rat *et al.*, 1974).

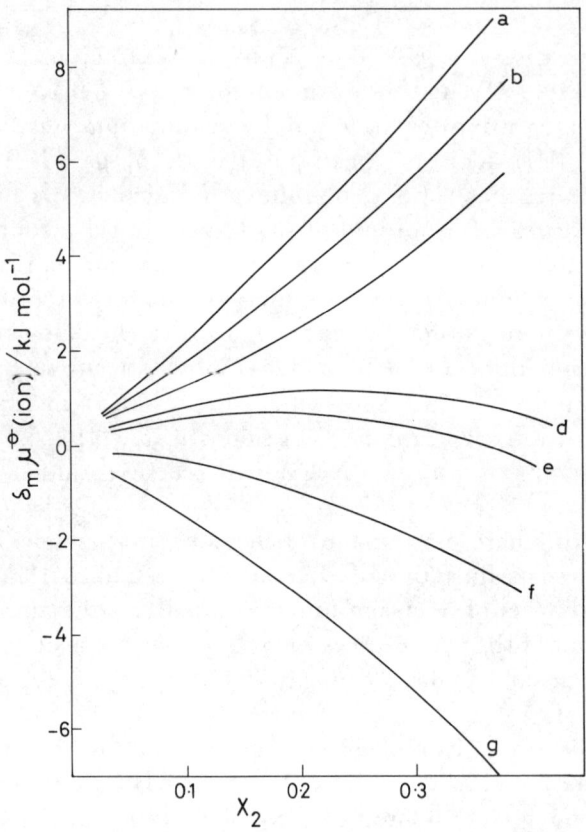

Figure 44. Dependence of single ion chemical potentials on mole fraction of methyl alcohol, $\delta_m \mu^\ominus$ (ion), for various ions in methyl alcohol + water mixtures; (a) Cl^-, (b) Br^-, (c) I^-, (d) K^+, (e) Na^+, (f) Li^+ and H^+ (Wells, 1973).

The empirical relationship (38) proposed by Criss (Criss *et al.*, 1968; Criss, 1974) connects the partial molar entropy of an ion in a

$$S^\ominus (\text{ion}; X) = a + b \cdot S^\ominus (\text{ion}; \text{water}) \tag{38}$$

given solvent X to the entropy of the ion in water through two empirical constants. Franks and Reid (1969) discovered that (38) is also valid for ionic entropies (mole fraction scale) in dioxan + water and methyl alcohol + water mixtures. However, they noted a more interesting observation. When $S^\ominus$(ion) is plotted against the mole fraction of methyl alcohol in the aqueous mixture, the points deviate from a linear dependence towards less negative values than predicted (Fig. 45). The deviation becomes more marked as the size of the ion decreases. Nevertheless, in each case the maximum deviation from the straight line occurs at the mole fraction where S^E for the mixture

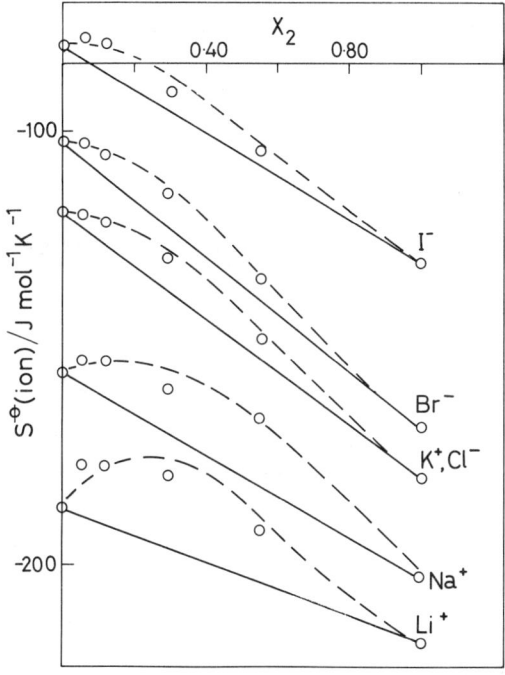

Figure 45. Single ion entropies for various ions in methyl alcohol + water mixtures at 298 K; full lines represent idealized dependence and dotted lines show observed dependence (Franks and Reid, 1969).

is a minimum, and, therefore, where the ions have the most disruptive effect on the mixture. It would be interesting to know if this pattern holds for other aqueous mixtures.

An insight into the dynamic aspects of ionic solvation in mixed solvents is obtained from nmr data. For example, in methyl alcohol + water mixtures, Rb^+ ions appear to avoid the neighbourhood of methyl alcohol molecules in contrast to Cl^- ions (Neggia et al., 1974).

Salts of Alkylammonium ions in TA mixtures

The properties of alkylammonium and related ions in these mixtures are complex and still not completely understood (Sarma and Ahluwalia, 1973). For example, partial molar enthalpies and heat capacities vary considerably as the mole fraction x_2 is changed. However, a link between these variations and the mole fractions indicated in Table 10 is often clearly seen.

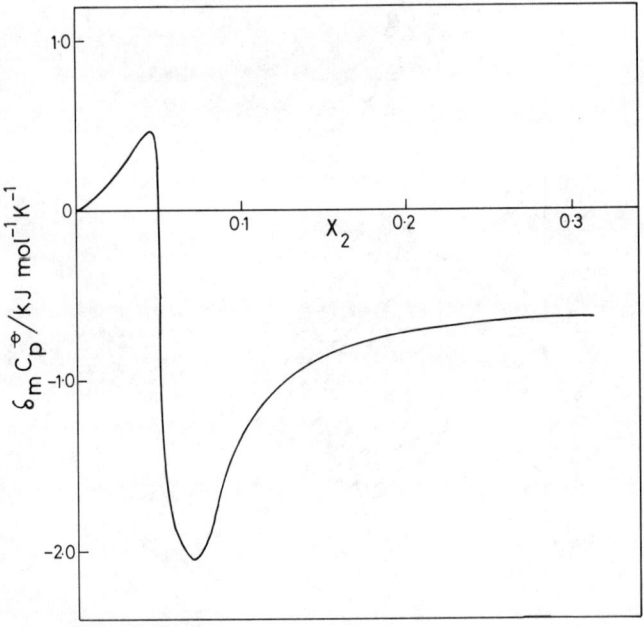

Figure 46. Solvent dependence of the partial molar heat capacity of sodium tetraphenylborate in t-butyl alcohol + water mixtures at 298 K as a function of alcohol mole fraction, x_2 (Arnett and McKelvey, 1966b).

Thus the partial molar enthalpy of $Bu_4N^+Br^-$ increases sharply when t-butyl alcohol is added to aqueous solutions until $x_2 = 0.1$ and then decreases slowly (Mohanty et al., 1971). Similar complex patterns emerge in the enthalpies of these salts when amide co-solvents are added, e.g. formamide (de Visser and Somsen, 1974a,b). Striking changes are observed in the partial molar heat capacities of salts in TA mixtures when x_2 is varied (Avedikian et al., 1975). Thus for $Am_4N^+Br^-$ in t-butyl alcohol + water mixtures (Mohanty et al., 1972), the partial molar heat capacity increases as x_2 increases to a maximum near $x_2 \simeq 0.04$, drops sharply to a minimum and then increases again when $x_2 > 0.1$. The size of these variations is indicated for $Na^+BPh_4^-$ in this same solvent mixture in Fig. 46; on going from $x_2 = 0.04$ to $x_2 = 0.07$ $\delta_m C_p^\ominus$ is ca. 2·5 kJ mol^{-1} K^{-1} (Arnett and McKelvey, 1965; 1966b). Such a complicated double functional dependence of a property of a solute indicates that marked changes must occur in the solvent medium as x_2 is increased (Table 10). Extrema are generally marked in t-butyl alcohol + water mixtures but similar though less marked behaviour is also observed in other aqueous mixtures (Mohanty and Ahluwalia, 1972; Arnett et al., 1972).

The increase in $C_p^\ominus$ (salt) for these 'onium salts can be understood in terms of mutual structure-forming or long-range hydrophobic interaction. The hydrocarbon side chains occupy voids in the "solvent-lattice", which is stabilized by co-solvent but which collapses when $x_2 > x_2^*$. It is interesting therefore to note that the partial molar volume of R_4N^+ ions decreases as x_2 increases in ethyl alcohol + water mixtures, reaching a minimum near $x_2 \simeq 0.1$ which becomes more clearly defined with increase in size of the R-groups (Lee and Hyne, 1970). A similar trend is observed in acetone + water mixtures (Macdonald and Hyne, 1970). By way of contrast, the partial molar volumes, $V^\ominus$(ion), for I^-, Br^-, and Cl^- ions decrease steadily to minima near $x_2 = 0.3$ in ethyl alcohol + water mixtures where $V_2^\ominus$ (R_4N^+) is a maximum (Lee, 1972; 1973).

An explanation similar to that advanced above may account for the minima in $V_2^\ominus$ observed for potassium benzoate and p-bromobenzoate at low x_2 in t-butyl alcohol + water mixtures (Dollet and Juillard, 1973).

Ionic mobilities

The electrical mobilities of ions in aqueous mixtures are difficult to interpret because the ionic mobility may change as a result of changes in water-structure brought about by the co-solvent and it may also change because the additive simply obstructs the passage of ions. Nevertheless, several studies have shown that the first of these two contributions can be identified. For example, structure enhancement by added hexamethylene tetramine leads to a fall in ionic mobilities for KCl in water (Barone et al., 1968). Ionic mobilities have been measured for a variety of salt solutions in ethyl alcohol + water (Lee, 1974), THF + water (Renard and Justice, 1974) and for the hydrogen ion in various alcohol + water mixtures (Tourky and Abdel-Hamid, 1971; Tourky et al., 1973). The complexity of the phenomena is illustrated by the dependence of the Walden product, $\Lambda^0 \eta$, for isobutyrate ions on the mole fraction of t-butyl alcohol. Two maxima (0·05 and 0·18) and one minimum (0·10) are observed (Juillard et al., 1972), and it is possible that these extrema reflect the bifunctional nature, hydrophobic and hydrophilic, of these ions. Maxima in Walden products, λ^0(ion) η, for alkali metal and halide ions are observed near $x_2 = 0.05$ for water + t-butyl alcohol mixtures (Broadwater and Kay, 1970). Moreover, the extrema cannot be accounted for in terms of classical "ion-in-solvent-continuum" models but are shown by Kay and Broadwater (1971) to reflect the changes in solvent structure as x_2 increases.

The mobilities of H^+ and OH^- ions in these aqueous mixtures are of some importance but their understanding presents problems. It is recalled that in ice the mobility of H^+ dramatically exceeds that of H^+ in water, and that in water the proton mobility is dependent on the ability of water molecules to rotate into a configuration which allows ready proton transfer (Hills et al., 1965). It is therefore noteworthy that addition of t-butyl alcohol to H^+ in water lowers the proton diffusion coefficient, a trend expected if this co-solvent enhances water–water interactions (Lannung et al., 1974).

Other aspects of transport phenomena including activation energies for viscosity parameters and Washburn transport numbers (Feakins, 1974a; Feakins and Lorimer, 1974) have also been measured to probe ion–solvent interactions in mixed solvent systems.

Equilibria in Typically Aqueous Mixtures

As a preliminary to a discussion of kinetics of reactions in aqueous mixtures, it is interesting to review briefly the behaviour of equilibrium quantities as a function of co-solvent mole fraction. Interpretation of the data is necessarily complex because, for example, in the case of acid dissociation constants, the quantity $\delta_m \Delta X^{\ominus}$ represents the result of the individual variations of the partial molar quantities for acid, conjugate base and hydrogen ion. Nevertheless patterns of behaviour are observed which demonstrate the impact of co-solvent on water structure and on solute properties along the lines discussed in the previous section.

In general $\delta_m \Delta G^{\ominus}$ is small, being a smooth function of x_2, a result of compensation effects (p. 247). Examples are acetic and benzoic acids in water + t-butyl alcohol mixtures (Arnaud et al., 1970) and acetic acid in acetone + water mixtures (Morel, 1966). However, the solvent does play an important role here. This is shown in the solvent dependence of the double functions $\delta_R \delta_m pK$ which for phenols, benzoic acid and anilinium derivatives show extrema in

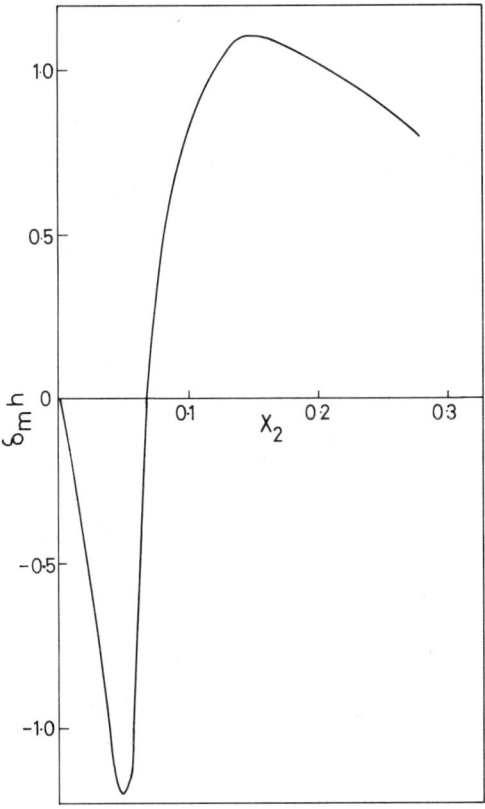

Figure 47. Solvent dependence of the enthalpy of proton transfer between benzoic acid and acetate anion in t-butyl alcohol + water mixtures; $\delta_m h = \delta_m \Delta H^\ominus / 2 \cdot 3 RT$ (Avedikian and Morel, 1971).

ethyl alcohol + water mixtures, three regions being identified, $0 \leqslant x_2 < 0 \cdot 25$, $0 \cdot 25 \leqslant x_2 \leqslant 0 \cdot 50$, and $0 \cdot 50 \leqslant x_2 \leqslant 1 \cdot 0$ (Thuaire, 1972). The protonation constants for bases, e.g. ammonia, show minima in methyl alcohol–water mixtures near $w_2\% = 60$ (Rorabacher et al., 1971).

More marked dependences on mole fraction are observed in the related enthalpy and entropy functions, $\delta_m \Delta H^\ominus$ and $\delta_m \Delta S^\ominus$, e.g. acetic and chloracetic acids in ethyl alcohol + water mixtures (Millero et al., 1969). Extrema in these quantities are observed for the ionization of anilinium ions in ethyl alcohol + water mixtures (Van de Poel, 1971; Van de Poel and Slootmaekers, 1970). One way of probing the effects of the solvent on acid strength has been to examine the thermodynamic functions for proton exchange between two acids, viz., $HA_1 + A_2^- \rightleftharpoons A_1^- + HA_2$ where, for example, HA_1 = benzoic and HA_2 = acetic acid. The function $\delta_m \Delta H^\ominus$ is remarkably sensitive to co-solvent and mole fraction (Avedikian et al., 1970; Avedikian and Morel, 1971). Thus in t-butyl alcohol + water, $\delta_m \Delta H^\ominus$ for butyric acid-acetic acid shows a sharp minimum near x_2 = 0·05 (Fig. 47), as does $\delta_m \Delta S^\ominus$. Moreover enthalpy-entropy compensation plots yield good straight lines with slope ca. 200 K in water-rich mixtures, but as x_2 increases this correlation is lost, consistent with a collapse of water structure beyond a certain value of x_2 (cf. Table 10) (Morel et al., 1974).

Kinetics of Reactions in typically Aqueous Mixtures

There is an extensive literature concerned with kinetics of reactions in aqueous mixtures and it is impossible to cite every example. We have indicated many of the relevant points in previous sections and so here we shall use selected examples to illustrate the key points at issue. Hyne (1968) has reviewed some of the relevant material with particular emphasis on $\Delta V^\ddagger$-quantities. As noted on p. 210, the complex pattern reported almost 20 years ago for the dependence on solvent composition of $\delta_m \Delta H^\ddagger$ and $\delta_m \Delta S^\ddagger$ for the hydrolysis of t-butyl chloride (Fig. 48) has attracted and still attracts comment and speculation (Winstein and Fainberg, 1957). For aqueous mixtures, $\delta_m \Delta G^\ddagger$ reflects to a considerable degree the changes in the initial state, whereas, on going from a solution in ethyl alcohol to one in methyl alcohol, the change in the chemical potential of the transition state is more important. Indeed, on going from an aqueous mixture containing 40 v_2% ethyl alcohol to a solution in ethyl alcohol, $\delta_m \mu^\ominus$ for t-butyl chloride is almost twice as large as $\delta_m \mu^\ddagger$. The $S_N 1$ transition state for t-butyl chloride is destabilized when either methyl or ethyl alcohol is added to water. Nevertheless, quite dramatic changes in initial states often contribute the major part to the changes in $\Delta G^\ddagger$. In the reaction between $HgCl_2$ and Et_4Sn, stabilization of reactants by addition of methyl alcohol to the reaction in water is responsible for the large increase in $\Delta G^\ddagger$ (Abraham, 1971). The complex patterns shown by $\delta_m \Delta H^\ddagger$ in Fig. 48 probably reflect a complex dependence of $\delta_m H^\ominus$ for the initial state on solvent composition. The dramatic change in $\Delta H^{\ddagger}$ over the range $0 \leqslant x_2 < 0.15$ is twice that for the remainder of the range, $0.15 \leqslant x_2 \leqslant 1.0$. By combining values of the enthalpy of solution and activation enthalpies, Arnett and co-workers were able to calculate the dependence of partial molar enthalpies of both initial and transition states on mole fraction of alcohol (Fig. 49; Arnett et al., 1965, 1963). The dependence of $\delta_m H^{\ominus}$ for the initial state follows the pattern for apolar solutes discussed on p. 303. The most

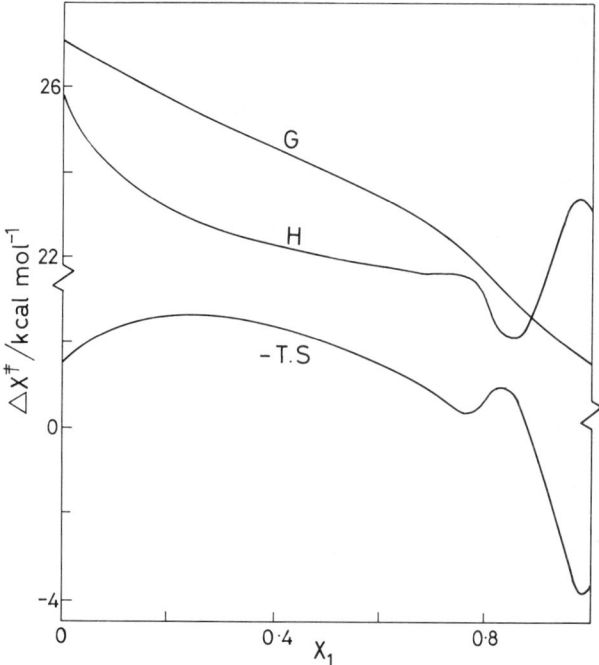

Figure 48. Dependence of activation parameters for solvolysis of t-butyl chloride on mole fraction of water in ethyl alcohol + water mixtures at 298 K (Winstein and Fainberg, 1957), 4·184 J = 1 cal.

striking observation is the small variation in the enthalpy of the transition state, $\delta_m H^{\ddagger}$ over the range $0 \leqslant x_2 < 0.4$. This does not imply that the transition state is not solvated nor that the nature of its solvation is invariant. Rather, the results show how a marked change in $\delta_m \Delta H^{\ddagger}$ can stem from a comparable variation in the initial state properties. It also does not follow that a similar pattern emerges for all systems. Indeed in the solvolysis of t-butyl dimethylsulphonium ions in ethyl alcohol + water, the slight dependence of $\Delta H^{\ddagger}$ on x_2 masks much more dramatic variations in initial and transition state enthalpies.

In view of the points made in Section 11 (cf. Table 10), it is not unexpected that the minimum in $\Delta H^{\ddagger}$ for the hydrolysis of t-butyl chloride should occur at progressively lower mole fractions of co-solvent on going from ethyl alcohol to isopropyl and t-butyl alchols (Robertson and Sugamori, 1969). The extrema arise because the apolar initial state enhances water structure to a smaller extent

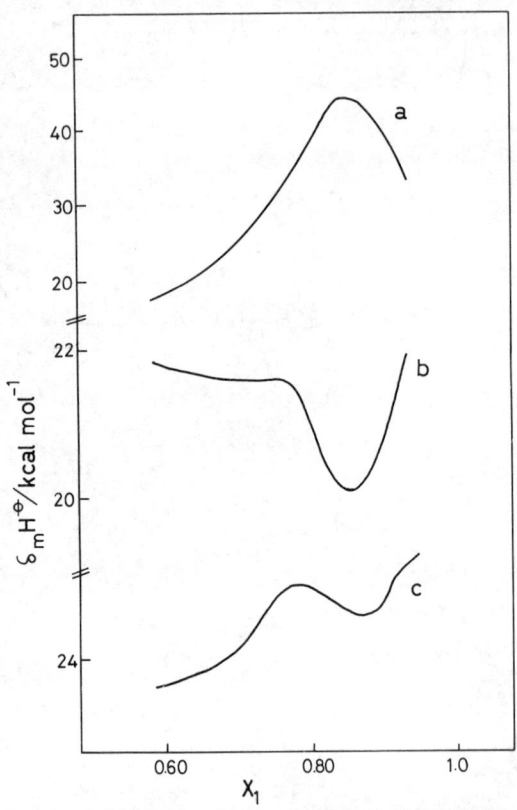

Figure 49. Dependence of enthalpy of activation for the hydrolysis of t-butyl chloride on the mole fraction of water in ethyl alcohol + water mixtures (curve b), together with dependence of partial molar enthalpies of initial (curve a) and transition states (curve c) (Arnett et al., 1965).

as x_2 increases $(x_2 < x_2^*)$ since this enhancement is also being accomplished by the co-solvent. This enhanced structure is increasingly sensitive to temperature as indicated in Fig. 50 by the tendency for $\Delta C_p^{\ddagger}$ to become more negative (cf. p. 293). The collapse of solvent structure when $x_2 > x_2^*$ is indicated by sharp extrema in $\delta_m \Delta C_p^{\ddagger}$ and minima in $\Delta H^{\ddagger}$ when $x_2 = x_2^*$ where the co-solvent is a monohydric alcohol or THF (Robertson and Sugamori, 1972). These features are particularly marked in t-butyl alcohol + water mixtures. A similar pattern is observed in $\delta_m \Delta C_p^{\ddagger}$ and $\delta_m \Delta H^{\ddagger}$ for the hydrolysis of chloromethyl dichloroacetate and related compounds in acetone + water mixtures (Cleve, 1972a, b). Further, for t-butyl chloride solvolysis, the plot of $\Delta H^{\ddagger}$ against $\Delta S^{\ddagger}$ generates two straight lines containing data points for the ranges $0 \leq x_2 \leq x_2^*$ and $x_2 > x_2^*$, thus reinforcing the idea that the solvent properties of the mixture change considerably when $x_2 \simeq x_2^*$.

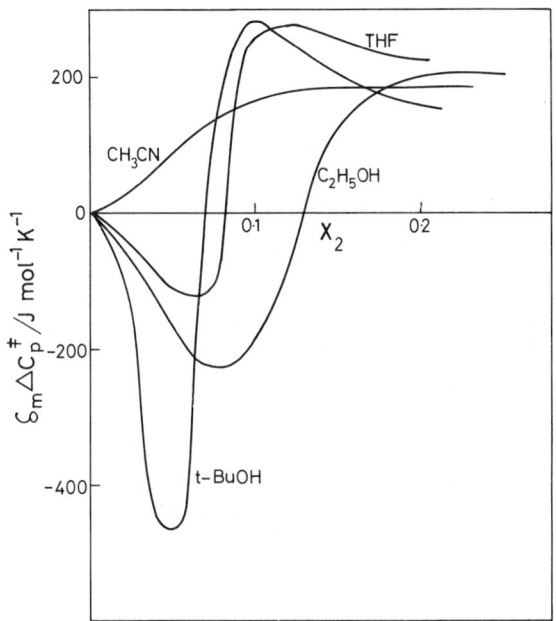

Figure 50. Dependence of heat capacity of activation for solvolysis of t-butyl chloride on co-solvent mole fraction in aqueous mixtures at 283 K (Robertson and Sugamori, 1969; 1972).

The importance of the entropy of activation in these reactions is clear when it is realized that if the rate constant were solely a function of $\Delta H^{\ddagger}$, the rate constant would increase initially as x_2 increases. Instead, the fall in rate constant is determined by the decrease in $\Delta S^{\ddagger}$. Thus, as the co-solvent enhances water-structure, so the partial molar entropy of the apolar initial state increases, the solute entering the structure-enhanced solvent, possibly as a guest in a preformed cavity. In any event, the change in this entropy term is probably more important than the endothermic shift in $\delta_m H^{\ominus}$, so that $\Delta G^{\ddagger}$ increases. Nevertheless, the balance of changes in $T \cdot \delta_m \Delta S^{\ddagger}$ and $\delta_m \Delta H^{\ddagger}$ are subtle and in a few cases $\delta_m \Delta G^{\ddagger}$ does actually decrease when x_2 increases, e.g. the hydrolysis of sulphonyl chlorides in aqueous dioxan (Foon and Hambly, 1962) and the decarboxylation of 2-cyano-2-phenylacetate ions in aqueous ethyl alcohol (Thomson, 1970). However, with few exceptions, the key quantity in the kinetics of hydrolysis of apolar solutes in TA mixtures is the entropy change; examples include the hydrolysis of benzyl chloride derivatives in ethyl alcohol + water mixtures (Hyne and Wills, 1963), hydrolysis of aromatic sulphonyl chlorides in dioxan + water mixtures (Tonnet and Hambly, 1971) and of aliphatic and substituted aliphatic sulphonyl halides in aqueous acetone and aqueous dioxan (Foon and Hambly, 1962). This control of the kinetics by the activation entropy was noted by Tommila, Tiilikainen and Voipo (1955) who examined the hydrolysis of t-butyl chloride, t-butyl bromide and ethyl chloride in acetone + water mixtures. Here the activation parameters, e.g. $\Delta H^{\ddagger}$, show the "roller coaster" pattern. It is interesting to note that Tommila *et al.* emphasized in their discussion the importance of changes in the properties of the transition state; in the intervening years, attention has turned to the changes in the initial state, at least for the water-rich mixtures. Actually acetone + water appears to be an important system and it would be interesting to know how $\Delta C_p^{\ddagger}$ depends on x_2 for one or more of these reactions.

It is interesting to compare the patterns shown by the activation parameters, $\delta_m \Delta X^{\ddagger}$, with those calculated using the endostatic approach discussed on p. 289. Unfortunately, only one set of data has so far been subjected to this analysis, namely the solvolysis of t-butyl chloride in ethyl alcohol + water mixtures (Grunwald and Effio, 1974), but this reveals some interesting trends. Thus $\delta_m \Delta G_\alpha^{\ddagger}$ is more sensitive to x_2 than is $\delta_m \Delta G^{\ddagger}$, the endostatic quantity increasing rapidly over the range $0 < x_2 < 0.2$, then remaining almost steady when $0.2 < x_2 < 0.8$, before increasing rapidly again. The quantity $\delta_m \Delta H_\alpha^{\ddagger}$ has a broad minimum near $x_2 = 0.2$. The minimum occurs at higher mole fraction, x_2, than the minimum for $\delta_m \Delta H^{\ddagger}$, although the minima in $\delta_m \Delta C_p^{\ddagger}$ and $\delta_m \Delta C_{p\alpha}^{\ddagger}$ occur at the same mole fraction (0.075). How these patterns should be interpreted is not clear. It would be valuable to convert enthalpy data for solutes to endostatic partial molar enthalpies, and then to see whether these quantities showed extrema near $x_2 = 0.2$. The sensitivity shown by $\delta_m \Delta G_\alpha^{\ddagger}$ to x_2 is also noteworthy and it would again be interesting to know whether this sensitivity is a common feature of reactions in aqueous mixtures.

An important feature of reactions in solvent mixtures is the effect of substituents in the reactant on the kinetic parameters. As Foon and Hambly (1962) note with reference to the hydrolysis of a series of sulphonyl halides, it is possible to obtain almost any order for the rate constants by selection of the appropriate solvent mixture. The substituent effects also depend on the temperature and, in the case of bimolecular reactions, on the nature of the attacking group, e.g. OH^- instead of H_2O (Tommila, 1967).

With reference to the dependence on co-solvent of the kinetics of one reaction, the hydrolysis of benzyl chloride (Golinkin and Hyne, 1968), the extrema in $\delta_m \Delta H^{\ddagger}$ and $\delta_m \Delta S^{\ddagger}$ become more marked along the series methyl alcohol, ethyl alcohol, isopropyl alcohol, t-butyl alcohol. Moreover the extrema in $\delta_m \Delta H^{\ddagger}$ do not occur at the mole fraction at which H^E for a given mixture is a minimum, nor at precisely the same mole fraction as the extrema in $\delta_m \Delta S^{\ddagger}$.

The dependence of volumes of activation, $\Delta V^{\ddagger}$, on solvent composition for the hydrolysis of benzyl chloride in aqueous mixtures (Golinkin et al., 1966; Hyne et al., 1966) has been analysed using the solvent dependence of the partial molar volume of the initial state (Golinkin et al., 1967). In aqueous alcohols, as x_2 increases, $\Delta V^{\ddagger}$ decreases to a minimum value at a mole fraction characteristic of the alcohol, e.g. $x_2 = 0.1$ for t-butyl alcohol, the intensity of the extremum increasing on going from methyl alcohol to t-butyl alcohol. The dependence of $\Delta V^{\ddagger}$ on x_2 is a consequence of quite marked changes in the quantities $\delta_m V_3^{\ominus}$ and $\delta_m V^{\ddagger}$ (Dickson and Hyne, 1971). For example, in ethyl alcohol + water mixtures, $\delta_m V^{\ominus}$ for benzyl chloride in the initial state has a maximum near $x_2 = 0.3$, whereas $\delta_m V^{\ddagger}$ has a shallow minimum near $x_2 = 0.1$ (cf. $V_2^{\ominus}$ for $Me_4N^+Cl^-$ in these mixtures; Lee and Hyne, 1969). In particular, minima in plots of $\Delta V^{\ddagger}$ against x_2 are not necessarily due to the trend in the initial state quantity, and volumetric data can provide some indication of the details of reaction mechanism. Thus for the hydrolysis of p-chlorobenzyl chloride (S_N2), a shallow minimum in $\delta_m \Delta V^{\ddagger}$ for reaction in aqueous ethyl alcohol stems from a more intense maximum in $\delta_m V^{\ominus}$ than for $\delta_m V^{\ddagger}$. At the other end of the scale, a sharp minimum in $\delta_m \Delta V^{\ddagger}$ for t-butyl chloride solvolysis (S_N1) results from a sharp minimum in $\delta_m V^{\ddagger}$, $\delta_m V_3^{\ominus}$ changing only gradually as x_2 is increased. The behaviour of the volumetric properties for the benzyl chloride system (Fig. 51) places this hydrolysis reaction as borderline S_N1-S_N2 (Mackinnon et al., 1970; Baliga and Whalley, 1970). A similar conclusion is indicated by the analysis of volume quantities for benzyl chloride in acetone + water mixtures (Macdonald and Hyne, 1970a).

So far we have considered only activation quantities calculated at constant pressure. A related set of constant volume quantities, e.g.

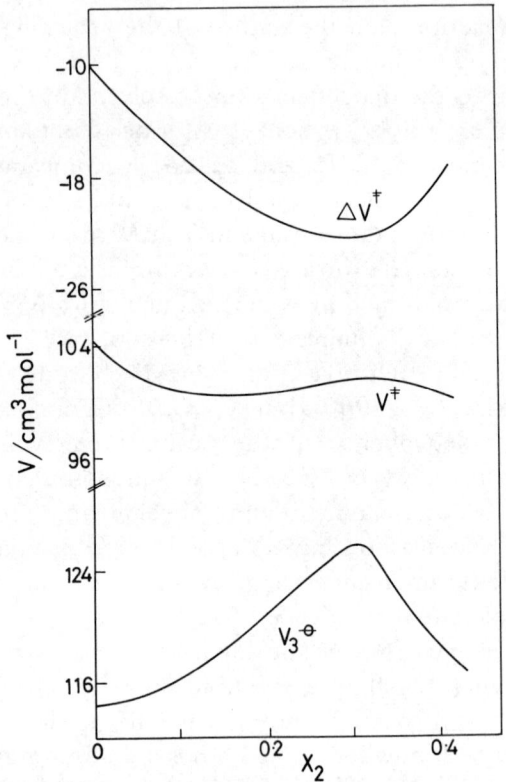

Figure 51. Dependence of volume parameters for the solvolysis of benzyl chloride on mole fraction of alcohol in ethyl alcohol + water mixtures at 323 K (Mackinnon et al., 1970).

$\Delta U_v^\ddagger$ and $\Delta S_v^\ddagger$ can be obtained (p. 214). Baliga and Whalley (1967) suggest that these quantities are more fundamental and that minima in $\delta_m \Delta H^\ddagger$ are a consequence of the dependence on x_2 of the thermal expansivity, α, of the solvent mixture [eqn (5)]. Thus for the acid catalysed hydrolysis of methyl acetate in acetone + water mixtures, a minimum in $\delta_m \Delta H^\ddagger$ is observed but not in $\delta_m \Delta U_v^\ddagger$ (Whalley, 1966). However, the task of determining the properties of aqueous systems at constant volume and repeating the analyses outlined here and in previous sections is awesome.

The activation parameters for more complex reactions show similar patterns to that outlined above. For example, in the alkaline hydrolysis of ethyl acetate in aqueous mixtures (Tommila et al., 1952), the second-order rate constant decreases with increasing mole fraction of ethyl alcohol, while $\Delta G^{\ddagger}$ increases, and $\Delta H^{\ddagger}$ falls to a minimum near $x_2 = 0.1$ (Fig. 52), $\delta_m \Delta S^{\ddagger}$ having a minimum in the same region. A similar trend is observed in aqueous acetone and in

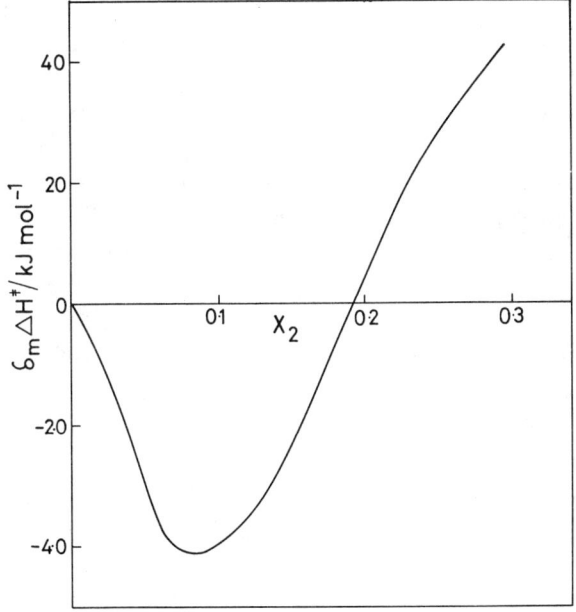

Figure 52. Dependence of enthalpy of activation for the alkaline hydrolysis of ethyl acetate on alcohol mole fraction in ethyl alcohol + water mixtures (Tommila et al., 1952)..

aqueous dioxan, while in aqueous t-butyl alcohol the extremum in $\delta_m \Delta H^{\ddagger}$ near $x_2 = 0.05$ is marked. Indeed $\Delta H^{\ddagger}$ decreases so rapidly as t-butyl alcohol is added that there is a slight increase in rate constant when this alcohol is first added at 298 K. The extremum in $\delta_m \Delta H^{\ddagger}$ becomes more marked as the hydrophobic character of the co-solvent increases (Tommila, 1952).

Because the rate constants are second order, the solvent dependence of the calculated activation parameters for these reactions, $\delta_m \Delta X^{\ddagger}$, can be represented in terms of three functions [eqn (39); cf. p. 213]. Based on our previous discussion it is possible to understand qualitatively the patterns which emerge in the kinetic parameters. One general trend will be for the chemical potential of the ester to

$$\delta_m \Delta X^\ddagger = \delta_m X^\ddagger - \delta_m X^\ominus (\text{ester}) - \delta_m X^\ominus (\text{OH}^-) \qquad (39)$$

decrease with increase in x_2 i.e. $\delta_m \mu^\ominus (\text{ester}) < 0$ (cf. Fig. 30). Consequently, the rate constant should tend to decrease. This is opposed by the tendency for $\delta_m \mu^\ominus (\text{OH}^-)$ to increase (p. 309) contributing towards an increase in rate constant. In t-butyl alcohol + water mixtures, this destabilization of the anion may be responsible for the initial increase in rate constant. The destabilization of OH^- may also account for the observation that in acetone + water mixtures, the fall in rate constant with increasing x_2 is more dramatic for the spontaneous hydrolysis of benzyl chloride (Macdonald and Hyne, 1970a) than for the alkaline hydrolysis (Tommila and Pitkanen, 1966). A similar pattern emerges in a comparison with the kinetics of the reaction between benzyl chloride and thiosulphate ions (Kalliorinne and Tommila, 1969). In at least one case, the alkaline hydrolysis of ethyl p-nitrobenzoate, the rate constant increases $[k/k(x_2 = 0) > 1]$ in acetone-rich aqueous mixtures (Tommila, 1964).

The endothermic maxima observed for apolar solutes and salts in water-rich mixtures must also contribute towards the minima in $\Delta H^\ddagger$ for alkaline ester hydrolysis in these mixtures. As before, the tendency for the rate constant to decrease is determined by the behaviour of $\delta_m \Delta S^\ddagger$. Plots of $\Delta H^\ddagger$ against $\Delta S^\ddagger$ are complicated but in mixtures for which $x_2 < x_2^*$ the data points generally fall on a straight line. Of course, there are new problems in this class of reactions. For example, the possibility arises that the rate constant is a function of quantities describing the equilibrium between, say, RO^- and OH^-. However, the patterns which emerge indicate that this may not usually be an important consideration in water-rich mixtures. One exception may be the alkaline hydrolysis of ethyl acetate and methyl acetate (Tommila and Maltamo, 1955) in methyl alcohol + water mixtures for which $\Delta H^\ddagger$ increases gradually as x_2 increases.

The effect of added co-solvent on the initial state is also important in more complicated reactions. For example, in the α-chymotrypsin-catalysed hydrolysis of p-nitrophenyl acetate and of N-acetyl-L-tryptophan methyl ester, the difference in the pattern of rates of hydrolysis when the solvent composition is varied can be attributed to the variation in the properties of the initial states of the esters (Bell et al., 1974).

12. TYPICALLY NON-AQUEOUS MIXTURES WITH G^E NEGATIVE (TNAN MIXTURES)

Properties of the Mixtures

The TNAN aqueous mixtures are characterized by the following conditions for the excess molar thermodynamic functions of mixing; $G^E < 0$ and $|H^E| > |T \cdot S^E|$.

The simplest example of this class is the hydrogen peroxide + water system (Fig. 53; Giguère and Carmichael, 1962). Thus G^E is negative (Scatchard et al., 1952), the mixing being strongly exothermic (Giguère et al., 1958). The molar excess volumes of mixing

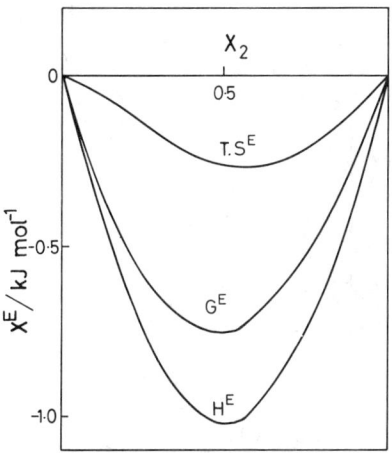

Figure 53. Excess thermodynamic functions of mixing for hydrogen perioxide + water mixtures at 298 K (Giguère and Carmichael, 1962).

are negative (Phibbs and Giguère, 1951) but, in contrast to the TA systems (Fig. 39), the partial molar volume of H_2O_2 does not pass through a minimum value in water-rich mixtures (Mitchell and Wynne-Jones, 1953).

The mixture dimethyl sulphoxide + water has attracted a great deal of interest. The excess function H^E is negative for this mixture at 298 K (Clever and Piggott, 1971; Fox and Whittingham, 1975), as also are G^E (Lam and Benoit, 1974; Philippe and Jambon, 1974) and V^E-quantities (Lau et al., 1970). A set of smoothed thermodynamic excess functions is shown in Fig. 54 (Kenttamaa and Lindberg, 1960). The dependence on x_2 of the isothermal compressibilities of DMSO + water mixtures is quite different from that for the TA monohydric alcohols + water mixtures. The curves for the latter systems show sharp minima in the water-rich mixtures, but the compressibility for DMSO + water system decreases gradually over the same range, reaching a broad minimum near $x_2 = 0.4$ (Jung and Hyne, 1970). Viscosity (Cowie and Toporowski, 1961; Schichman and Amey, 1971) and dielectric data (Lindberg and Kenttamaa, 1960; Tommila and Pajunen, 1968) are consistent with extensive intercomponent association, and this conclusion is borne out by nmr spectra (Glasel, 1970), phase diagram (Rasmussen and MacKenzie, 1968) and

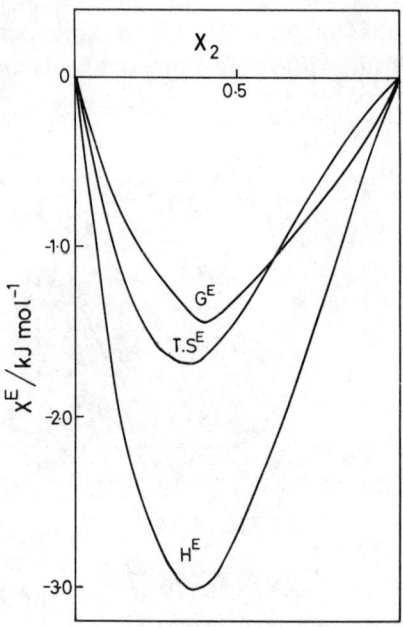

Figure 54. Excess thermodynamic functions of mixing for dimethyl sulphoxide + water mixtures at 298 K (Kenttamaa and Lindberg, 1960).

energy-volume coefficient data (Macdonald and Hyne, 1971). Symons (1971) suggests that DMSO breaks the water structure and that in the range $0.25 < x_2 < 0.4$ the properties of the mixture can be explained in terms of associates of composition DMSO . $3H_2O$ and DMSO . $2H_2O$. Infrared studies yield little evidence of extensive water-structure formation in these mixtures (Brink and Falk, 1970) and nmr studies show a minimum in molecular mobility for either component when $x_2 = 0.35$ (Packer and Tomlinson, 1970). As the hydrophobic character of this sulphoxide is increased by, for example, replacing the CH_3 groups to form diethyl sulphoxide

(DESO), so the properties change and DESO enhances water–water interactions (Macdonald et al., 1971).

The co-solvent ethylene glycol is probably borderline between a TA and TNAN component. The vapour pressure-composition curves for aqueous mixtures below 390 K show slight negative deviations from ideal behaviour consistent with a negative value for G^E (Trimble and Potts, 1935). The mixing is exothermic (Konneke et al., 1957) and the phase diagram shows that a 1 : 1 solid hydrate is formed (Orr et al., 1972). While these observations place the ethylene glycol + water mixtures in the TNAN class, the observation that the partial molar volume of ethylene glycol passes through a minimum value at $x_2 = 0.06$ (Ray and Nemethey, 1973) indicates that this mixture belongs to the TA class.

Hexamethylphosphoramide + water also belongs to this class of mixtures, the negative G^E and other excess functions show maximum deviations from ideality near $x_2 = 0.3$ consistent with intense intercomponent interaction (Jose et al., 1975; Jambon and Philippe, 1975).

Solutes in TNAN Aqueous Mixtures

At 298 K, argon is only slightly more soluble in ethylene glycol than in water. When ethylene glycol is added to water, the solubility of argon decreases ($\delta_m \mu_3^\ominus > 0$ using molar standard states) and does not become more soluble in the mixture than in water until $x_2 > 0.7$ (Ben-Naim, 1968). When glycol is added to water, the enthalpy of solution for argon increases (i.e. becomes less negative) as does the entropy change, but the change with mole fraction of the enthalpy and entropy parameters is more rapid than required by a linear relationship. There is some discrepancy between the enthalpies of solution calculated by Ben-Naim for argon in ethylene glycol and data reported by Gjaldbaek and Niemann (1958) which indicate that, on the mole fraction scale, solution is endothermic when $x_2 = 1$. However, Ben-Naim suggests that the solution is exothermic. The transfer of a wide range of solutes from water to ethylene glycol + water mixtures where $x_2 = 0.33$ is endothermic (Brower et al., 1969).

The impact of intercomponent association is also revealed in the solubilities of hydrogen in DMSO + water mixtures (Symons, 1971). At low temperatures, e.g. 298 K, the solubility of hydrogen first decreases and then increases when $x_2 > 0\cdot3$, but the minimum is lost at higher temperatures (e.g. 353 K) and the solubility increases steadily with increase in x_2. In general, small neutral solutes such as sulphur are more soluble in DMSO than in water (Kawakami et al., 1971).

A similar pattern to that outlined above is shown (Fig. 55) by the transfer quantities for ethyl acetate on going from water to DMSO + water mixtures (Fuchs et al., 1974). When DMSO is added to ethyl acetate in water, the ester is destabilized until $x_2 > 0\cdot35$ when $\delta_m \mu^\ominus$ (ester) < 0. At the same time the quantity $T\delta_m S^\ominus$ increases rapidly at first before levelling off, while the enthalpy of transfer shows a maximum near $x_2 = 0\cdot3$. Thus in these mixtures, as for the

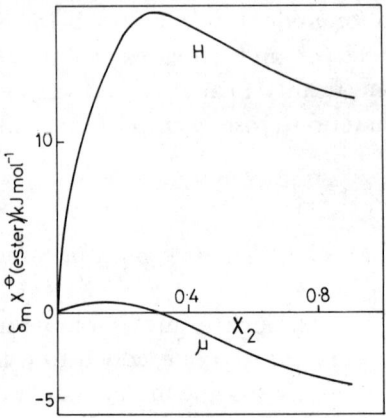

Figure 55. Dependence of the thermodynamic properties of ethyl acetate on co-solvent mole fraction in dimethyl sulphoxide + water mixtures at 298 K (Fuchs et al., 1974).

TA systems, there is a close link between the properties of the mixture and of solutes in the mixtures. Also the impact of the solvent variation on the properties of a neutral solute cannot be overlooked. For example, the dissociation of benzoic acid in DMSO + water mixtures becomes increasingly endothermic as x_2 increases, and this can be accounted for to a large extent by the exothermic transfer of undissociated benzoic acid from water to the mixtures. (Rodante et al., 1974).

Single ion transfer properties have been estimated for ions in DMSO + water mixture (Fuchs et al., 1971; Fuchs and Hagan, 1973). Assuming $\delta_m H^\ominus$ ($Ph_4 P^+$) = $\delta_m H^\ominus$ ($Ph_4 B^-$), the transfer of large hydrophobic ions, e.g. $Bu_4 N^+$, from water to DMSO + water mixtures is endothermic, the data showing a maximum when $x_2 \simeq$

0·3. In contrast, the transfer of halide and hydroxide ions is exothermic in water-rich mixtures, but, as more DMSO is added, endothermic for OH^-, Br^-, and Cl^- ions, the change in sign for $\delta_m H^\ominus$ (ion) occurring near $x_2 = 0·2$. Thus it appears that anions are better solvated at low than at high DMSO mole fractions. In these mixtures, NO_3^- ions are preferentially solvated by DMSO, Li^+ ions, by water (Covington et al., 1974). There is no preferential solvation for chloride ions (Covington and Thain, 1974), although it is noteworthy that the partial molar volume data for chloride ions in DMSO + water mixtures show a maximum in $V_2^\ominus$ (Cl^-) when $x_2 \simeq 0·2$ whereas $V_2^\ominus(R_4N^+)$ shows a minimum. This minimum for R_4N^+ becomes increasingly well defined as the size of the group R increases (Macdonald and Hyne, 1970). Extrema in Washburn transport numbers are observed for salt solutions in DMSO + water mixtures when $x_2 \simeq 0·05$ (Feakins et al., 1974b).

Arnett and McKelvey (1966a) have reported the enthalpies of transfer for a wide range of solutes between water and DMSO. For most non-polar solutes, the transfer is endothermic, benzene being an exception. As Arnett notes, the small exothermic transfer for benzene must be set against a large negative change in chemical potential, as deduced from the relative solubilities of benzene in water and in DMSO. This illustrates the danger of predicting trends in rate and equilibrium data on the basis of enthalpy data.

In the context of the kinetics of ester hydrolysis in DMSO + water mixtures, the properties of OH^- and H^+ ions are important (Gillett et al., 1975). It is noteworthy therefore that Das and Kundu (1973) conclude that the transfer of OH^- ions from water to water + DMSO mixtures is unfavourable over the range $0 \leqslant x_2 < 0·25$, but that of H^+ ions is favourable ($\delta_m \mu^\ominus$ (H^+) < 0).

In ethylene glycol + water mixtures, the enthalpies of transfer for HCl show evidence for three mole fraction ranges. Thus $\delta_m H^\ominus$ (HCl) is endothermic in water-rich mixtures, exothermic over a middle range but again endothermic glycol-rich mixture (Fig. 56). At the same time, however, the $\delta_m \mu^\ominus$ (HCl) function increases gradually (Stern and Nobilione, 1968a). A similar trend is observed for the transfer functions of HCl in acetic acid + water mixtures at 298 K except that $\delta_m H^\ominus$ (HCl) is always endothermic (Stern and Nobilione, 1968b).

Wells (1975) has analysed transfer function in ethylene glycol + water mixtures and shown that H^+ ions are stabilized and Cl^- ions are destabilized when x_2 increases, the curve for $\delta_m \mu^\ominus$ (H^+) having an extremum near $x_2 = 0.2$ at 298 K. The enthalpies of transfer for NaCl from water to hydrogen peroxide + water mixtures becomes more exothermic as x_2 increases (Stern and Bottenberg, 1971), consistent with water-structure breaking by added H_2O_2 as a result of intercomponent association.

For transfer of several alkali metal fluorides from water to aqueous hydrogen peroxide, values of $\delta_m \mu^{\ominus}$ (salt) are negative. Although

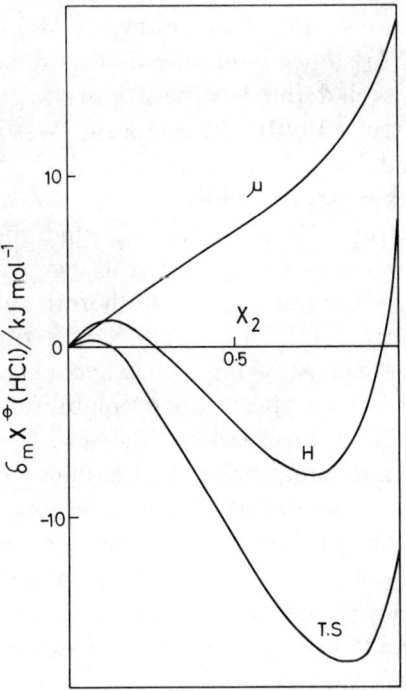

Figure 56. Dependence of the thermodynamic properties of HCl on co-solvent mole fraction in ethylene glycol + water mixtures at 298 K (Stern and Nobilione, 1968).

these mixtures are almost isodielectric, when $x_2 = 0.73$, $\delta_m \mu^{\ominus}$ (KF) is -15.1 kJ mol^{-1} (Covington and Thain, 1975), confirming the importance of factors other than solvent permittivity in the quantitative treatment of ionic solvation. In the $H_2O + H_2O_2$ mixtures, the dependence of $\delta_m \mu^{\ominus}$ (salt) on x_2 is not linear for CsF, whereas it is close to linear for CsCl in methyl alcohol + water mixtures. It is interesting to speculate whether compensation in Gibbs functions and chemical potentials is stronger in TA than in other aqueous mixtures.

Kinetics of Reactions in TNAN Aqueous Mixtures

Although a great deal of interest has been shown in the kinetics of alkaline hydrolysis in DMSO + water mixtures (see below), rather less interest has been shown in the kinetics of spontaneous hydrolysis reactions in TNAN mixtures.

The rate constant for the hydrolysis of t-butyl chloride at 298 K decreases as x_2 increases in DMSO + water mixtures (Heinonen and Tommila, 1965). A clear-cut contrast between TA and TNAN mixtures is shown by the volumes of activation and related parameters for the solvolysis of benzyl chloride in acetone + water (TA) and DMSO + water mixtures (Fig. 57). Thus, in the latter system, the curves show no marked extrema but there is a shallow minimum in $\Delta V^{\ddagger}$ near $x_2 = 0.4$. Extrema in $\delta_m \Delta H^{\ddagger}$ and $T \cdot \delta_m \Delta S^{\ddagger}$ for the hydrolysis of benzyl chloride are also smoothed out when the co-solvent is changed from acetone to DMSO (Tommila, 1966). A similar trend is observed in the kinetic parameters for the hydrolysis of chloromethyl and methyl trifluoroacetates (Cleve, 1972a). For example, in the case of the chloro derivative, $\delta_m \Delta C_p^{\ddagger}$ decreases gradually over the range $0 \leqslant x_2 < 0.2$ for DMSO + water mixtures, whereas a minimum is observed in this range for acetone + water mixtures.

The rate constant for the hydrolysis of t-butyl chloride in water increases when hydrogen peroxide is added, $\Delta H^{\ddagger}$ decreasing more rapidly than $T\Delta S^{\ddagger}$ so that $\Delta G^{\ddagger}$ decreases (Blandamer and Membrey, 1974). This rate increase may stem from a destabilization of the initial state, intercomponent hydrogen bonding between H_2O_2 and H_2O "salting-out" the apolar t-butyl chloride.

Interest in the kinetics of alkaline hydrolysis of esters in DMSO + water mixtures was stimulated by the observation that the rate constant often increased gradually as x_2 increased. This is observed, for example, in the alkaline hydrolysis of ethyl acetate. For higher esters, e.g. ethyl p-nitrobenzoate, the rate constant drops slightly at low x_2 but then rises again until $k/k(x_2 = 0) > 1$ (Tommila, 1964). The rate of alkaline hydrolysis of esters of benzoic acid is accelerated when DMSO is added (Tommila and Palenius, 1963), as also is the rate of alkaline hydrolysis of 2,4-dinitrofluorobenzene. In the latter case the effect is less dramatic because the rate constant for spontaneous hydrolysis also increases (Murto and Hiiro, 1964). The rate constants also increase when DMSO is added to aqueous solution for reactions between hydroxide ions and benzyl chloride (Tommila and Pitkänen, 1966), methyl iodide (Murto, 1961) and trimethylamine sulphur trioxide (Krueger and Johnson, 1968), and for reaction between thiosulphate ions and benzyl chloride and derivatives of benzyl chloride (Kalliorinne and Tommila, 1969). It is noteworthy that in the reaction between hydroxide ion and benzyl

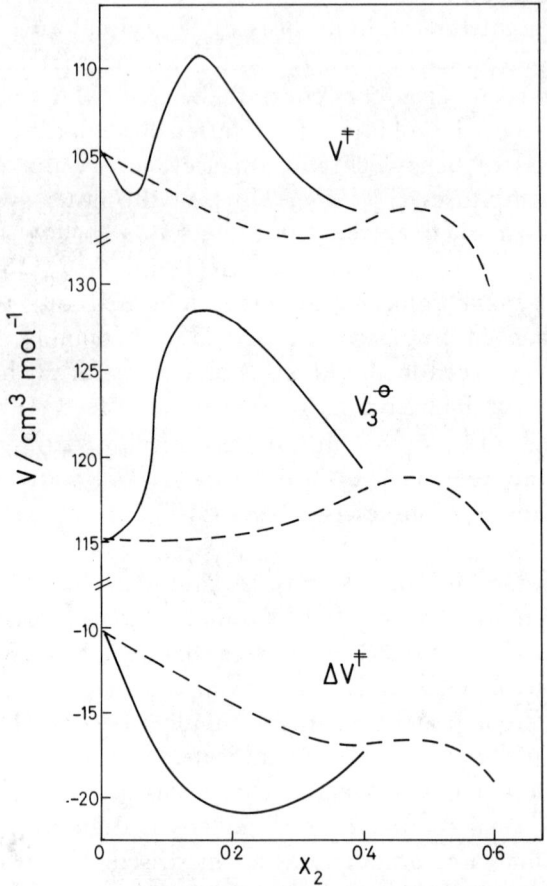

Figure 57. Comparison of the volumes of activation for solvolysis of benzyl chloride and related quantities as a function of co-solvent mole fraction at 323 K for acetone + water mixtures (full line) and dimethyl sulphoxide + water mixtures (Macdonald and Hyne, 1970a).

chloride, the activation energy is larger in DMSO + water mixtures than in acetone + water mixtures even though the reaction is roughly ten times faster in the DMSO system when $0.15 \leqslant x_2 \leqslant 0.28$. Presumably this increase in $\Delta H^{\ddagger}$ arises because there is no marked endothermic destabilization of the apolar benzyl chloride in DMSO mixtures as is observed in TA mixtures (p. 306).

A careful analysis has been reported of the alkaline hydrolysis of ethyl acetate in these mixtures (Fuchs et al., 1974). The thermodynamic transfer quantities for ethyl acetate on going from water to DMSO + water mixtures together with $\delta_m H^\ominus$ (OH$^-$) show that the increase in rate constant over the range $0 \leqslant x_2 < 0.3$ is due to the change in $\Delta S^\ddagger$ which offsets an increase in $\Delta H^\ddagger$. Transfer of ethyl acetate and hydroxide ions from water to the aqueous mixtures is generally endothermic but $\delta_m \Delta H^\ddagger$ is not negative when $x_2 > 0.15$ because $\delta_m H^\ddagger$ is more positive than for $\delta_m H^\ominus$ (ester) + $\delta_m H^\ominus$ (OH$^-$).

The kinetics of acid-catalysed ester hydrolysis in DMSO + water mixtures (Tommila and Murto, 1963) is similarly complex, reflecting the importance of intercomponent association. Nevertheless the analysis is made more difficult because the mechanism of reaction involves at least two stages (Wolford, 1964).

13. TYPICALLY NON-AQUEOUS MIXTURES WITH G^E POSITIVE (TNAP MIXTURES)

Properties of the Mixtures

For TNAP systems, the thermodynamic excess functions of mixing follow the general pattern: $G^E > 0$ and $|H^E| > |T \cdot S^E|$.

A clear-cut example of this class of systems is provided by acetonitrile + water mixtures. G^E is positive (Vierk, 1950) and the mixture separates into phases with a UCST at 272 K when $x_2 = 0.38$ (Armitage et al., 1968). The excess enthalpy of mixing is positive over most of the range, although at low temperatures and low mole fractions (e.g. $x_2 < 0.03$ at 298 K) the mixing is exothermic (Fig. 58; Morcom and Smith, 1969). This exothermic mixing at low x_2 does not stem from enhancement of water structure by this co-solvent. For example, there is no minimum in the partial molar volume of acetonitrile in this region. Instead, the overall trend appears to be for this aprotic co-solvent to disrupt water–water interactions comparable to the effect which added carbon tetrachloride has on methyl alcohol. Dielectric and viscosity data for acetonitrile + water mixtures have been documented by Cunningham et al. (1967).

Sulpholane + water mixtures belong to this class of systems because $G^E > 0$ and $G^E \simeq H^E$ at 298 K (Tommila et al., 1969;

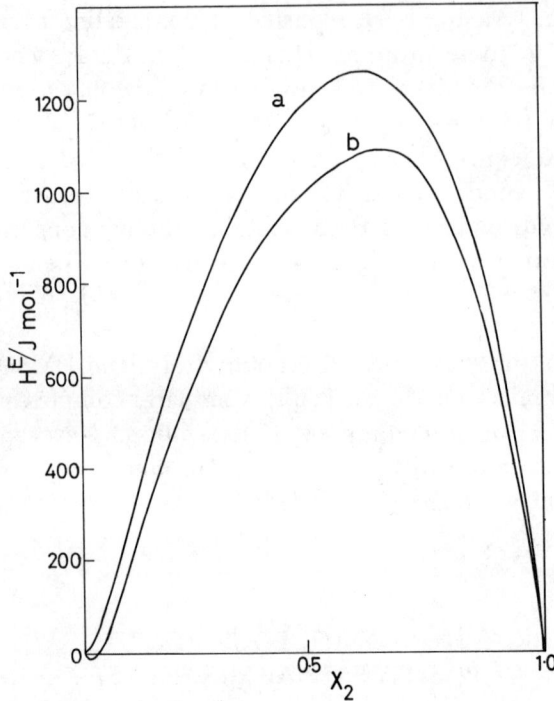

Figure 58. Enthalpies of mixing for acetonitrile + water mixtures at (a) 318 and (b) 298 K (Morcom and Smith, 1969).

Benoit and Choux, 1968). Similarly, propylene carbonate + water mixtures belong to this class, having a UCST at 334 K when $x_2 \simeq 0.2$ (Catherall and Williamson, 1971) with $G^E > 0$ at 298 K (Lam and Benoit, 1974). Propylene carbonate is an interesting solvent because its relative permittivity, 65·1, is close to that for water (Krishnan and Friedman, 1969).

Solutes in TNAP Aqueous Mixtures

Compared to the extensive information concerning the properties of solutes in TA mixtures, there is little information for this class of systems and most of it refers to acetonitrile + water mixtures.

The UCST for acetonitrile + water mixtures is sensitive to added salt (Renard and Heichelheim, 1967; Benter and Schneider, 1973; Moolel and Schneider, 1971), some salts (e.g. sodium benzoate) increasing, others (e.g. silver benzoate) lowering the UCST. Coetzee and Campion (1967) have shown that chloride, bromide and iodide ions are destabilized on going from solutions in water to solutions in acetonitrile. The enthalpies of transfer for alkali metal cations from water to acetonitrile are exothermic but endothermic for small anions. Thus acetonitrile is poor as an anion-solvating solvent but a good one for cations (Hedwig and Parker, 1974; Cox et al., 1974). Covington and Thain (1974), in an analysis of spectroscopic data, conclude that in the mixtures Na^+, Cl^-, Br^- and I^- ions are preferentially hydrated.

The properties of ions in the aqueous mixtures, $x_2 < 0.64$ show no extrema comparable to those observed in TA mixtures, compensation effects also being absent (Moreau and Douheret, 1974). These observations are consistent with the lack of any water-structure enhancement by this co-solvent.

Some emf studies have been reported for salt solutions in sulpholane + water mixtures (Tommila and Belinskij, 1969). It is noteworthy that pK_a for m-nitroanilinium ions in these mixtures does not change regularly with change in solvent permittivity (Ang, 1972). Covington et al. (1974) have suggested that, in these mixtures, Cs^+ ions are preferentially solvated by sulpholane. The Walden product, $\Lambda^0 \eta$, for $KClO_4$ has a minimum near $x_2 = 0.2$ in sulpholane + water mixtures at 298 K (D'Aprano et al., 1972).

Kinetics of Reactions in TNAP Mixtures

There have been relatively few kinetic studies of reactions in this class of mixtures and those that have been reported generally concern acetonitrile + water mixtures. Nevertheless these investigations have proved extremely valuable, affording a comparison with the kinetics in TA mixtures. Thus we saw (p. 256) that in the spontaneous hydrolysis of apolar alkyl halides $\Delta C_p^{\ddagger} < 0$. Addition of acetonitrile generally makes $\Delta C_p^{\ddagger}$ less negative (Fig. 50). Examples of such behaviour are shown in the kinetics of hydrolysis of t-butyl chloride (Robertson and Sugamori, 1972), sulphamoyl chlorides (Ko and Robertson, 1972a, b), and chloromethyl dichloroacetate (Cleve, 1972a). In the last case, $\delta_m \Delta C_p^{\ddagger}$ shows a maximum at low x_2 in these mixtures, whereas it shows a minimum at low x_2 in acetone + water mixtures. This behaviour is accounted for if added acetonitrile lowers the partial molar heat capacity of the apolar initial states in water by a structure breaking action. However, the kinetic data still present problems of interpretation. Thus for hydrolysis of t-butyl chloride, $\delta_m \Delta H^{\ddagger}$ shows a minimum at low values of x_2 which is not markedly different from that observed in TA mixtures. Clearly a great deal more needs to be learnt and understood about the kinetics of reactions in aqueous mixtures although the progress made so far has been impressive.

ACKNOWLEDGEMENTS

I should like to express my appreciation to M. C. R. Symons for many valuable discussions and to Ross Robertson who stimulated my interest in the kinetics of reactions in water and aqueous mixtures.

REFERENCES

Abraham, F. F. (1974). *J. Chem. Phys.* 61, 1221.
Abraham, M. H. (1971). *J. Chem. Soc.* (A) 1061.
Abraham, M. H. (1973). *J. C. S. Faraday I* 69, 1375.
Adams, D. M., Blandamer, M. J., Symons, M. C. R., and Waddington, D. (1971). *Trans. Faraday Soc.* 67, 611.
Adams, P. A., Sheppard, J. G., and Swart, E. R. (1973). *Chem. Comm.* 663.
Ahluwalia, J. C., and Chawla, B. (1973). *J. C. S. Faraday I* 69, 434.
Albery, W. J., and Curran, J. S. (1972). *Chem. Comm.* 425.
Albery, W. J., and Robinson, B. H. (1969). *Trans. Faraday Soc.* 65, 980.
Alexander, D. M. (1959). *J. Chem. Eng. Data* 4, 252.
Alexander, D. M., and Hill, D. J. T. (1969). *Aust. J. Chem.* 22, 347.
Allen, A. T., Wood, R. M., and McDonald, M. P. (1974). *Sugar Technology Reviews* 2, 165.
Allen, K. W., and Jeffrey, G. A. (1963). *J. Chem. Phys.* 38, 2304.
Allnatt, A. R. (1964). *Mol. Phys.* 8, 533.
Amis, E. S., and Hinton, J. F. (1973). "Solvent Effects in Chemical Phenomena", Academic Press, New York, Volume 1.
Anderson, H. L., and Wood, R. H. (1973). "Water—A Comprehensive Treatise" (F. Franks, ed.), Plenum Press, New York, Vol. 3, p. 119.
Anderson, R. G., and Symons, M. C. R. (1969). *Trans. Faraday Soc.* 65, 2550.
Andrews, A. L., Bennetto, H. P., Feakins, D., Lawrence, K. G., and Tomkins, R. P. T. (1968). *J. Chem. Soc.* (A) 1486.
Ang, K. P. (1972). *J. Soln. Chem.* 1, 369.
Angell, C. A. (1971). *J. Phys. Chem.* 75, 3698.
Armitage, D. A., Blandamer, M. J., Foster, M. J., Hidden, N. J., Morcom, K. W., Symons, M. C. R., and Wootten, M. J. (1968). *Trans. Faraday Soc.* 64, 1193.
Arnaud, R., Avedikian, L., and Morel, J.-P. (1972). *J. Chim. Phys.* 45.
Arnaud, R., Morin, J., and Morel, J.-P. (1970). *C.R. Acad. Sci. Ser. C.* 271, 927.
Arnett, E. M. (1967). "Physico-Chemical Processes in Mixed Aqueous Systems" (F. Franks, ed.), Heinemann, London, p. 105.

Arnett, E. M., Bentrude, W. G., Burke, J. J., and Duggleby, P. McC. (1965). *J. Amer. Chem. Soc.* **87**, 1541.
Arnett, E. M., Duggleby, P. McC., and Burke, J. J. (1963). *J. Amer. Chem. Soc.* **85**, 1350.
Arnett, E. M., Ho, M., and Schaleger, L. L. (1970). *J. Amer. Chem. Soc.* **92**, 7039.
Arnett, E. M., Ko, H. C., and Minasz, R. J. (1972). *J. Phys. Chem.* **76**, 2474.
Arnett, E. M., Kover, W. B., and Carter, J. V. (1969). *J. Amer. Chem. Soc.* **91**, 4028.
Arnett, E. M., and McKelvey, D. R. (1965). *Record Chem. Progr.* **26**, 185.
Arnett, E. M., and McKelvey, D. R. (1966). *J. Amer. Chem. Soc.* **87**, 1393.
Arnett, E. M., and McKelvey, D. R. (1966a). *J. Amer. Chem. Soc.* **88**, 2598.
Arnett, E. M., and McKelvey, D. R. (1966b). *J. Amer. Chem. Soc.* **88**, 5031.
Arnett, E. M., and McKelvey, D. R. (1969). "Solute-Solvent Interactions", (J. F. Coetzee and C. D. Ritchie, eds.), M. Dekker, New York and London, p. 343.
Assarsson, P., and Eirich, F. R. (1968). A. C. S. Adv. in Chem. Series, No. 84, p. 1.
Assarsson, P., and Eirich, F. R. (1968b). *J. Phys. Chem.* **72**, 2710.
Avedikian, L., Juillard, J., Morel, J.-P., and Ducros, M. (1973). *Thermochimica Acta* **6**, 283.
Avedikian, L., and Morel, J.-P. (1971). *J. Chim. Phys.* 1201.
Avedikian, L., Morin, J., and Morel, J.-P. (1970). *C.R. Acad. Sci. Ser. C* **271**, 988.
Avedikian, L., Perron, G., and Desnoyers, J. E. (1975). *J. Soln. Chem.* **4**, 331.
Aveyard, R., and Heselden, R. (1974). *J. C. S. Faraday I* **70**, 1953.
Bahe, L. W. (1972). *J. Phys. Chem.* **76**, 1062, 1608.
Bale, C. W., and Pelton, A. D. (1974). *Metallurgical Trans.* **5**, 2323.
Bale, H. D., Shepler, R. E., and Sorgen, D. K. (1970). *Phys. Chem. Liquids* **1**, 181.
Baliga, B. T., and Whalley, E. (1967). *J. Phys. Chem.* **71**, 1166.
Baliga, B. T., and Whalley, E. (1970). *Canad. J. Chem.* **48**, 528.
Barker, J. A. (1953). *Aust. J. Chem.* **6**, 207.
Barone, G., Crescenzi, V., and Vitagliano, V. (1968). *J. Phys. Chem.* **72**, 2588.
Barone, G., Rizzo, E., and Vitagliano, V. (1970). *J. Phys. Chem.* **74**, 2230.
Bateman, L. C., Church, M. G., Hughes, E. D., Ingold, C. K., and Taher, N. A. (1940). *J. Chem. Soc.* 979.
Bates, R. G. (1968). "Hydrogen-Bonded Solvent Systems", (A. K. Covington and P. Jones, eds.), Taylor and Francis, London, p. 49.
Battino, R. (1971). *Chem. Rev.* **71**, 5.
Battino, R., and Clever, H. L. (1966). *Chem. Rev.* **66**, 395.
Baumgartner, E. K., and Atkinson, G. (1971). *J. Phys. Chem.* **75**, 2336.
Bax, D., de Ligny, C. L., and Remijnse, A. G. (1972). *Rec. Trav. Chim. Pays-Bas* **91**, 1225.
Bell, R. P., Critchlow, J. E., and Page, M. I. (1974). *J. C. S. Perkin II* 66.
Benjamin, L., and Benson, G. C. (1963). *J. Phys. Chem.* **67**, 858.
Ben-Naim, A. (1965). *J. Phys. Chem.* **69**, 3245.
Ben-Naim, A. (1967). *J. Phys. Chem.* **71**, 4002.
Ben-Naim, A. (1968). *J. Phys. Chem.* **72**, 2998.
Ben-Naim, A. (1972). *Mol. Phys.* **24**, 705.
Ben-Naim, A. (1972a). *Chem. Phys. Lett.* **13**, 406.
Ben-Naim, A. (1972b). *J. Chem. Phys.* **57**, 3605, 5257, 5266.
Ben-Naim, A. (1972c). *Mol. Phys.* **24**, 723.
Ben-Naim, A. (1973). "Water—A Comprehensive Treatise" (F. Franks, ed.), Plenum Press, New York, vol. 1, p. 413.

Ben-Naim, A. (1973a). *J. Stat. Phys.* 7, 3.
Ben-Naim, A. (1973b). *J. Chem. Phys.* 59, 6535.
Ben-Naim, A. (1975). *J. Phys. Chem.* 79, 1268.
Ben-Naim, A., and Baer, S. (1964). *Trans. Faraday Soc.* 60, 1736.
Ben-Naim, A., and Egel-Thal, M. (1965). *J. Phys. Chem.* 69, 3250.
Ben-Naim, A., and Moran, G. (1965). *Trans. Faraday Soc.* 61, 821.
Ben-Naim, A., and Stillinger, F. H. (1971). "Structure and Transport Processes in Water and Aqueous Solutions" (R. A. Horne, ed.), Wiley, New York.
Ben-Naim, A., Wilf, J., and Yaacobi, M. (1973). *J. Phys. Chem.* 77, 95.
Ben-Naim, A., and Yaacobi, M. (1974). *J. Phys. Chem.* 78, 170, 175.
Bennetto, H. P. (1973). *Ann. Rep. Chem. Soc.* 70A, 223.
Bennetto, H. P., and Caldin, E. F. (1971). *J. Chem. Soc.* (*A*), 2130, 2198.
Bennetto, H. P., Feakins, D., and Lawrence, K. G. (1968). *J. Chem. Soc.* (*A*) 1493.
Bennetto, H. P., and Letcher, J. W. (1972). *Chem. and Ind.* 847.
Benoit, R. L., and Choux, G. (1968). *Canad. J. Chem.*, 46, 3215.
Benter, G., and Schneider, H. (1973). *Ber. Bunsenges. Phys. Chem.* 77, 997.
Berendsen, H. J. C. (1967). "Theoretical and Experimental Biophysics" (A. Cole, ed.), Arnold, London, Vol. 1, p. 1.
Berendsen, H. J. C. (1973). "Water—A Comprehensive Treatise" (F. Franks, ed.), Plenum Press, New York, Vol. 5, p. 293.
Bernal, J. D., and Fowler, R. H. (1933). *J. Chem. Phys.* 1, 515.
Bertie, J. E. (1968). *Appl. Spectroscopy* 22, 634.
Bertie, J. E., and Whalley, E. (1964). *J. Chem. Phys.* 40, 1637, 1646.
Bertrand, G. L., Larson, J. W., and Hepler, L. G. (1968). *J. Phys. Chem.* 72, 4194.
Bertrand, G. L., Millero, F. J., Wu, C.-H., and Hepler, L. G. (1966). *J. Phys. Chem.* 70, 699.
Bett, K. E., and Cappi, J. B. (1965). *Nature* 207, 620.
Beurskens, P. T., and Jeffrey, G. A. (1964). *J. Chem. Phys.* 41, 924.
Bezboruah, C. P., Filomena, M., Camoes, G. F. C., Covington, A. K., and Dobson, J. V. (1973). *J. C. S. Faraday I* 69, 949.
Blandamer, M. J. (1970). *Quart. Rev.* 24, 169.
Blandamer, M. J. (1973). "Water—A Comprehensive Treatise" (F. Franks, ed.), Plenum Press, Volume 2, p. 495.
Blandamer, M. J., and Burgess, J. (1974). *Chem. Soc. Rev.* 4, 55.
Blandamer, M. J., Burgess, J., and Dupree, M. (1975b). unpublished work.
Blandamer, M. J., Burgess, J., and Morris, S. H. (1975a), in press.
Blandamer, M. J., Foster, M. J., and Waddington, D. (1970a). *Trans. Faraday Soc.* 66, 1369.
Blandamer, M. J., and Fox, M. F. (1970). *Chem. Rev.* 70, 59.
Blandamer, M. J., and Fox, M. F. (1973). "Water—A Comprehensive Treatise" (F. Franks, ed.), Plenum Press, New York, Volume 2, p. 459.
Blandamer, M. J., Fox, M. F., Symons, M. C. R., and Wootten, M. J. (1970b). *Trans. Faraday Soc.* 66, 1574.
Blandamer, M. J., Godfrey, E., and Membrey, J. R. (1974). *J. Soln. Chem.* 3, 885.
Blandamer, M. J., Golinkin, H. S., and Robertson, R. E. (1969a). *J. Amer. Chem. Soc.* 91, 2678.
Blandamer, M. J., Hidden, N. J., Morcom, K. W., Smith, R. W., Treloar, N. C., and Wootten, M. J. (1969b). *Trans. Faraday Soc.* 65, 2633.
Blandamer, M. J., and Membrey, J. R. (1974). *J. C. S. Perkin II* 1400.

Blandamer, M. J., and Symons, M. C. R. (1968). "Hydrogen-Bonded Solvent Systems" (A. K. Covington and P. Jones, ed.), Taylor and Francis, London, p. 211.
Blandamer, M. J., and Waddington, D. (1970). *J. Chem. Phys.* 52, 6247.
Blyth, C. A., and Knowles, J. R. (1971). *J. Amer. Chem. Soc.* 93, 3017, 3021.
Bockris, J. O'M., and Saluja, P. P. S. (1972). *J. Phys. Chem.* 76, 2140.
Boettger, G., Harders, H., and Luck, W. A. P. (1967). *J. Phys. Chem.* 71, 459.
Bolton, P. D. (1970). *J. Chem. Educ.* 47, 638.
Bolton, P. D., Fleming, K. A., and Hall, F. M. (1972). *J. Amer. Chem. Soc.* 94, 1033.
Bonamico, M., Jeffrey, G. A., and McMullan, R. K. (1962). *J. Chem. Phys.* 32, 2219.
Bose, K., Das, A. K., and Kundu, K. (1975). *J. C. S. Faraday I* 71, 1838.
Boyne, J. A., and Williamson, A. G. (1967). *J. Chem. Eng. Data* 12, 318.
Brandts, J. F., and Hunt, L. (1967). *J. Amer. Chem. Soc.* 89, 4826.
Brink, G., and Falk, M. (1970). *J. Mol. Struct.* 5, 27.
Broadwater, T. L., and Kay, R. L. (1970). *J. Phys. Chem.* 74, 3802.
Brooks, C. A. G., Davis, K. M. C., and Blandamer, M. J. (1974). *J. Soln. Chem.* 3, 247.
Brower, K. R., Peslak, J., and Elrod, J. (1969). *J. Phys. Chem.* 73, 207.
Brower, K. R., and Wu, T.-L. (1970). *J. Amer. Chem. Soc.* 92, 5303.
Buijs, K., and Choppin, G. R. (1963). *J. Chem. Phys.* 39, 2035.
Bunton, C. A., Fuller, N. A., Perry, S. G., and Pitman, I. H. (1962). *J. Chem. Soc.* 4478.
Cabani, S., Conti, G., and Lepori, L. (1971). *Trans. Faraday Soc.* 67, 1943.
Cabani, S., Conti, G., and Lepori, L. (1972). *J. Phys. Chem.* 76, 1338.
Cabani, S., Conti, G., and Lepori, L. (1974). *J. Phys. Chem.* 78, 1030.
Cabani, S., Conti, G., Martinelli, A., and Matteoli, E. (1973). *J. C. S. Faraday I* 69, 2112.
Caldin, E. F., and Bennetto, H. P. (1973). *J. Soln. Chem.* 2, 217.
Caldin, E. F., and Godfrey, P. (1974). *J. C. S. Faraday I* 70, 2260.
Caldin, E. F., and Hasinoff, B. B. (1975). *J. C. S. Faraday I* 71, 515.
Carbonnel, L., and Rosso, J.-C. (1972). *Rev. Chim. Minerale* 9, 771.
Carbonnel, L., Rosso, J.-C., and Caranoni, C. (1973). *C.R. Acad. Sci. Ser. C* 276, 619.
Cargill, R. W., and Morrison, T. J. (1975). *J. C. S. Faraday I* 71, 618.
Cassel, R. B., and Wood, R H. (1974). *J. Phys. Chem.* 78, 1924, 2460, 2465.
Catherall, N. F., and Williamson, A. G. (1971). *J. Chem. Eng. Data* 16, 335.
Cayley, G. R., and Margerum, D. W. (1974). *Chem. Comm.* 1002.
Chawla, B., and Ahluwalia, J. C. (1972). *J. Phys. Chem.* 76, 2582.
Clarke, E. C. W., and Glew, D. N. (1966). *Trans. Faraday Soc.* 62, 539.
Clarke, E. C. W., and Glew, D. N. (1971). *Canad. J. Chem.* 49, 691.
Clarke, G. A., and Taft, R. W. (1962). *J. Amer. Chem. Soc.* 84, 2295.
Claussen, W. F. (1951). *J. Chem. Phys.* 19, 259, 662.
Cleve, N. J. (1972a). *Suomen Kemistilehti* 46B, 385.
Cleve, N. J. (1972b). *Ann. Acad. Sci. Fennicae, Ser. A II* No. 167.
Clever, H. L., and Holland, C. J. (1968). *J. Chem. Eng. Data.* 13, 411.
Clever, H. L., and Piggott, S. P. (1971). *J. Chem. Thermodynamics* 3, 221.
Coetzee, J. F., and Campion, J. J. (1967). *J. Amer. Chem. Soc.* 89, 2517.
Coles, M. J., Walrafen, G. E., and Wecht, K. W., (1970). *Chem. Phys. Lett.* 4, 621.
Conway, B. E., and Laliberte, L. H. (1968). *J. Phys. Chem.* 72, 4317.

Conway, B. E., and Novak, D. M. (1975). Aqueous Gas Solutions (W. A. Adams, G. Greer and J. E. Desnoyers, ed.), Electrochem. Soc. Inc., Princeton, N.J. p. 115.
Conway, B. E., Verrall, R. E., and Desnoyers, J. E. (1965). *Z. phys. Chem. (Leipzig)* 230, 157.
Conway, B. E., Verrall, R. E., and Desnoyers, J. E. (1966). *Trans. Faraday Soc.* 62, 2738.
Copp, J. L. (1955). *Trans. Faraday Soc.* 51, 1056.
Copp, J. L., and Everett, D. H. (1953). *Disc. Faraday Soc.* 15, 174.
Covington, A. K., Lantzke, I. R., and Thain, J. M. (1974). *J. C. S. Faraday I* 70, 1869.
Covington, A. K., Newmann, K. E., and Wood, M. (1972). *Chem. Comm.* 1234.
Covington, A. K., Lilley, T. H., Newman, K. E., and Porterhouse, G. A. (1973). *J. C. S. Faraday I* 69, 963.
Covington, A. K., and Thain, J. M. (1974). *J. C. S. Faraday I* 70, 1879.
Covington, A. K., and Thain, J. M. (1975). *J. C. S. Faraday I* 71, 78.
Cowie, J. M. G., and Toporowski, P. M. (1961). *Canad. J. Chem.* 39, 2240.
Cox, B. G. (1973). *Ann. Rep. Chem. Soc.* 70A, 249.
Cox, B. G. (1973a). *J. C. S. Perkin II*, 607.
Cox, B. G., Hedwig, C. R., Parker, A. J., and Watts, D. W. (1974). *Aust. J. Chem.* 27, 477.
Crescenzi, V., Quadrifoglio, F., and Vitagliano, V. (1967). *J. Phys. Chem.* 71, 2313.
Criss, C. M. (1974). *J. Phys. Chem.* 78, 1000.
Criss, C. M., Held, R. P., and Luksha, E. (1968). *J. Phys. Chem.* 72, 2970.
Cunningham, G. P., Vidulich, G. A., and Kay, R. L. (1967). *J. Chem. Eng. Data* 12, 336.
D'Aprano, A., Donato, I. D., and Palombo, R. (1972). *J. Phys. Chem.* 76, 2923.
Darling, B. T., and Denison, D. M. (1940). *Phys. Rev.* 57, 128.
Das, A. K., and Kundu, K. K. (1973). *J. C. S. Faraday I* 69, 730.
Davidson, D. W. (1973). "Water—A Comprehensive Treatise" (F. Franks, ed.), Plenum Press, New York, Volume 2, p. 115.
Davies, J., Ormondroyd, S., and Symons, M. C. R. (1972). *J. C. S. Faraday I* 68, 686.
Davis, C. M., and Jarzynski, J. (1967). *Adv. Mol. Relax. Proc.* 1, 155.
Davis, C. M., and Jarzynski, J. (1973). "Water—A Comprehensive Treatise" (F. Franks, ed.), Plenum Press, New York, Vol. 1, p. 443.
Dawe, R. A., Newsham, D. M. T., and Ng, S. B. (1973). *J. Chem. Eng. Data.* 18, 44.
Del Bene, J. E., and Pople, J. A. (1970). *J. Chem. Phys.* 52, 4858.
Del Bene, J. E., and Pople, J. A. (1973). *J. Chem. Phys.* 58, 3605.
de Ligny, C. L., and Alfenaar, M. (1965). *Rec. Trav. Chim. Pays-Bas* 84, 82.
Desnoyers, J. E., Arel, M., Perron, G., and Jolicoeur, C. (1969). *J. Phys. Chem.* 73, 3346.
Desnoyers, J. E., and Conway, B. E. (1964). *J. Phys. Chem.* 68, 2305.
Desnoyers, J. E., and Jolicoeur, C. (1969). "Modern Aspects of Electrochemistry" (J. O'M. Bockris and B. E. Conway, ed.), Plenum Press, New York, Vol. 5, p. 1.
Desnoyers, J. E., Pelletier, G. E., and Jolicoeur, C. (1965). *Canad. J. Chem.* 43, 3232.
Desnoyers, J. E., and Perron, G. (1972). *J. Soln. Chem.* 1, 199.
Desrosiers, N., Perron, G., Mathieson, J. G., Conway, B. E., and Desnoyers, J. E. (1974). *J. Soln. Chem.* 3, 789.

Devine, W., and Lowe, B. M. (1971). *J. Chem. Soc. (A)* 2113.
De Nooijer, B., Spencer, D., Whittington, S. G., and Franks, F. (1971). *Trans. Faraday Soc.* 67, 1315.
De Visser, C., and Somsen, G. (1973). *J. C. S. Faraday I* 69, 1440.
De Visser, C., and Somsen, G. (1974a). *J. Soln. Chem.* 3, 847.
De Visser, C., and Somsen, G. (1974b). *J. Phys. Chem.* 78, 1719.
Diamond, R. M. (1963). *J. Phys. Chem.* 67, 2513.
Dickson, S. J., and Hyne, J. B. (1971). *Canad. J. Chem.* 49, 2394.
Dobson, H. J. E. (1925). *J. Chem. Soc.* 2866.
Doherty, D. G., and Vaslow, F. (1952). *J. Amer. Chem. Soc.* 74, 931.
Dollet, N., and Juillard, J. (1973). *J. Chim. Phys.* 1733.
Dorsey, N. E. (1940). "Properties of Ordinary Water Substance", Reinhold Publishing Co., New York.
Drost-Hansen, W. (1968). *Chem. Phys. Lett.* 2, 647.
Drost-Hansen, W. (1972). "Symposium of the Society for Experimental Biology", Cambridge University Press, No. 26, p. 61.
Drost-Hansen, W. (1973). *Ann. N. Y. Acad. Sci.* 204, 100.
Drost-Hansen, W., and Curry, R. W. (1970). *J. Coll. Int. Sci.* 32, 464.
Duncan, A. B. F., and Pople, J. A. (1953). *Trans. Faraday Soc.* 49, 217.
Duttachoudhury, M. K., and Mathur, H. B. (1974). *J. Chem. Eng. Data* 19, 145.
Eagland, D. (1973). "Water—A Comprehensive Treatise" (F. Franks, ed.), Plenum Press, New York, Vol. 4, Chapt. 5.
Egan, E. P., and Luff, B. B. (1966). *J. Chem. Eng. Data* 11, 192.
Eisenberg, D., and Kauzmann, W. (1969). "The Structure and Properties of Water", Oxford University Press, Oxford.
Eley, D. D. (1939). *Trans. Faraday Soc.* 35, 1281.
Ellis, C. M. (1967). *J. Chem. Educ.* 44, 405.
Endo, H. (1972). *Bull. Chem. Soc. Japan* 45, 3270.
Engels, P., and Schneider, G. M. (1972). *Ber. Bunsenges. Physik. Chem.* 76, 1239.
Evans, A. G. (1946). *Trans. Faraday Soc.* 42, 719.
Ewing, G. J., and Ionescu, L. G. (1974). *J. Chem. Eng. Data* 19, 367.
Exner, O. (1972). *Coll. Czech. Chem. Comm.* 37, 1425.
Exner, O. (1973a). *Prog. Phys. Org. Chem.* 10, 411.
Exner, O. (1973b). *Coll. Czech. Chem. Comm.* 38, 781.
Falcone, J. S., Levine, A. S., and Wood, R. H. (1973). *J. Phys. Chem.* 77, 2137.
Falcone, J. S., and Wood, R. H. (1974). *J. Soln. Chem.* 3, 233.
Falk, M. (1971). *Canad. J. Chem.* 49, 1137.
Falk, M., and Ford, T. A. (1966). *Canad. J. Chem.* 44, 1699.
Falk, M., and Wyss, H. R. (1969). *J. Chem. Phys.* 51, 5727.
Fanning, R. J., and Kruus, P. (1970). *Canad. J. Chem.* 48, 2052.
Farrow, L. A., and Edelson, D. (1974). *Int. J. Chem. Kinetics* 6, 787.
Feakins, D. (1967). "Physico-Chemical Processes in Mixed Aqueous Solvents" (F. Franks, ed.), Heinemann, London, p. 71.
Feakins, D., Freemantle, D. J., and Lawrence, K. G. (1974a). *J. C. S. Faraday I* 70, 795.
Feakins, D., Hickey, B. E., Lorimer, J. P., O'Shaughnessy, D. A., and Voice, P. M. (1974b). *Electroanalyt. Chem. Interfacial Electrochem.* 54, 443.
Feakins, D., Hickey, D., Lorimer, J. P., and Voice, P. J. (1975). *J. C. S. Faraday I* 71, 780.
Feakins, D., Lawrence, K. G., and Tomkins, R. P. T. (1967). *J. Chem. Soc. (A)* 753.
Feakins, D., and Lorimer, J. P. (1974). *J. C. S. Faraday I* 70, 1888.

Feakins, D., and Turner, D. J. (1965). *J. Chem. Soc.* 4986.
Feakins, D., and Voice, P. J. (1972). *J. C. S. Faraday I* 68, 1390.
Feakins, D., and Voice, P. J. (1973). *J. C. S. Faraday I* 69, 1711.
Feil, D., and Jeffrey, G. A. (1961). *J. Chem. Phys.* 35, 1863.
Feillolay, A., and Lucas, M. (1972). *J. Phys. Chem.* 76, 3068.
Fendler, E. J., and Fendler, J. H. (1970). *Adv. Phys. Org. Chem.* 8, 271.
Fendler, J. H., and Fendler, E. J. (1975). "Catalysis in Micellar and Macromolecular Systems", Academic Press, New York.
Finer, E. G., Franks, F., and Tait, M. J. (1972). *J. Amer. Chem. Soc.* 94, 4424.
Fixman, M. (1962). *J. Chem. Phys.* 36, 1957.
Foon, R., and Hambly, A. N. (1962). *Aust. J. Chem.* 15, 668, 684.
Ford, T. A., and Falk, M. (1968). *Canad. J. Chem.* 46, 3579.
Fortier, J.-L., Leduc, P.-A., and Desnoyers, J. E. (1974a). *J. Soln. Chem.* 3, 323.
Fortier, J.-L., Philip, P. R., and Desnoyers, J. E. (1974b). *J. Soln. Chem.* 3, 523.
Fowler, D. L., Loebenstein, W. L., Pall, D. B., and Kraus, C. A. (1940). *J. Amer. Chem. Soc.* 62, 1140.
Fox, M. F., and Whittingham, K. P. (1975). *J. C. S. Faraday I* 71, 1407.
Franck, E. U. (1970). *Rev. Pure Appl. Chem.* 24, 13.
Frank, H. S. (1958). *Proc. Roy. Soc.* 247A, 481.
Frank, H. S. (1963). *J. Phys. Chem.* 67, 1554.
Frank, H. S. (1973). "Water—A Comprehensive Treatise" (F. Franks, ed.), Plenum Press, New York, Volume 1, p. 515.
Frank, H. S., and Evans, M. W. (1945). *J. Chem. Phys.* 13, 507.
Frank, H. S., and Franks, F. (1968). *J. Chem. Phys.* 48, 4746.
Frank, H. S., and Thompson, P. T. (1959). "The Structure of Electrolyte Solutions" (W J. Hamer, ed.), Wiley, New York, Chapt. 8.
Frank, H. S., and Quist, A. S. (1961). *J. Chem. Phys.* 34, 604.
Frank, H. S., and Robinson, A. L. (1940). *J. Chem. Phys.* 8, 933.
Frank, H. S., and Wen, W.-Y. (1957). *Disc. Faraday Soc.* 24, 133.
Franklin, J. F. (1952). *Trans. Faraday Soc.* 48, 443.
Franks, F. (1968b). *Chem. and Ind.* 560.
Franks, F. (1968a). "Hydrogen-Bonded Solvent Systems" (A. K. Covington and P. Jones, ed.), Taylor and Francis, London, p. 31.
Franks, F. (1973). "Water—A Comprehensive Treatise" (F. Franks, ed.), Plenum Press, New York, vol. 1, p. 115.
Franks, F. (1973a). "Water—A Comprehensive Treatise" (F. Franks, ed.), Plenum Press, New York, vol. 2, p. 1.
Franks, F. (1973b). "Water—A Comprehensive Treatise" (F. Franks, ed.), Plenum Press, New York, vol. 4, p. 1.
Franks, F., and Ives, D. J. G. (1966). *Quart. Rev.* 20, 1.
Franks, F., Pedley, M., and Reid, D. S. (1975). Private communication.
Franks, F., Quikenden, M. A. J., Reid, D. S., and Watson, B. (1970a). *Trans. Faraday Soc.* 66, 582.
Franks, F., Ravenhill, J. R., Egelstaff, P. A., and Page, D. I. (1970b). *Proc. Roy. Soc.* 319A, 189.
Franks, F., Ravenhill, J. R., and Reid, D. S. (1972). *J. Soln. Chem.* 1, 3.
Franks, F., and Reid, D. S. (1969). *J. Phys. Chem.* 73, 3152.
Franks, F., and Reid, D. S. (1973). "Water—A Comprehensive Treatise" (F. Franks, ed.), Plenum Press, New York, vol. 2, p. 323.
Franks, F., Reid, D. S., and Suggett, A. (1973). *J. Soln. Chem.* 2, 99.
Franks, F., and Smith, H. T. (1968). *Trans. Faraday Soc.* 64, 2962.
Franks, F., and Watson, B. (1967). *Trans. Faraday Soc.* 63, 329.
Friedman, H. L. (1960). *J. Chem. Phys.* 32, 1351.

Friedman, H. L. (1971). "Modern Aspects of Electrochemistry" (J. O'M. Bockris and B. E. Conway, ed.), Plenum Press, New York, Vol. 6, Chapter 1.
Friedman, H. L. and Krishnan, C. V. (1973a). *Ann. N.Y. Acad. Sci.* 204, 79.
Friedman, H. L., and Krishnan, C. V. (1973b). "Water—A Comprehensive Treatise" (F. Franks, ed.), Plenum Press, New York, Vol. 3, p. 1.
Friedman, H. L., and Krishnan, C. V. (1973c). *J. Soln. Chem.* 2, 119.
Friedman, H. L., and Krishnan, C. V. (1974). *J. Phys. Chem.* 78, 1927.
Friedman, H. L., Smitherman, A., and De Santis, R. (1973). *J. Soln. Chem.* 2, 59.
Friedman, M. E., and Scheraga, H. A. (1965). *J. Phys. Chem.* 69, 3795.
Fromm, J., Clementi, E., and Watts, R. O. (1975). *J. Chem. Phys.* 62, 1388.
Fuchs, R., and Hagan, C. P. (1973). *J. Phys. Chem.* 77, 1797.
Fuchs, R., Hagan, C. P., and Rodewald, R. F. (1974). *J. Phys. Chem.* 78, 1509.
Fuchs, R., Plumlee, D. S., and Rodewald, R. F. (1971). *Thermochimica Acta* 2, 515.
Gaboriaud, R. (1971). *Bull. Soc. Chim. France* 1605.
Gammell, P. M., and Angell, C. A. (1974). *J. Chem. Phys.* 60, 584.
Garnjost, von H. (1974). *Ber. Bunsenges. Phys. Chem.* 78, 1002.
Gear, C. W. (1971). "Numerical Initial Value Problems in Ordinary Differential Equations", Prentice Hall, N.J.
Ghosh, J. C. (1918). *J. Chem. Soc.* 113, 449, 707.
Giguère, P. A., and Carmichael, J. L. (1962). *J. Chem. Eng. Data* 7, 526.
Giguère, P. A., Knop, O., and Falk, M. (1958). *Canad. J. Chem.* 36, 883.
Gill, S. J., Nichols, N. F., and Wadso, I. (1975). *J. Chem. Thermodyn.* 7, 125.
Gillet, H., Avedikian, L., and Morel, J.-P. (1975). *Canad. J. Chem.* 53, 455.
Gjaldbaek, J. C., and Niemann, H. (1958). *Acta Chem. Scand.* 12, 1015.
Glasel, J. A. (1970). *J. Amer. Chem. Soc.* 92, 372.
Glasel, J. A. (1973). "Water—A Comprehensive Treatise" (F. Franks, ed.), Plenum Press, New York, Vol. 1, p. 215.
Glasstone, S., Laidler, K. J., and Eyring, H. (1941). "The Theory of Rate Processes", McGraw-Hill, New York.
Glew, D., Mak, H. D., and Rath, N. S. (1968). "Hydrogen-Bonded Solvent Systems", (A. K. Covington and P. Jones, ed.), Taylor and Francis, London, p. 195.
Glew, D. N. (1962). *J. Phys. Chem.* 66, 605.
Glew, D. N., and Moelwyn-Hughes, E. A. (1952). *Proc. Roy. Soc.* 211A, 254.
Glew, D. N., and Moelwyn-Hughes, E. A. (1953). *Disc. Faraday Soc.* 15, 150.
Glew, D. N., and Watts, H. (1971). *Canad. J. Chem.* 49, 1830.
Glew, D. N., and Watts, H. (1973). *Canad. J. Chem.* 51, 1933.
Goel, A., Murthy, A. S. N., and Rao, C. N. R. (1971). *J. Chem. Soc. (A)* 190.
Gold, V. (1969). *Adv. Phys. Org. Chem.* 7, 259.
Goldammer, E. V., and Hertz, H. G. (1970). *J. Phys. Chem.* 74, 3734.
Goldammer, E. V., and Zeidler, M. D. (1969). *Ber. Bunsenges. Phys. Chem.* 73, 4.
Golinkin, H. S., and Hyne, J. B. (1968). *Canad. J. Chem.* 46, 125.
Golinkin, H. S., Laidlaw, W. G., and Hyne, J. B. (1966). *Canad. J. Chem.* 44, 2193.
Golinkin, H. S., Lee, I., and Hyne, J. B. (1967). *J. Amer. Chem. Soc.* 89, 1307.
Good, W., Ingham, D. B., and Stone, J. (1975). *Tetrahedron* 31, 257.
Good, W., and Ingham, D. B. (1975). *Electrochimica Acta* 20, 57.
Grant, N. H., and Alburn, H. E. (1967). *Arch. Biochem. Biophys.* 118, 292.
Greyson, J., and Snell, H. (1969). *J. Phys. Chem.* 73, 3208.
Gross, G. W. (1968). A. C. S. *Advances in Chem. Series* 73, 27.

Gross, G. W., McKee, C., and Wu, C.-H. (1975). *J. Chem. Phys.* **62**, 3080.
Grunwald, E., and Butler, A. F. (1960). *J. Amer. Chem. Soc.* **82**, 5647.
Grunwald, E., and Effio, A. (1974). *J. Amer. Chem. Soc.* **96**, 423.
Grunwald, E., and Winstein, S. (1948). *J. Amer. Chem. Soc.* **70**, 846.
Guggenheim, E. A. (1937a). *Trans. Faraday Soc.* **33**, 151.
Guggenheim, E. A. (1937b). *Trans. Faraday Soc.* **33**, 607.
Gurney, R. W. (1953). "Ionic Processes in Solution", McGraw-Hill, New York.
Guthrie, J. P., and Ueda, Y. (1974). *Chem. Comm.* 991.
Hafemann, D. R., and Miller, S. L. (1969). *J. Phys. Chem.* **73**, 1392.
Halberstadt, E. S., and Prue, J. E. (1952). *J. Chem. Soc.* 2234.
Hamann, S. D., and Smith, F. (1971). *Aust. J. Chem.* **24**, 2431.
Hankins, D., Moskowitz, J., and Stillinger, F. (1970) *J. Chem. Phys.* **53**, 4544.
Hargreaves, W. A., and Kresheck, G. C. (1969). *J. Phys. Chem.* **73**, 3249.
Hasted, J. B. (1973). "Water—A Comprehensive Treatise" (F. Franks, ed.), Plenum Press, New York, vol. 1, p. 255.
Hawkins, R. E., and Davidson, D. W. (1966). *J. Phys. Chem.* **70**, 1889.
Hayduk, W., Laudie, H., and Smith, O. H. (1973). *J. Chem. Eng. Data* **18**, 373.
Hedwig, G. R., and Parker, A. J. (1974). *J. Amer. Chem. Soc.* **96**, 6589.
Heinonen, K., and Tommila, E. (1965). *Suomen Kemistilehti* **B38**, 9.
Heinzinger, K., and Vogel, P. C. (1974). *Z. Naturforsch.* **29A**, 1164.
Hepler, L. G. (1969). *Canad. J. Chem.* **47**, 4613.
Hepler, L. G. (1971). *Canad. J. Chem.* **49**, 2803.
Hepler, L. G., and O'Hara, W. F. (1961). *J. Phys. Chem.* **65**, 811.
Herskovits, T. T., and Harrington, J. P. (1972). *Biochemistry* **11**, 4800.
Herskovits, T. T., and Kelly, T. M. (1973). *J. Phys. Chem.* **77**, 381.
Hertz, H. G. (1964). *Ber. Bunsenges. Phys. Chem.* **68**, 907.
Hertz, H. G. (1970). *Angew. Chem. Int. Ed.* **9**, 174.
Hertz, H. G., Holz, M., and Mills, R. (1974). *J. Chim. Phys.* 1355.
Hertz, H. G., and Rädle, C. (1973). *Ber. Bunsenges. Phys. Chem.* **77**, 521.
Hertz, H. G., Tutsch, R., and Versmold, H. (1971). *Ber. Bunsenges. Phys. Chem.* **78**, 1177.
Hertz, H. G., and Wen, W. Y. (1974). *Z. phys. Chem. (Frankfurt)* **93**, 313.
Hibbert, F., and Long, F. A. (1972). *J. Amer. Chem. Soc.* **94**, 7637.
Hills, G. J., Ovenden, P. J., and Whitehouse, D. R. (1965). *Disc. Faraday Soc.* **39**, 207.
Hills, G., and Viana, C. A. N. (1971). *Nature* **229**, 194.
Höjer, G., and Keller, J. (1974). *J. Amer. Chem. Soc.* **96**, 3746.
Hoover, T. (1969). *J. Phys. Chem.* **73**, 57.
Hornig, D. F., White, H. F., and Reding, F. P. (1958). *Spectrochim. Acta* **12**, 338.
Horvath, A. L. (1972). *J. Chem. Documentation* **12**, 163.
Hoyland, J. R., and Kier, L. B. (1969). *Theor. Chim. Acta* **15**, 1.
Hsieh, Y. C., Inglefield, P. T., and Wen, W.-Y. (1974). *J. Soln. Chem.* **3**, 351.
Hyne, J. B. (1968). "Hydrogen-Bonded Solvent Systems" (A. K. Covington and P. Jones, ed.), Taylor and Francis, London, p. 99.
Hyne, J. B., Golinkin, H. S., and Laidlaw, W. G. (1966). *J. Amer. Chem. Soc.* **88**, 2104.
Hyne, J. B., and Wills, R. (1963). *J. Amer. Chem. Soc.* **85**, 3650.
Ingold, C. K. (1953). "Structure and Mechanism in Organic Chemistry", Bell, London, p. 347.
Ives, D. J. G., and Lemon, T. H. (1968). *R.I.C. Reviews* **1**, 62.
Ives, D. J. G., and Marsden, P. D. (1965). *J. Chem. Soc.* 649.
Ives, D. J. G., and Moseley, P. G. N. (1970). *J. Chem. Soc. (B)* 1655.

Izatt, J. R., Sakai, H., and Benedict, W. S. (1969). *J. Opt. Soc. Amer.* **59**, 19.
Jambon, C., and Philippe, R. (1975). *J. Chem. Thermodynamics* **7**, 479.
Jeffrey, G. A., and McMullan, R. K. (1967). *Prog. Inorg. Chem.* **8**, 43.
Jencks, W. P. (1969). "Catalysis in Chemistry and Enzymology", McGraw-Hill, New York, p. 393.
Johnson, D. A., and Martin, J. F. (1973). *J. C. S. Dalton* 1585.
Jolicoeur, C., and Boileau, J. (1974). *J. Soln. Chem.* **3**, 889.
Jolicoeur, C., and Friedman, H. L. (1974). *J. Soln. Chem.* **3**, 15.
Jolicoeur, C., and Philip, P. R. (1975). *J. Soln. Chem.* **4**, 3.
Jolicoeur, C., Philip, P. R., Perron, G., Leduc, P. A., and Desnoyers, J. E. (1972). *Canad. J. Chem.* **50**, 3167.
Jose, J., Philippe, R., and Clechet, P. (1975). *Canad. J. Chem. Eng.* **53**, 88.
Juillard, J., Morel, J.-P., and Avedikian, L. (1972). *J. Chim. Phys.* 787.
Jung, K. H., and Hyne, J. B. (1970). *Canad. J. Chem.* **48**, 2423.
Kalliorinne, K., and Tommila, E. (1969). *Acta. Chem. Scand.* **23**, 2567.
Kamb, B. (1965). *Science* **150**, 205.
Kamb, B. (1968). "Structural Chemistry and Molecular Biology" (A. Rich and N. Davidson, ed.), Freeman, San Francisco, p. 507.
Kamb, B., and Davis, B. L. (1964). *Proc. Nat. Acad. Sci.* **52**, 1433.
Kaminsky, M. (1957). *Disc. Faraday Soc.* **24**, 171.
Kaulgud, M. V., and Patil, K. J. (1974). *J. Phys. Chem.* **78**, 714.
Kauzmann, W. (1959). *Adv. Protein Chem.* **14**, 1.
Kawakami, T., Kubota, N., and Terui, H. (1971). *Tech. Report. Iwate Univ.* **5**, 77.
Kay, R. L. (1968). A.C.S. Advances in Chemistry Series, **73**, 1.
Kay, R. L. (1973). "Water—A Comprehensive Treatise" (F. Franks, ed.), Plenum Press, New York, Volume 3, p. 173.
Kay, R. L., and Broadwater, T. L. (1971). *Electrochimica Acta* **16**, 667.
Kebarle, P. (1974). "Modern Aspects of Electrochemistry" (B. E. Conway and J. O'M Bockris, eds.), Plenum Press, Vol. 9, p. 1.
Kell, G. S. (1967). *J. Chem. Eng. Data* **12**, 66.
Kell, G. S. (1970). *J. Chem. Eng. Data* **15**, 119.
Kell, G. S. (1973). "Water—A Comprehensive Treatise" (F. Franks, ed.), Plenum Press, New York, Vol. 1, p. 363.
Kell, G. S. (1975). *J. Chem. Eng. Data* **20**, 97.
Kenttamaa, J., and Lindberg, J. J. (1960). *Suomen Kemistilehti*, **B33**, 98.
Kenttamaa, J., Tommila, E., and Martti, E. (1959). *Ann. Acad. Sci. Fennicae*, Ser. A **11** (93), 3.
Kern, C. W., and Karplus, M. (1973). "Water—A Comprehensive Treatise" (F. Franks, ed.), Plenum Press, New York, Vol. 1, p. 21.
Kingston, B., and Symons, M. C. R. (1973). *J. C. S. Faraday II* **69**, 978.
Kistenmacher, H., Popkie, H., and Clementi, E. (1974). *J. Chem. Phys.* **60**, 799, 4455
Ko, E. C. F., and Robertson, R. E. (1972). *Canad. J. Chem.* **50**, 434.
Ko, E. C. F., and Robertson, R. E. (1972a). *J. Amer. Chem. Soc.* **94**, 573.
Ko, E. C. F., and Robertson, R. E. (1972b). *Canad. J. Chem.* **50**, 946.
Kohnstam, G. (1967). *Adv. Phys. Org. Chem.* **5**, 121.
Kollman, P. A., and Allen, L. C. (1970). *J. Amer. Chem. Soc.* **92**, 753.
Kollman, P. A., and Allen, L. C. (1972). *Chem. Rev.* **72**, 283.
Konicek, J., and Wadso, I. (1971). *Acta. Chem. Scand.* **25**, 1571.
Konneke, H.-G., Steinert, H., and Leibnitz, E. (1957). *Z. phys. Chem. (Leipzig)* **208**, 148.
Koren, R., and Perlmutter-Hayman, B. (1971). *J. Phys. Chem.* **75**, 2372.

Korson, L., Drost-Hansen, W., and Millero, F. J. (1969). *J. Phys. Chem.* 73, 34.
Koshy, K. M., and Robertson, R. E., (1974). *Canad. J. Chem.* 52, 2485.
Koshy, K. M., Robertson, R. E., and Strachan, W. M. J. (1973). *Canad. J. Chem.* 51, 2958.
Kozak, J. J., Knight, W. S., and Kauzmann, W. (1968). *J. Chem. Phys.* 48, 675.
Kresheck, G. C., Schneider, H., and Scheraga, H. A. (1965). *J. Phys. Chem.* 69, 3132.
Kretschmer, C. B., and Wiebbe, R. (1951). *J. Amer. Chem. Soc.* 73, 3778.
Krishnan, C. V., and Friedman, H. L. (1969). *J. Phys. Chem.* 73, 1572.
Krishnan, C. V., and Friedman, H. L. (1970). *J. Phys. Chem.* 74, 2356, 3900.
Krishnan, C. V., and Friedman, H. L. (1971). *J. Phys. Chem.* 75, 3598.
Krishnan, C. V., and Friedman, H. L. (1973). *J. Soln. Chem.* 2, 37.
Krishnan, C. V., and Friedman, H. L. (1974). *J. Soln. Chem.* 3, 727.
Krueger, J. H., and Johnson, W. A. (1968). *Inorg. Chem.* 7, 679.
Kundu, K. K., and Mazumdar, K. (1975). *J. C. S. Faraday I* 71, 1422.
Laidler, K. J. (1956). *Disc. Faraday Soc.* 22, 88.
Laiken, N., and Nemethy, G. (1970). *J. Phys. Chem.* 74, 3501.
Lam, S. Y., and Benoit, R. L. (1974). *Canad. J. Chem.* 52, 718.
Lama, R. F. and Lu, B. C-Y. (1965). *J. Chem. Eng. Data* 10, 216.
La Mer, V. K. (1933). *J. Chem. Phys.* 1, 289.
Lanning, J. A., Pikal, M. J., and Chambers, J. Q. (1974). *J. Phys. Chem.* 78 70.
Lannung, A., and Gjaldbaek, J. C. (1960). *Acta. Chem. Scand.* 14, 1124.
Larkin, J. A. (1975). *J. Chem. Thermodynamics* 7, 137.
Larson, J. W., and Hepler, L. G. (1969). "Solute–Solvent Interactions" (J. F. Coetzee and C. D. Ritchie, eds.), M. Dekker, New York and London, p. 1.
Lau, C. F., Wilson, P. T., and Fenby, D. V. (1970). *Aust. J. Chem.* 23, 1143.
Laughton, P. M., and Robertson, R. E. (1969). "Solute–Solvent Interactions", (J. F. Coetzee and C. D. Ritchie, eds.), M. Dekker, New York and London, p. 400.
LeDuc, P.-A., and Desnoyers, J. E. (1973). *Canad. J. Chem.* 51, 2993.
LeDuc, P.-A., Fortier, J.-L., and Desnoyers, J. E. (1974). *J. Phys. Chem.* 78, 1217.
Lee, I., and Hyne, J. B. (1968). *Canad. J. Chem.* 46, 2333.
Lee, I., and Hyne, J. B. (1969). *Canad. J. Chem.* 47, 1437.
Lee, I., and Hyne, J. B. (1970). *Canad. J. Chem.* 46, 2333.
Lee, J.-J. (1974). *J. Korean Inst. Chem. Eng.* 12, 203.
Lee, M. D. (1972). *J. Korean Inst. Chem. Eng.* 10, 289.
Lee, M. D. (1973). *J. Korean Inst. Chem. Eng.* 11, 232.
Lee, Y., and Jones, J. (1973). *J. Chem. Phys.* 59, 5845.
Leffler, J. E., and Grunwald, E. (1963). "Rates and Equilibria of Organic Reactions", Wiley, New York.
Lemberg, H. L., and Stillinger, F. H. (1975). *J. Chem. Phys.* 62, 1677.
Lentz, B. R., Hagler, A. T., and Scheraga, H. A. (1974). *J. Phys. Chem.* 78, 1531.
Levine, A. S., and Wood, R. H. (1973). *J. Phys. Chem.* 77, 2380.
Levine, S., and Perram, J. W. (1968). "Hydrogen-Bonded Solvent Systems", (A. Covington and P. Jones, eds.), Taylor and Francis, London, p. 115.
Lietzke, M. H., Stoughton, R. W., and Fuoss, R. M. (1968). *Proc. Nat. Acad. Sci.* 59, 39.
Lindberg, J. J., and Kenttamaa, J. (1960). *Suomen Kemistilehti* B33, 104.
Lindenbaum, S. (1966). *J. Phys. Chem.* 70, 814.
Lindenbaum, S. (1970). *J. Phys. Chem.* 74, 3027.
Lindenbaum, S. (1971). *J. Phys. Chem.* 75, 3733.

Linderstrøm-Lang, C. U., and Vaslow, F. (1968). *J. Phys. Chem.* 72, 2645.
Long, F. A., and McDevit, W. F. (1952). *Chem. Rev.* 51, 119.
Lowe, B. M., and Rendall, H. M. (1971). *Trans. Faraday Soc.* 67, 2318.
Lowe, B. M., and Rubienski, G. A. (1974). *Electrochimica Acta* 19, 393.
Lucas, M., and de Trobriand, A. (1971). *J. Phys. Chem.* 75, 1803.
Luck, W. A. P. (1973). "Water—A Comprehensive Treatise" (F. Franks, ed.), Plenum Press, New York, Vol. 2, p. 235.
Luck, W. A. P., and Ditter, W. (1967). *J. Mol. Struct.* 1, 339.
Luck, W. A. P., and Ditter, W. (1969). *Z. Naturforsch* 24b, 482.
Lumry, R. (1974). *Ann. N.Y. Acad. Sci.* 227, 471.
Lumry, R., and Biltonen, R. (1969). "Structure and Stability of Biological Macromolecules" (S. Timasheff and G. Fasman, eds.), M. Dekker, New York, Vol. 2, p. 65.
Lumry, R., and Rajender, S. (1970). *Biolpolymers* 9, 1125.
Lumry, R., and Rajender, S. (1971). *J. Phys. Chem.* 75, 1387.
Macdonald, D. D., and Hyne, J. B. (1970). *Canad. J. Chem.* 48, 2416.
Macdonald, D. D., and Hyne, J. B. (1970a). *Canad. J. Chem.* 48, 2494.
Macdonald, D. D., and Hyne, J. B. (1971). *Canad. J. Chem.* 49, 611, 2636.
Macdonald, D. D., Smith, M. D., and Hyne, J. B. (1971). *Canad. J. Chem.* 49, 2817.
MacDonald, J. C., and Guerrera, J. J. (1970). *J. Chem. Eng. Data* 15, 546.
Mackinnon, M. J., Lateef, A. B., and Hyne, J. B. (1970). *Canad. J. Chem.* 48, 2025.
Mak, T. C. W. (1965). *J. Chem. Phys.* 43, 2799.
Mak, T. C. W., and McMullan, R. K. (1965). *J. Chem. Phys.* 42, 2732.
Malcolm, G. N., and Rowlinson, J. S. (1957). *Trans. Faraday Soc.* 53, 921.
Masterson, W. L., Bolocofsky, D., and Lee, T. P. (1971). *J. Phys. Chem.* 75, 2809.
Mathieson, J. G., and Curthoys, G. (1975). *Aust. J. Chem.* 28, 975.
Mathieson, J. G., and Conway, B. E. (1974a). *J. C. S. Faraday I* 70, 752.
Mathieson, J. G., and Conway, B. E. (1974b). *J. Soln. Chem.* 3, 455.
Mathieson, J. G., and Conway, B. E. (1975). *J. Soln. Chem.* 4, 17.
Matous, J., Hrncirik, J., Novak, J. P., and Sobr, J. (1970). *Coll. Czech. Chem. Comm.* 35, 1904.
Matous, J., Novak, J. P., Sobr, J., and Pick, J. (1972). *Coll. Czech. Chem. Comm.* 37, 2653.
McDowell, J. M., and Vincent, C. A. (1974). *J. C. S. Faraday I* 70, 1862.
McMillan, W. G., and Mayer, J. E. (1945). *J. Chem. Phys.* 13, 276.
McMullan, R. K., Bonamico, M., and Jeffrey, G. A. (1963). *J. Chem. Phys.* 39, 3295.
McMullan, R. K., and Jeffrey, G. A. (1965). *J. Chem. Phys.* 42, 2725.
McMullan, R. K., Jeffrey, G. A., and Jordan, T. H. (1967). *J. Chem. Phys.* 47, 1229.
Maugh, T., and Bruice, T. C. (1971). *J. Amer. Chem. Soc.* 93, 6584.
Menninga, L., and Engberts, J. B. F. N. (1973). *J. Phys. Chem.* 77, 1271.
Meranda, D., and Furter, W. F. (1974). *A. I. Chem. Eng. J.* 20, 103.
Mikhailov, V. A. (1968). *J. Struct. Chem.* 9, 332.
Mikhailov, V. A., and Ponomarova, L. I. (1968). *J. Struct. Chem.* 9, 8.
Millero, F. J. (1969). *J. Phys. Chem.* 73, 2417.
Millero, F. J. (1971). *Chem. Rev.* 71, 147.
Millero, F. J., Wu, C.-H., and Hepler, L. G. (1969). *J. Phys. Chem.* 73, 2453.
Mitchell, A. G., and Wynne-Jones, W. F. K. (1953). *Disc. Faraday Soc.* 15, 161.
Moelwyn-Hughes, E. A. (1934). *Z. phys. Chem.* 26B, 281.

Moelwyn-Hughes, E. A. (1936). *Proc. Roy. Soc.* **157A**, 667.
Moelwyn-Hughes, E. A. (1938). *Proc. Roy. Soc.* **164A**, 295.
Moelwyn-Hughes, E. A. (1953). *Proc. Roy. Soc.* **220A**, 386.
Moelwyn-Hughes, E. A. (1971). "The Chemical Statics and Kinetics of Solutions", Academic Press, London, 1971, p. 113.
Moelwyn-Hughes, E. A., Robertson, R. E., and Sugamori, S. E. (1965). *J. Chem. Soc.* 1965.
Mohanty, R. K., and Ahluwalia, J. C. (1972). *J. Soln. Chem.* **1**, 531.
Mohanty, R. K., Sarma, T. S., Subramanian, S., and Ahluwalia, J. C. (1971). *Trans. Faraday Soc.* **67**, 305.
Mohanty, R. K., Sunder, S., and Ahluwalia, J. C. (1972). *J. Phys. Chem.* **76**, 2577.
Moolel, M., and Schneider, H. (1971). *Z. phys. Chem. (Frankfurt)* **74**, 237.
Morcom, K. W., and Smith, R. W. (1969). *J. Chem. Thermodynamics* **1**, 503.
Morcom, K. W., and Smith, R. W. (1970). *Trans. Faraday Soc.* **66**, 1073.
Moreau, C., and Douheret, G. (1974). *J. Chim. Phys.* **71**, 1313.
Morel, J.-P. (1966). *Bull. Soc. Chim. France* 2112.
Morel, J.-P., Fauve, J., Avedikian, L., and Juillard, J. (1974). *J. Soln. Chem.* **3**, 403.
Morel, J.-P., and Morin, J. (1970). *J. Chim. Phys.* **67**, 2018.
Morokuma, K., and Winick, J. (1970). *J. Chem. Phys.* **52**, 1301.
Murakami, S., Tanaka, R., and Fujishiro, R. (1974). *J. Soln. Chem.* **3**, 71.
Murto, J. (1961). *Suomen Kemistilehti* **B34**, 92.
Murto, J., and Hiiro, A. M. (1964). *Suomen Kemistilehti* **B37**, 177.
Nakajima, T., Komatsu, T., and Nakagawa, T. (1975). *Bull. Chem. Soc. Japan* **48**, 783, 788.
Nakanishi, K., Kato, N., and Maruyamo, M. (1967). *J. Phys. Chem.* **71**, 814.
Nakayama, H. (1972). *Bull. Chem. Soc. Japan* **45**, 1371.
Nakayama, H., and Shinoda, K. (1971). *J. Chem. Thermodynamics* **3**, 401.
Narten, A. H., Danford, M. D., and Levy, H. A. (1967). *Disc. Faraday Soc.* **43**, 97.
Narten, A. H., and Levy, H. A. (1973). "Water—A Comprehensive Treatise" (F. Franks, ed.), Plenum Press, New York, Vol. 1, p. 311.
Neggia, P., Holz, M., and Hertz, H. G. (1974). *J. Chim. Phys.* **56**.
Neilson, G. W., Howe, R. A., and Enderby, J. E. (1975). *Chem. Phys. Lett.* **33**, 284.
Nelson, H. D., and de Ligny, C. L. (1968). *Rec. Trav. Chim. Pays Bas* **87**, 528, 623.
Nemethy, G., and Scheraga, H. A. (1962a). *J. Chem. Phys.* **36**, 3382.
Nemethy, G., and Scheraga, H. A. (1962b). *J. Chem. Phys.* **36**, 3401.
Nemethy, G., and Scheraga, H. A. (1962c). *J. Phys. Chem.* **66**, 1773.
Nemethy, G., and Scheraga, H. A. (1964). *J. Chem. Phys.* **41**, 680.
Nichols, N., and Wadso, I. (1975). *J. Chem. Thermodynamics* **7**, 329.
Oakenfull, D. G. (1971). *Aust. J. Chem.* **24**, 2547.
O'Connell, J. P. (1971). *A. I. Chem. Eng. J.* **17**, 658.
Onsager, L., and Runnels, L. K. (1969). *J. Chem. Phys.* **50**, 1089.
Orr, J. B., Goates, J. R., and Lamb, J. D. (1972). *J. Chem. Thermodynamics* **4**, 123.
Packer, K. J., and Tomlinson, D. J. (1970). *Trans. Faraday Soc.* **66**, 1302.
Page, D. I. (1973). "Water—A Comprehensive Treatise" (F. Franks, ed.), Plenum Press, New York, Vol. 1, p. 333.
Parker, A. J. (1969). *Chem. Rev.* **69**, 1.
Pauling, L. (1959). "Hydrogen Bonding" (D. Hadzi and H. W. Thompson, eds.), Pergamon, New York, p. 1.

Pauling, L. (1960). "The Nature of the Chemical Bond", Cornell Univ. Press, Ithaca, 3rd. edn.
Perlmutter-Hayman, B. (1971). *Prog. Reaction Kinetics* 6, 239.
Peron, J.-J., Bourderon, C., and Sandorfy, C. (1971). *Canad. J. Chem.* 49, 3901.
Perram, J. W., and Levine, S. (1974). *Adv. Mol. Relax. Proc.* 6, 85.
Perron, G., and Desnoyers, J. E. (1972). *J. Soln. Chem.* 1, 537.
Peterson, S. W., and Levy, H. A. (1957). *Acta Crystallog.* 10, 70.
Phibbs, M. K., and Giguère, P. A. (1951). *Canad. J. Chem.* 36, 883.
Philip, P. R., Perron, G., and Desnoyers, J. E. (1974). *Canad. J. Chem.* 52, 1709.
Philippe, R., and Jambon, C. (1974). *J. Chim. Phys.* 1041.
Picker, P., Leduc, P.-A., Philip, P. R., and Desnoyers, J. E. (1971). *J. Chem. Thermodyn.* 3, 631.
Pierrotti, R. A. (1965). *J. Phys. Chem.* 69, 281.
Pinder, K. L. (1973). *J. Chem. Eng. Data* 18, 275.
Pitzer, K. S., and Kim, J. J. (1974). *J. Amer. Chem. Soc.* 96, 5701.
Pointud, Y., Juillard, J., Avedikian, L., Morel, J.-P., and Ducros, M. (1974a). *Thermochimica Acta* 8, 423.
Pointud, Y., Juillard, J., Morel, J.-P., and Avedikian, L. (1974b). *Electrochimica Acta* 19, 229.
Polanyi, M., and Ogg, R. (1935). *Trans. Faraday Soc.* 31, 604.
Pople, J. A. (1950). *Proc. Roy. Soc.* 202A, 323.
Powles, J. G. (1974). *Contemp. Phys.* 15, 409.
Pratt, K. C., and Wakeham, W. A. (1975). *Proc. Roy. Soc.* 342A, 401.
Prigogine, I., and Defay, R. (1954). "Chemical Thermodynamics", trans. D. H. Everett, Longmans, London.
Queen, A. (1967). *Canad. J. Chem.* 45, 1619.
Queen, A., and Robertson, R. E. (1966). *J. Amer. Chem. Soc.* 88, 1363.
Rao, C. N. R. (1973). "Water—A Comprehensive Treatise" (F. Franks, ed.), Plenum Press, New York, Vol. 1, p. 93.
Rahman, A., and Stillinger, F. H. (1971). *J. Chem. Phys.* 55, 3336.
Rahman, A., and Stillinger, F. H. (1973). *J. Amer. Chem. Soc.* 95, 7943.
Rahman, A., and Stillinger, F. H. (1974). *Phys. Rev.* 10A, 368.
Ramanthan, P. S., and Friedman, H. L. (1971). *J. Chem. Phys.* 54, 1086.
Ramanthan, P. S., Krishnan, C. V., and Friedman, H. L. (1972). *J. Soln. Chem.* 1, 237.
Raridon, R. J., and Kraus, K. A. (1965). *J. Coll. Sci.* 20, 1000.
Rasaiah, J. C. (1970). *J. Chem. Phys.* 52, 704.
Rasaiah, J. C. (1973). *J. Soln. Chem.* 2, 301.
Rasmussen, D. H., and MacKenzie, A. P. (1968). *Nature* 220, 1315.
Rat, J. C., Villermaux, S., and Delpuech, J. J. (1974). *Bull. Soc. Chim. France* 815.
Ray, A., and Nemethy, G. (1973). *J. Chem. Eng. Data* 18, 309.
Rehfeld, S. J. (1973). *J. Amer. Chem. Soc.* 95, 4489.
Reichardt, C. (1965). *Angew. Chem. Int. Ed.* 4, 29.
Reid, D. S., Quickenden, M. A. J., and Franks, F. (1969). *Nature* 224, 1293.
Renard, E., and Justice, J.-C. (1974). *J. Soln. Chem.* 3, 633.
Renard, J. A., and Heichelheim, H. R. (1967). *J. Chem. Eng. Data* 12, 33.
Robertson, R. E. (1960). *Suomen Kemistilehti*, 33A, 63.
Robertson, R. E. (1967). *Prog. Phys. Org. Chem.* 4, 213.
Robertson, R. E., Heppolette, R. L., and Scott, J. M. W. (1959). *Canad. J. Chem.* 37, 803.
Robertson, R. E., and Sugamori, S. E. (1969). *J. Amer. Chem. Soc.* 91, 7254.
Robertson, R. E., and Sugamori, S. E. (1972). *Canad. J. Chem.* 50, 1353.

Rochester, C. H. (1972). *J. C. S. Dalton* 5.
Rochester, C. H., and Symonds, J. R. (1973). *J. C. S. Faraday I* 69, 1577.
Rochester, C. H., and Symonds, J. R. (1974). *J. Fluorine Chem.* 4, 141.
Rodante, F., Rallo, F., and Fiordiponti, P. (1974). *Thermochimica Acta* 9, 269.
Rorabacher, D. B., MacKellar, W. J., Shu, F. R., and Bonavita, M. (1971). *Ann. Chem.* 43, 561.
Rosso, J.-C., and Carbonnel, L. (1968). *C.R. Acad. Sci. Ser. C* 267, 1.
Rosso, J.-C., and Carbonnel, L. (1971). *C.R. Acad. Sci. Ser. C* 272, 713.
Rosso, J.-C., and Carbonnel, L. (1972). *C.R. Acad. Sci. Ser. C* 274, 1108.
Rosso, J.-C., and Carbonnel, L. (1973). *C.R. Acad. Sci. Ser. C* 276, 1389.
Rowlinson, J. S. (1951). *Trans. Faraday Soc.* 47, 120.
Rowlinson, J. S. (1969). "Liquids and Liquid Mixtures", Butterworths, London, 2nd. edn.
Rowlinson, J. S. (1974). *Pure Appl. Chem.* 38, 495.
Roy, R. N., and Bothwell, A. L. M. (1971). *J. Chem. Eng. Data* 16, 347.
Sada, E., Ohno, T., and Kito, S. (1972). *J. Chem. Eng. Japan* 5, 215.
Samoilov, O. Ya. (1965). "Structure of Aqueous Electrolyte Solutions and the Hydration of Ions", trans. D. J. G. Ives, Consultants Bureau, New York.
Safford, G. J., and Leung, P. S. (1973). "Electrochemistry" (E. Yeager and A. J. Salkind, eds.), Wiley, New York, Vol. 2, p. 173.
Sarma, T. S., and Ahluwalia, J. C. (1973). *Chem. Soc. Rev.* 2, 203.
Sarma, T. S., Mohanty, R. K., and Ahluwalia, J. C. (1969). *Trans. Faraday Soc.* 65, 2333.
Sarnowski, M., and Baranowski, B. (1965). *Z. phys. Chem. (Leipzig)* 228, 310.
Scatchard, G. (1949). *Chem. Rev.* 44, 7.
Scatchard, G., Kavanagh, G. M., and Ticknor, L. B. (1952). *J. Amer. Chem. Soc.* 74, 3715.
Schichman, S. A., and Amey, R. L. (1971). *J. Phys. Chem.* 75, 98.
Schiffer, J., and Hornig, D. F. (1968). *J. Chem. Phys.* 49, 4150.
Schneider, G. (1966). *Ber. Bunsenges. Phys. Chem.* 70, 497.
Schneider, G. (1973). "Water—A Comprehensive Treatise" (F. Franks, ed.), Plenum Press, New York, Vol. 2, p. 381.
Schukarev, S. A., and Tolmacheva, T. A. (1968). *Zh. Strukt. Khim.* 9, 221.
Scott, J. M. W., and Robertson, R. E. (1972). *Canad. J. Chem.* 50, 167.
Senior, W. A., and Verrall, R. E. (1969). *J. Phys. Chem.* 73, 4242.
Shipman, L. L., Owicki, J. C., and Scheraga, H. A. (1974). *J. Phys. Chem.* 78, 2055.
Stein, A., and Allen, G. F. (1973). *J. Chem. Phys.* 59, 6079.
Stein, A., Allen, G. F. (1974). *J. Phys. Chem. Data* 2, 443.
Stern, J. H., and Bottenberg, W. R. (1971). *J. Phys. Chem.* 75, 2229.
Stern, J. H., and Hansen, S. L. (1971). *J. Chem. Eng. Data* 16, 360.
Stern, J. H., and Hermann, A. (1967). *J. Phys. Chem.* 71, 309.
Stern, J. H., and Hermann, A. (1968). *J. Phys. Chem.* 72, 364.
Stern, J. H., and Nobilione, J. M. (1968a). *J. Phys. Chem.* 72, 3937.
Stern, J. H., and Nobilione, J. M. (1968b). *J. Phys. Chem.* 72, 1064.
Stern, J. H., and O'Connor, M. E. (1972). *J. Phys. Chem.* 76, 3077.
Stern, J. H., Sandstrom, J. P., and Hermann, A. (1967). *J. Phys. Chem.* 71, 3623.
Stern, K. H., and Amis, E. S. (1959). *Chem. Rev.* 59, 1.
Stillinger, F. H. (1970). *J. Phys. Chem.* 74, 3677.
Stillinger, F. H. (1973). *J. Soln. Chem.* 2, 141.
Stillinger, F. H., and Ben-Naim, A. (1969). *J. Phys. Chem.* 73, 900.
Stillinger, F. H., and Lemberg, H. L. (1975). *J. Chem. Phys.* 62, 1340.

Stillinger, F. H., and Rahman, F. H. (1972). *J. Chem. Phys.* 57, 1281.
Stillinger, F. H., and Rahman, A. (1974). *J. Chem. Phys.* 60, 1545.
Stimson, E. R., and Schrier, E. E. (1974). *J. Chem. Eng. Data* 19, 354.
Streitweiser, A. (1956). *Chem. Rev.* 56, 571.
Stokes, R. H. (1967). *Aust. J. Chem.* 20, 2087.
Suggett, A., Jr. (1973). "Water—A Comprehensive Treatise" (F. Franks, ed.), Plenum Press, New York, Vol. 4, Chapt. 6.
Sunder, S., Chawla, B., and Ahluwalia, J. C. (1974). *J. Phys. Chem.* 79, 738.
Symons, E. A. (1971). *Canad. J. Chem.* 49, 3940.
Symons, M. C. R. (1975). *Proc. Roy. Soc.* 272B, 13.
Szent-Gyorgi, A. (1957). "Bioenergetics", Academic Press, New York, p. 39.
Tait, M. J., Suggett, A., Franks, F., Ablett, S., and Quikenden, P. A. (1973). *J. Soln. Chem.* 1, 131.
Tamaki, K., Isomura, Y., and Ohara, Y. (1972). *Bull. Chem. Soc. Japan* 45, 2939.
Tamaki, K., Ohara, Y., and Isomura, Y. (1973). *Bull. Chem. Soc. Japan* 46, 1551.
Thomson, A. (1970). *J. Chem. Soc. (B)* 1198.
Thuaire, R. (1972). *J. Chim. Phys.* 30.
Tiepel, E. W., and Gubbins, K. E. (1972). *Canad. J. Chem. Eng.* 50, 361.
Timimi, B. A. (1974). *Electrochim. Acta* 19, 149.
Tödheide, K. (1973). "Water—A Comprehensive Treatise" (F. Franks, ed.), Plenum Press, New York, Vol. 1, p. 463.
Tommila, E. (1952). *Suomen Kemistilehti, Ser. B.* No. 9.
Tommila, E. (1964). *Suomen Kemistilehti* B37, 117.
Tommila, E. (1966). *Acta. Chem. Scand.* 20, 923.
Tommila, E. (1967). *Ann. Acad. Sci. Fenn. A. II, Chemica*, 139.
Tommila, E., and Belinskij, I. (1969). *Suomen Kemistilehti* B42, 185.
Tommila, E., Koivisto, A., Lyrra, J. P., Antell, K., and Heimo, S. (1952). *Ann. Acad. Sci. Fennicae, Ser. A. II. Chemica*, No. 47.
Tommila, E., Lindell, E., Virtalaine, M.-L., and Laakso, R. (1969). *Suomen Kemistilehti* B42, 95.
Tommila, E., and Maltamo, S. (1955). *Suomen Kemistilehti* B28, 118.
Tommila, E., and Murto, M. (1963). *Acta. Chem. Scand.* 17, 1957.
Tommila, E., and Pajunen, A. (1968). *Suomen Kemistilehti* B41, 172.
Tommila, E., and Palenius, I. (1963). *Acta. Chem. Scand.* 17, 1980.
Tommila, E., and Pitkänen, I. P. (1966). *Acta. Chem. Scand.* 20, 937.
Tommila, E., Tiilikainen, M., and Voipo, A. (1955). *Ann. Acad. Sci. Fennicae, Ser. A, II Chemica*, No. 65.
Tonnet, M. L., and Hambly, A. N. (1971). *Aust. J. Chem.* 24, 703.
Tourky, A. R., and Abdel-Hamid, A. A. (1971). *Z. phys. Chem. (Leipzig)* 248, 9.
Tourky, A. R., Mikhail, S. Z., and Abdel-Hamid, A. A. (1973). *Z. phys. Chem. (Leipzig)* 252, 289.
Treiner, C. (1973). *J. Chim. Phys.* 1183.
Treiner, C., Bocquet, J. F., and Chemla, M. (1973). *J. Chim. Phys.* 72.
Trimble, H. M., and Potts, W. (1935). *Ind. Eng. Chem.* 27, 66.
Van de Poel, W. (1971). *Bull. Soc. Chim. Belges* 80, 401.
Van de Poel, W., and Slootmaekers, P. J. (1970). *Bull. Soc. Chim. Belges* 79, 223.
Vaslow, F. (1969). *J. Phys. Chem.* 73, 3745.
Vaslow, F. (1971). *J. Phys. Chem.* 75, 3317.
Vaslow, F., and Doherty, D. G. (1953). *J. Amer. Chem. Soc.* 75, 928.
Vierk, A. L. (1950). *Z. anorg. Chem.* 261, 283.

Villermaux, S., Baudot, V., and Delpuech, J. J. (1972). *Bull. Soc. Chim. France*, 1781.
Vuks, M. F., and Shurupova, L. V. (1972). *Optics Comm.* 5, 277.
Walrafen, G. (1966). *J. Chem. Phys.* 44, 3726.
Walrafen, G. (1972). *Adv. Mol. Relax. Proc.* 3 43.
Walrafen, G. (1973). "Water—A Comprehensive Treatise" (F. Franks, ed.), Plenum Press, New York, Volume 1, p. 151.
Wang, J. H. (1965). *J. Phys. Chem.* 69, 4412.
Warner, D. T. (1965). *Ann. N.Y. Acad. Sci.* 125, 605.
Watts, R. O., Clementi, E., and Fromm, J. (1974). *J. Chem. Phys.* 61, 2550.
Wauchope, R. D., and Haque, R. (1972). *Canad. J. Chem.* 50, 133.
Wells, C. F. (1973). *J. C. S. Faraday I* 69, 984.
Wells, C. F. (1974). *J. C. S. Faraday I* 70, 694.
Wells, C. F. (1975). *J. C. S. Faraday I* 71, 1868.
Wells, P. R. (1963). *Chem. Rev.* 63, 171.
Wen, W.-Y. (1973). *J. Soln. Chem.* 2, 253.
Wen, W.-Y., and Chen, C. L. (1969). *J. Phys. Chem.* 73, 2895.
Wen, W.-Y., and Hertz, H. G. (1972). *J. Soln. Chem.* 1, 17.
Wen, W.-Y., and Hung, J. H., (1970). *J. Phys. Chem.* 74, 170.
Wen, W.-Y., and Saito, S. (1964). *J. Phys. Chem.* 68, 2639.
Wen, W.-Y., Takeguchi, N., and Wilson, D. P. (1974). *J. Soln. Chem.* 3, 103.
Weres, O., and Rice, S. A. (1972). *J. Amer. Chem. Soc.* 94, 8083.
Wetlaufer, D. B., Malik, S. K., Stoller, L., and Coffin, R. I. (1964). *J. Amer. Chem. Soc.* 86, 509.
Whalley, E. (1966). *Ber. Bunsenges. Phys. Chem.* 70, 958.
Whalley, E., Davidson, D. W., and Heath, J. B. R. (1966). *J. Chem. Phys.* 45, 3976.
Wheeler, J. C. (1972). *Ber. Bunsenges. Phys. Chem.* 76, 308.
Wheeler, J. C. (1975). *J. Chem. Phys.* 62, 433.
Wilhelm, E., and Battino, R. (1973). *Chem. Rev.* 73, 1.
Wilson, G. J., Chan, R. K., Davidson, D. W., and Whalley, E. (1965). *J. Chem. Phys.* 43, 2384.
Winstein, S., and Fainberg, A. H. (1957). *J. Amer. Chem. Soc.* 79, 5937.
Wirth, H. E., and Lo Surdo, A. (1968). *J. Phys. Chem.* 72, 751.
Wold, S. (1970). *Acta. Chem. Scand.* 24, 2321.
Wold, S. (1972). *J. Phys. Chem.* 76, 369.
Wold, S., and Exner, O. (1973). *Chimica Scripta* 3, 5.
Wolford, R. K. (1964). *J. Phys. Chem.* 68, 3392.
Worley, J. D., and Klotz, I. M. (1966). *J. Chem. Phys.* 45, 2868.
Wu, Y.-C., and Friedman, H. L. (1966). *J. Phys. Chem.* 70, 166.
Yaacobi, M., and Ben-Naim, A. (1973). *J. Soln. Chem.* 2, 425.
Zana, R., and Yeager, E. (1966). *J. Phys. Chem.* 70, 954.
Zeidler, M. D. (1973). "Water—A Comprehensive Treatise" (F. Franks, ed.), Plenum Press, New York, Vol. 2, p. 529.

Author Index

Numbers in italics refer to the pages on which references are listed at the end of each article.

A

Aaron, J. J., 119, *128*
Abdel-Hamid, A. A., 314, *351*
Abidaud, A., 143, *200*
Ablett, S., 261, *351*
Abraham, F. F., 236, *336*
Abraham, M. H., 28, 61, *62*, 137, 141, 162, *196*, 288, 316, *336*
Abraham, R. J., 61, *63*
Adams, P. A., 231, 234, 258, *336*
Adsetts, J. R., 191, *196*
Afanas'ev, I. B., 123, *127*
Ahluwalia, J. C., 264, 266, 267, 312, 313, *336*, *339*, *348*, *350*, *351*
Albagli, A., 143, 146, 148, 169, *196*
Albery, W. J., 84, 85, *127*, 212, 258, *336*
Alburn, H. E., 225, *344*
Alcais, P., 118, *128*
Alexander, D. M., 251, 252, *336*
Alexander, R., 76, *127*, 162, *196*
Alfenaar, M., 309, *340*
Allen, A. T., 262, *336*
Allen, G. F., 297, *350*
Allen, K. W., *336*
Allen, L. C., 220, 221, 222, *345*
Allerhand, A., 135, *196*
Allnatt, A. R., 245, *336*
Alunni, S., 187, *196*
Amberger, E., 190, *202*
Amey, R. L., 326, *350*
Amis, E. S., 135, *196*, 206, 218, *336*, *350*
Anbar, M., 191, *199*
Anderson, H. G., 301, *336*
Anderson, H. L., 243, 275, *336*
Ando, T., 37, *67*
Andrews, A. L., 288, 309, *336*
Andrews, L-J., 39, *63*
Ang, K. P., 335, *336*
Angell, C. A., 235, 297, *336*, *343*
Antell, K., 323, *351*

Apeloig, Y., 77, *131*
Appel, B., 99, *132*
Arel, M., M., 267, *340*
Armitage, D. A., 333, *336*
Arnaud, R., 293, 297, 315, *336*
Arnett, E. M., 58, 61, *62*, 136, 147, 148, *196*, 210, 251, 275, 306, 307, 312, 313, 317, 318, *336*, *337*
Assarsson, P., 298, *337*
Atkinson, G., 300, *337*
Avedikian, L., 293, 294, 297, 301, 307, 308, 312, 314, 315, 316, 329, *336*, *337*, *343*, *345*, *348*, *349*
Ayediran, D., 177, 178, 179, *196*
Aveyard, R., 275, *337*

B

Bach, R. D., 43, *62*
Baciocchi, E., 187, *196*
Badger, R. C., 23, *63*
Baer, S., 303, *338*
Bahe, L. W., 241, *337*
Bailey, T. H., 107, 108, 109, *127*
Baker, A. D., 50, *67*
Baker, C., 50, *67*
Baker, R., 99, *132*
Balakrishnan, M., 163, 164, 165, 166, *196*
Bale, C. W., 281, *337*
Bale, H. D., 302, *337*
Baliga, B. T., 322, *337*
Bamkole, T. O., 177, 178, 179, *196*
Baranowski, B., 261, *350*
Bares, J. E., 144, 145, 146, 191, *200*
Barker, J. A., 281, *337*
Barnett, R., 89, *127*
Barone, G., 260, 261, 313, *337*
Bartmess, J. E., 144, 145, 146, 147, 191, *197*, *200*
Bartsch, R. A., 185, 186, 187, 188, *196*
Basch, H., 50, *62*

Bateman, L. C., 279, *337*
Bates, R. G., 308, *337*
Battino, R., 216, 281, *337, 352*
Baudot, V., 310, *352*
Baugham, G., *199*
Baughman, E. H., 148, *200*
Baumgartner, E. K., 300, *337*
Bax, d., 309, *337*
Becker, A., 98, 99, 100, *129*
Belinskij, I., 335, *351*
Bell, H. M., 102, *128*
Bell, R. P., 72, 82, 83, 85, 86, 88, 89, 94, 95, *127, 128*, 150, 151, 152, 153, 154, *196*, 324, *337*
Belloli, R., 9, 20, 31, *65*
Bender, M. L., 150, *196*
Benedict, W. S., 219, *345*
Benjamin, L., 290, *337*
Ben-Naim, A., 235, 236, 239, 254, 255, 273, 303, 305, 306, 327, *337, 338, 350, 352*
Bennetto, H. P., 206, 210, 241, 263, 277, 288, 309, 310, *336, 338, 339*
Benoit, R. L., 148, 149, 195, *200*, 325, 334, *346, 388*
Benson, G. C., 290, *337*
Benson, S. W., 122, *128*
Benter, G., 334, *338*
Bentley, M. O., 47, *62*
Bentley, T. W., 8, 9, 10, 11, 12, 15, 19, 21, 24, 27, 30, 34, 35, 37, 38, 39, 45, 46, 47, 48, 51, 53, 54, 55, 57, 58, 59, *62, 66*, 98, *128*
Bentrude, W. G., 62, *62*, 136, *196*, 306, 307, 317, 318, *337*
Berendson, H. J. C., 216, 263, *338*
Berger, J. E., 39, *63*
Bernal, J. D., 207, *338*
Bernasconi, C. F., 175, 177, 179, 180, *196*
Berrigan, P. J., 189, *200*
Berson, J. A., 135, *196*
Bertie, J. E., 223, 224, *338*
Bertini, F., 125, 126, *130*
Bertrand, G. L., 290, 295, *338*
Bethell, D., 32, 77, *62, 128*
Bett, K. E., 230, *338*
Beurskens, P. T., 229, *338*
Bezboruah, C. P., 212, *338*
Bhowmik, S., 148, *200*
Bienvenue, A., 135, *198*
Biltonen, R., 247, *347*
Bingham, R. C., 3, 5, 8, 9, 12, 34, *63, 65, 66*, 110, *131*
Blanchard, E. P., 116, *131*
Blandamer, M. J., 210, 218, 231, 233, 234, 243, 247, 257, 263, 269, 280, 289, 293, 300, 301, 302, 331, 333, *336, 338, 339*
Bloom, A., 162, *199*
Blyth, C. A., 259, *339*
Bockris, J. O.' M., 218, *339*
Bocquet, J. F., 281, 291, *351*

Boettger, G., 232, *339*
Bohlmann, F., 39, 41, 42, 44, *63*, 135, *198*
Bohme, H., 44, *62*
Boileau, J., 270, *345*
Bolocofsky, D., 272, 275, *347*
Bolton, P. D., 212, *339*
Bonamico, M., 229, *339, 347*
Bonavita, M. 315, *350*
Bone, J. A., 6, 13, 17, *62*
Bopp, R. J., 31, *65*
Borchardt, J. K., 184, 185, *196*
Borčić, S., 23, 25, *64, 66*, 102, *130*
Bordwell, F. G., 28, 30, *62*, 93, *128*, 144, 145, 146, 147, 152, 156, 191, *196, 197, 200*
Bose, K., 309, *339*
Bothwell, A. L. M., 308, *350*
Bottenberg, W. R., 330, *350*
Bourderon, C., 232, *349*
Bowden, K., 145, 147, 149, 160, 167, 168, 169, 170, 171, *197*
Bowen, C. 11, *62*
Bowen, D. E., 194, *197*
Bown, D. E., 39, *67*
Boyle, W. J. Jr. 93, *128*, 152, 156, *197*
Bradley, W., 28, *66*
Brandts, J. F., 305, *339*
Brauman, J. I., 146, *197, 201*
Brink, G., 326, *339*
Broadwater, T. L., 314, *339, 345*
Brokaw, M. L., 88, *131*
Brønsted, J. N., 79, 83, 88, *128*, 150, 151, *197*
Brooks, C. A. G., 247, *339*
Brower, K. R., 277, 327, *339*
Brown, C. A., 187, *197*
Brown, D. A., 44, *62*
Brown, F. J., 23, 24, 28, *65*
Brown, H. C., 4, 9, 19, *62, 66*, 102, 117, *128, 131, 197*
Brownstein, S., 135, *197*
Broxton, T. J., 76, *127*, 162, *196*
Bruice, T. C., 166, *197, 198*, 259, *347*
Brundle, C. R., 50, *62, 67*
Bryson, J. A., 146, *197*
Buck, P., 175, *197*
Buckley, A., 143, 148, *196*
Buijs, K., 232, *339*
Buisson, D. H., 62, *63*
Buncel, E., 146, 147, 150, 171, 172, 173, 175, 181, 189, *197, 201*
Bunnett, J. F., 45, *62*, 148, 149, 174, 176, 179, 182, *200*
Bunton, C. A., 279, *339*
Burfoot, G. D., *197*
Burgess, J., 210, 280, 289, *338*
Burke, J. J., 136, *196*, 306, 307, 317, 318, *337*
Burnett, J. F., *197*
Bushaw, B. A., 185, 186, 187. *196*
Buswell, R. L., 185, 186, *196*
Butler, A. F., 273, *344*

AUTHOR INDEX

C

Cabani, S., 252, 253, 298, *339*
Caldin, E. F., 83, *128*, 151, 154, 159, *197*, 206, 210, 277, *338*, *339*
Camoes, G. F. C., 212, *338*
Campbell, N. C. G., 16, *62*
Campbell-Crawford, A. N., 85, *127*
Campion, J. J., 334, *339*
Cappi, J. B., 230, *338*
Caputo, A., 166, *198*
Caranoni, 298, *339*
Carbonnel, L., 291, 298, *339*, *350*
Cargill, R. W., 303, 305, *339*
Carmichael, J. L., 325, *343*
Carter, J. V., 28, 58, *62*, *66*, 96, 107, *131*, 251, *337*
Casey, M. L., 70, 76, 91, *128*, *129*
Cassel, R. B., 255, 260, 275, *339*
Catherall, N. F., 334, *339*
Caveng, P., 190, *197*
Cayley, G. R., 258, *339*
Challis, B. C., 95, *128*
Chambers, J. Q., 314, *346*
Chan, R. K., 224, *352*
Chang, C. J., 169, *201*
Chantooni, M. K. Jr., 148, *199*, *200*
Chawla, B., 264, 267, *336*, *339*, *351*
Chaykovsky, M., 146, *198*
Chemla, M., 291, *351*
Chen, C. L., 261, 281, *352*
Chen, H. J., 86, 88, *130*
Chevli, D. M., 31, *65*
Chiang, Y., 86, 88, 94, 95, *130*, 191, *200*
Cholod, M. S., 112, 114, 115, *131*
Choppin, G. R., 232, *339*
Choux, G., 334, *338*
Chrisment, J., 146, *197*
Chu, W. K. C., 88, *131*
Church, M. G., 279, *337*
Cimino, G. M., 149, *200*
Ciuffarin, E., *201*
Clark, D. C., 98, 99, 100, *129*
Clarke, E. C. W., 249, *339*
Clarke, G. A., 136, 146, *197*, 212, 279, *339*
Claussen, W. F., 226. *339*
Clayman, L., 161, *199*
Clechet, P., 327, *345*
Clement, R. A., 60, *62*
Clementi, E., 236, 265, *343*, *345*, *352*
Clementi, S., 76, 117, 119, 120, *128*
Cleve, N. J., 319, 331, 335, *339*
Clever, H. L., 216, 273, 325, *337*, *339*
Clippinger, E., 99, *132*
Closs, G. L., 113, *128*
Cocivera, M., 36, *62*
Cockerill, A. F., 145, 147, 169, 182, *197*, *198*, *201*

Coetzee, J. F., 134 , *197*, 334, *339*
Coffin, R. I., 257, 260, *352*
Cohen, A., *197*
Coles, M. J., 234, *339*
Collins, C. J., 23, 24, 28, *65*
Constein, V. G., 189, *198*
Conti, G. 252, 253, 298, *339*
Conway, B. E., 218, 238, 239, 241, 260, 267, 272, *339*, *340*, *341*, *347*
Cook, D., 161, *197*
Cook, D. M., 185, 186, *196*
Cook, R. S., 160, 169, 170, 171, *197*
Cooper, J. S., 161, *199*
Cooper, T. K., 191, *198*
Copp, J. L., 284, 292, 296, *340*
Cordes, E. H., 79, 89, *128*
Corey, E. J., 146, *198*
Cornforth, F. J., 144, 145, 146, 191, *200*
Covington, A. K., 212, 219, 329, 330, 335, *338*, *340*
Cowie, J. M. G., 326, *340*
Cox, B. G., 137, 138, 139, 140, 142, 151, 152, 153, 154, 155, *196*, *198*, 219, 266, 288, 335, *340*
Cox, D., *128*
Cox, R. A., 144, 145, 148, 149, 167, *198*
Cram, D. J., 4, *64*, 87, *128*, 150, 154, 168, 169, 182, *198*, *199*, *202*
Crampton, M. R., 175, 181, *198*
Crescenzi, V. 261, 298, 313, *337*, *340*
Criss, C. M., 310, *340*
Critchlow, J. E., 324, *337*
Crooks, J. E., 85, *128*
Cunningham, G. P. 333, *340*
Curran, E. L., 31, *65*
Curran, J. S. 85, *127*, 258, *336*
Curry, R. W., 225, *341*
Curthoys, G., 264, *347*
Cvetanovic, R. J., 70, *128*

D

Dack, M. R. J., 32, *62*
Dafforn, G. A., 10, 26, *66*
Dais, P. J., 27, 33, *62*
Damrauer, R., 113, *131*
Danford, M. D., 230, *348*
Dannenburg, J. J., 49, *62*
D'Aprano, A., 335, *340*
Darling, B. T., 219, *340*
Da Roza, D. A., 39, *63*
Darwish, D., 32, 33, *63*, *67*
Das, A. K., 309, 329, *340*
Davidson, D. W., 224, 225, 227, *340*, *344*, *352*
Davies, J., *340*
Davis, B. L., 225, *345*
Davis, C. M., 234, *340*
Davis, K. M. C., 247, *339*

Dawe, R. A., 291, *340*
Dawson, L. R., 39, *63*
Dayton, J. C., *202*
De Boer, Th. J., 43, *66*
Dedieu, A., 30, *63*
Defay, R., 214, *349*
de la Mare, P. B. D., 32, *63*, 117, *128*
Del Bene, J. E., 221, 222, *340*
De Ligny, C. L., 249, 309, *337*, *340*, *348*
Delpuech, J. J., 146, 156, 160, 193, *197*, *198*, *202*, 310, *349*, *352*
De More, W. B., 122, *128*
Denison, D. M., 219, *340*
De Nooijer, B., 252, *341*
De Santis, R., 276, *343*
Desnoyers, J. E., 217, 218, 241, 242, 260, 265, 266, 267, 268, 269, 270, 275, 294, 301, *337*, *340*, *341*, *342*, *345*, *346*, *349*
Desrosies, N., 260, *340*
De Trobriano, A., 274, *347*
Devine, W., 261, *341*
De Visser, C., 266, 312, *341*
Dewar, M. J. S., 50, 58, *63*, 72, 73, 74, *128*
Deyrup, A. J., *199*
Diamond, R. M., 271, *341*
Diaz, A. F., 4, 21, 37, *63*, *65*, 99, *132*
Dickason, W. C., 9, 28, *66*
Dickason, W. R., *66*
Dickson, S. J., 214, 215, 321, *341*
Dillard, D. E., 31, *65*
Dillon, R. L., 88, *130*
Dimroth, K., 32, 39, 40, 41, 42, 43, 44, *63*, *65*, 135, *198*
Ditter, W., 233, *347*
Dittmer, D. C., *67*
Do Amaral, L., 89, *128*
Dobson, H. J. E., 290, *341*
Dobson, J. V., 212, *338*
Doering, W. von E., 6, 30, *63*, 116, *128*
Doherty, D. G., 247, *341*, *351*
Dollet, N., 313, *341*
Dolman, D., 147, 152, 168, 182, *197*, *198*
Donato, I. D., 335, *340*
Dorsey, N. J., 207, *341*
Doucet, J. P., 118, *128*
Dougherty, R. C., 30, 49, *63*
Douheret, G., 335, *348*
Dowd, W., 18, 19, 26, 35, 36, 39, *66*, 110, *131*
Drost-Hansen, W., 225, 230, 237, *341*, *346*
Drucker, G. E., 144, 145, 146, 147, 191, *197*, *200*
Dubois, J. E., 119, *128*, 135, *198*
Ducros, M., 307, 308, *337*, *349*
Duggleby, P. McC., 61, *62*, 136, *196*, 306, 307, 317, 318, *337*
Dukes, M. D., 36, 47, *64*, *65*
Duncan, A. B. F., 220, *341*
Dupree, M., 289, *338*
Durst, T., *198*
Duttachoudhury, M. K., *341*

E

Eaborn, C., 76, *129*
Eagland, D., 247, *341*
Earls, D. W., 145, 146, 147, 169, *197*, *198*
Eastman, M. P., 194, *197*
Eckstrom, H. C., 39, *63*
Edelson, D., 211, *341*
Edwards, J. O., 45, *63*
Effio, A., 33, *63*, 289, 290, 320, *344*
Egan, E. P., 260, *341*
Egelstaff, P. A., 234, *342*
Egel-Thal. M., 273, *338*
Eggimann, W., *181*, *197*
Ehrenson, S. J., 114, *129*
Eigen, M., 85, *129*, 151, 156, 180, *198*
Eirich, F. R., 298. *337*
Eisenberg, D., 219, 231, 234, *341*
Eley, D. D., 207, *341*
Ellis, C. M., 284, 285, *341*
Elrod, J., 327, *339*
Enderby, J. E., 241, *348*
Endo, H., 267, *341*
Engberts, J. B. F. N., 278, *347*
Engels, P., 298, *341*
Engler, E. M., 12, *63*
Euranto, E. K., 163, *198*
Evans, A. G., 205, *341*
Evans, D. F., 43, *63*, *65*
Evans, I. P., 161, *197*
Evans, J. C., 121, *130*
Evans, M. G., 72, 122, *129*
Evans, M. W., 207, 248, 263, *342*
Everett, D. H., 284, 292, *340*
Ewing, G. J., 251, *341*
Exner, J. H., 188, *198*
Exner, O., 248, *341*
Eyring, H., *343*
Eyring, H., 135, *200*

F

Fagan, J. F., 98, 99, 100, *129*
Fainberg, A. H., 32, 33, 34, 35, 36, 37, 39, 41, 45, 46, 47, 51, 52, 59, 60, 61, *63*, *66*, *67*, 99, 109, *129*, *132*, 135, 136, *201*, *202*, 210, 316, 317, *352*
Falcone, J. S., 269, 276, *341*
Falk, M., 232, 233, 325, 326, *339*, *341*, *342*, *343*
Fanning, R. J., 302, *341*
Farinacci, N. T., 43, *63*
Farrow, L. A., 211, *341*
Fauve, J., 316, *348*
Fawcett, W. R., 135, *200*
Feakins, D., 210, 288, 309, 310, 314, 329, *336*, *338*, *341*, *342*

Feil, D., 228, 229, *342*
Feillolay, A., 274, *342*
Feldman, H., 43, *65*
Felt, G. R., 28, *66*
Fenby, D. V., 324, *346*
Fendler, E. J., 189, *198*, 205, *342*
Fendler, J. H., 182, 189, *198*, 205, *342*
Feynman, R., 5, *63*
Fife, T. H., 166, *197*
Filomena, M., 212, *338*
Finer, E. G., 260, *342*
Finkelstein, M., 34, *63*
Fiordiponti, P., 193, 195, *198*, 328, *350*
Fischer, H. P., 154, *202*
Fischer, P. B., 190, *197*
Fisher, R. D., 9, 18, 19, 26, 35, 36, 39, *66*, 110, *131*
Fixman, M., 302, *342*
Fleischhauer, H., 103, *131*
Fleming, K. A., 212, *339*
Flood, S., 121, *130*
Flournoy, J. M., *202*
Foon, R., 320, 321, *342*
Ford, T. A., 232, *341*, *342*
Forsythe, D. A., 111, *129*
Fort, R. C., *63*
Fortier, J.-L., 242, 265, 266, 270, *342*, *346*
Foster, M. J., 302, 333, *336*, *338*
Foster, R., 175, *198*
Fowler, D. L., 229, *342*
Fowler, F. W., 43, *63*
Fowler, R. H., 207, *338*
Fox, J. R., 107, 108, 109, *127*
Fox, M. F., 210, 218, 325, *338*, *342*
Franck, E. U., 229, *342*
Frank, H. S., 207, 209, 217, 222, 236, 237, 241, 248, 250, 259, 260, 261, 263, *342*
Franklin, J. F., 205, *342*
Franks, F., 209, 217, 223, 230, 234, 238, 244, 246, 250, 251, 252, 253, 255, 260, 261, 262, 282, 283, 298, 299, 301, 310, 311, *341*, *342*, *343*, *345*, *346*, *349*, *352*
Frederich, E. C., 179, *202*
Freemantle, D. J., *341*
Frensdorff, H. K., 182, *201*
Friedman, H. L., 52, *64*, 208, 218, 238, 240, 241, 242, 243, 245, 246, 247, 251, 252, 254, 255, 261, 266, 270, 271, 275, 301, 304, *342*, *343*, *345*, *346*, *349*, *352*
Friedman, J., 163, *199*
Frisone, G. J., 35, *63*
Fromm, J., 265, *343*, *352*
Fry, J. L., 3, 8, 9, 10, 12, 23, 31, 34, *63*, *65*, *66*, 110, *131*
Fuchs, R., 137, 163, 166, 173, *198*, *199*, 328, 333, *343*
Fujishiro, R., 290, 307, *348*
Fuller, N. A., 279, *339*
Funderburk, L. H., 154, *200*
Fuoss, R. M., 241, *346*

Furter, W. F., 307, *347*
Fyfe, C. A., 175, *198*

G

Gaboriaud, R., 213, *343*
Gal, A., 108, 109, *131*
Gammell, P. M., 297, *343*
Gani, V., 189, *198*
Gardini, G. P., 125, 126, *130*
Garner, A. Y., 114, 116, *131*
Garnjost, Von H., 230, *343*
Gaylord, N. G., 190, *198*
Gear, C. W., 211, *343*
Gehriger, C. L., 180, *196*
Gelles, E., 85, 86, *128*
Ghosh, J. C., 241, *343*
Gibson, A., 154, 155, *198*
Giguère, P. A., 325, *343*, *349*
Gilbert, J. M., 146, *201*
Gilbert, T. J., 70, 95, *129*
Giles, D. E., 177, *198*
Gill, S. J., 217. *343*
Gilles, L., 191, *200*
Gillet, H., 329, *343*
Gillis, H. A., 191, *197*
Ginsberg, H., 162, *199*
Gjaldbaek, J. C., 286, 327, *343*, *346*
Glasel, J. A., 234, 326, *343*
Glasstone, S., *343*
Glew, D. N., 208, 209, 212, 249, 250, 251, 256, 281, 291, 292, 301, *339*, *343*
Goates, J. R., 327, *348*
Godfrey, E., 331, *338*
Godfrey, P., 210, *339*
Goel, A., 221, *343*
Goetz, E., 135, *198*
Goitein, R., *198*
Gold, V., 3, 8, 32, *62*, *63*, 76, 77, 83, 95, *128*, *129*, 175, 190, 191, *196*, *198*, 205, *343*
Goldammer, E. V., 246, 253, 302, *343*
Golinkin, H. S., 213, 321, *338*, *344*
Good, W., 248, 343
Goodall, D. M., 152, 153, *196*
Goodman, H., *197*
Graczyk, D. G., 28, *63*
Grant, N. H., 225, *344*
Gregoriou, G. A., 27, 33, *62*
Gregory, J., 47, *65*
Greyson, J., 266, *343*
Griffiths, T. R., 18, *63*
Gross, G. W., 225, *344*
Gruetzmacher, R. E., 9, 20, *65*
Grunwald, E., 3, 8, 32, 33, 45, 46, 47, 51, 59, *63*, *64*, 67, 77, 81, 109, *129*, *130*, 135, 140, *198*, *199*, 210, 215, 248, 273, 289, 290, 321, *344*, *346*
Gubbins, K. E., 287, *351*
Guerrera, J. J., 261, *347*

Guggenheim, E. A., 79, 88, *128*, 214, 281, *344*
Gurney, R. W., 207, 218, 238, 241, *344*
Guthrie, J. P., 259, *344*

H

Haake, P., 39, *63*
Haberfield, P., 137, 161, 162, 163, 182, *198*, *199*
Hafemann, D. R., 209, *344*
Hagan, C. P., 173, *198*, 328, 333, *343*
Hagler, A. T., 235, *346*
Halberstadt, E. S., 278, *344*
Hall, F. M., 212, *339*
Hall, R. E., 8, 9, 14, 26, *63*, *64*, *65*, 77, 97, *131*
Hall, T. N., 175, 199
Halle, J. C., 144, *199*
Hamann, S. D., 298, *344*
Hambly, A. N., 320, 321, *342*, *351*
Hamlet, Z., 135, *196*
Hammett, L. P., 2, 22, 43, 44, *63*, *66*, 79, *129*, 147, 148, *199*
Hammond, G. S., 71, *129*
Hammons, J. H., 147, *201*
Hankins, D., 222, *344*
Hanna, S. B., 95, *129*, 153, *199*
Hansen, S. L., 309, *350*
Haque, R., 212, 249, *352*
Harders, H., 232, *339*
Hargreaves, W. A., 260, *344*
Harper, J. J., 20, *63*
Harrington, J. P., 288, *344*
Harris, J. M., 3, 4, 5, 8, 9, 12, 17, 19, 22, 23, 26, 33, 34, 36, 39, *63*, *64*, *65*, 77, 97, 98, 99, 100, 110, *129*, *131*
Harriss, M. G., 39, *64*
Hart, E. J., 191, *199*
Hartman, S., 23, *64*
Hartshorn, S. R., 2, 23, 27, 35, 36, 39, 43, 44, 45, *64*, *66*
Harvey, J. T., 117, *128*
Hasinoff, B. B., 206, *339*
Hassid, A. I., 85, *129*
Hasted, J. B., 234, *344*
Hauck, F., 150, *198*
Hautala, J. A., 93, *128*
Hawkins, R. E., 227, *344*
Hayduk, W., 298, *344*
Heath, J. B. R., 224, *352*
Heck, R., 99, *132*
Hedwig, G. R., 137, 140, *198*, 219, 266, 288, 335, *340*, *344*
Hehre, W. J., 60, *64*
Heichelheim, H. R., 334, *349*
Heilbronner, E., 190, *197*
Heimo, S., 323, *351*
Heino, E., 39, *65*
Heinonen, K., 331, *344*
Heinzinger, K., 265, *344*

Held, R. P., 310, *340*
Henderson, W. A., Jr., 116, *128*
Hepler, L. G., 138, 143, *199, 202,* 212, 239, 247, 290, 295, 315, *338*, *344*, *346*, *347*
Heppolette, R. L., 136, *201*, 205, *349*
Herbrandson, H. F., 135, *199*
Hermann, A., 252, 273, 274, *350*
Herskovits, T. T. 261, 288, *344*
Hertz, H. G., 239, 246, 253, 264, 265, 267, 301, 302, 311, *343*, *344*, *348*, *352*
Heselden, R., 275, *337*
Hibbert, F., 278, *344*
Hickey, B. E., *341*
Hickey, D., 309, 310, 329, *341*
Hidden, N. J., 293, 333, *336*, *338*
Hiiro, A. M., 331, *348*
Hill, D. J. T., 251, *336*
Hill, E., 102, *129*
Hill, R. R., 16, *62*
Hills, G. J., 258, 314, *344*
Hine, J., 31, *64*, 114, *129*
Hinton, J. F., 206, *336*
Hinze, W. L., 182, *198*
Hiršl-Starčević, S., 17, *64*
Hirst, J., 177, 178, 179, *196*
Ho, M., 275, *337*
Hobbs, K. S., 85, *127*
Hockswender, T. R. Jr., 127, *130*
Hoffmann, H. M. R., 3, 9, *64*
Hogen-Esch, T. E., 3, 18, *64*
Höjer, G., 251, *344*
Hojo, M., 165, *199*
Holiday, E. R., 78, *130*
Holland, C. J., 273, *339*
Hollyhead, W. B., 88, *131*
Holz, M., 265, 311, *344*, *348*
Hoover, T., 298, *344*
Hopkins, H. P., Jr., 143, *200*
Hornig, D. F., 223, 232, *344*, *350*
Horvath, A. L., 249, *344*
Houminer, Y., 108, 109, *131*
House, H. O., 189, *199*
Howe, R. A., 241, *348*
Hoyland, A. R., 222, *344*
Hrncirik, J., 295, *347*
Hsieh, Y. C., 266, *344*
Huang, D., *131*
Hudson, R. F., 44, *62*, 92, *129*
Hughes, E. D., 3, 32, *64*, 135, *199*, 279, *337*
Humski, K., 17, 25, 26, 27, *64*
Hung, J. H., 261, 274, *352*
Hunt, L., 305, *339*
Hutchins, R. O., 190, *200*
Hutchison, J. D., 76, *132*
Hyne, J. B., 136, 167, *199*, *200*, 210, 213, 214, 215, 219, 302, 313, 316, 320, 321, 322, 324, 326, 327, 329, 322, *341*, *343*, *344*, *345*, *346*, *347*
Hyson, E., 14, *65*

I

Ibne-Rasa, K. M., 45, *64*
Imhoff, M. A., 19, 27, *62*
Imoto, T., 39, *67*
Ingham, D. B., 248, *343*
Inglefield, P. T., 266, *344*
Ingold, C. K., 3, 6, 32, *64*, *129*, 135, 136, 163, 199, 206, 279, 337, *344*
Ingold, K. U., 107, 125, 126, , *131*
Ionescu, L. G., 251, *341*
Isomura, Y., 267, *351*
Ives, D. J. G., 209, 212, 236, 247, 282, 283, *342*, *344*
Ivin, K. J., 159, *199*
Izatt, J. R., 219, *345*

J

Jackson, J., 107, 108, 109, *127*
Jacob, S. W., 134, *199*
Jaiswal, D. K., 137, *199*
Jambon, C., 325, 327, *345*, *349*
Jarzynski, J., 234, *340*
Jeffrey, G. A., 208, 225, 227, 228, 229, 336, 339, 342, 345, 347
Jencks, W. P., 79, 88, 89, 90, *127*, *128*, *129*, *130*, 150, 177, 181, *199*, 258, *345*
Jindal, S. P., 9, *65*
Jermini, C., 95, *129*, 153, *199*
Johnson, C. D., 70, 80, 81, 95, 96, 119, 120, 127, *128*, *129*
Johnson, D. A., 266, *345*
Johnson, W. A., 332, *346*
Johnston, D. E., 147, *196*
Jolicoeur, C., 246, 254, 267, 268, 270, 275, *340*
Jonczyk, A., 190, *197*
Jones, H. W., 8, 33, 51, *67*
Jones, J., 301, *346*
Jones, J. R., 137, 145, 146, 147, 152, 168, 169, *196*, *197*, *198*, *199*
Jordan, T. H., 228, *347*
Jose, J., 327, *345*
Juhlke, T., 23, 24, 28, *65*
Juillard, J., 307, 308, 313, 314, 316, *337*, *341*, *345*, *348*, *349*
Jung, K. H., 326, *345*
Justice, J. C., 313, *350*

K

Kahl, D. C., 146, *197*
Kaiser, L. E., *67*
Kalliorinne, K., 324, 332, *345*
Kamat, R. J., 31, *65*

Kamb, B., 223, 224, 225, *345*
Kamego, A. A., *131*
Kaminsky, M. 207, *345*
Kaplan, L. A., 190, *199*
Karplus, W., 220, *345*
Kaspi, 36, 39, 43, *65*
Kato, N. 252, *348*
Katritzky, A. R., 43, *63*, 76, 96, 119, 120, *128*, *129*
Kaufman, D. A., 108, 109, *130*
Kaulgud, M. V., 298, *345*
Kauzmann, W., 219, 231, 234, 254, *341*, *345*, *346*
Kavanagh, G. M., 325, *350*
Kawakami, T., 328, *345*
Kay, P. S., 28, *66*, 96, 107, *131*
Kay, R. L., 265, 266, 267, 314, 333, *339*, 340, *345*
Kayser, R. H., 185, *196*
Kebarle, P., 239, *345*
Keefer, R. M., 39, *63*
Keller, J., 251, *344*
Kell, G. S., 230, *345*
Kelley, R. E., Jr., 9, 20, 31, *65*
Kelley, W. J., 9, *65*
Kelly, T. M., 261, *344*
Kemp, D. S., 70, 76, 91, *128*, *129*
Kennan, S. L., 99, *129*
Kenttamaa, J., 291, 293, 325, 326, *345*, *346*
Kern, C. W., 220, *345*
Kerr, J. A., 121, 122, 129
Kessick, M. A., 35, 36, 39, *66*
Kevill, D. N., 13, 33, 35, 59, *64*, 81, 110, *129*
Kharasch, N., 134, *199*
Kice, J. L., 76, *129*
Kier, L. B., 222, *344*
Kim, C.-B., 13, 59, *64*, 81, *129*
Kim, C. J., 4, *62*
Kim, J. J., 276, *349*
Kim, S-G., 37, *67*
King, H. F., 172, *201*
Kingsbury, C. A., 182, *198*
Kingston, B., 301, *345*
Kinoshita, T., 6, *65*, 102, 103, *130*
Kirby, A. J., 177, *199*
Kistenmacher, H., 236, 265, , *345*
Kito, S., 307, *350*
Klassen, N. V., 191, *197*
Klopman, G., 92, *129*
Klotz, I. M., 233, *352*
Knight, W. S., 254, *346*
Knoefel, J., 43, *65*
Knop, O., 325, *343*
Knowles, J. R., 259, *339*
Ko, E. C. F., 76, *127*, 161, 162, *196*, *197*, 257, 335, *345*
Ko, H. C., 313, *337*
Kobayashi, S., 118, 121, *130*
Kochi, J. K., 122, *129*

Kohnstam, G., 107, 108, 109, *127*, *199*, 211, *345*
Koivisto, A., 323, *351*
Kollman, P. A., 220, 221, 222, *345*
Kollmeyer, W. D., 87, *128*, 168, 169, *199*
Kolthoff, I. M., 145, 148, *199*
Kolwyck, K. C., 13, 33, 34, 59, *64*, 81, 110, *129*
Komatsu, T., 252, 348
Konasewich, D. E., 85, *129*
Konicek, J., 252, *345*
Konneke, H. G., 327, *345*
Kopecky, K. R., 150, *198*
Koppel, I. A., 43, *64*, 135, *200*
Koren, R., 20, 21, *65*, 98, 101, 110, *130*, *131*, 212, *345*
Kornblum, N., 189, *200*
Korson, L., 230, *346*
Koshy, K. M., 257, 258, *346*
Koskikallio, J., 105, 106, *129*
Kosower, E. M., 32, 39, 40, 41, *64*, *129*, 135, *200*
Koulkes-Pujo, A. M., 191, *200*
Kovačević, D., 25, *64*
Kover, W. B., 58, *62*, 251, *337*
Kozak, J. J., 254, *346*
Kraus, C. A., 229, 342
Kraus, K. A., 229, 308, *349*
Kreevoy, M. M., 82, 85, *129*, 148, *200*
Kresge, A. J., 83, 85, 86, 87, 88, 93, 94, 95, *130*, 156, 157, 158, 191, *200*
Kresheck, G. C., 254, 260, *344*, *346*
Kretschmer, C. B., 286, *346*
Krishnan, C. V., 52, *64*, 208, 218, 238, 240, 246, 247, 251, 255, 261, 266, 270, 271, 275, 301, 334, *343*, *346*, *349*
Kroeger, D. J., 169, *200*
Kruegger, J. H., 332, *346*
Kruger, T. L., 88, *131*
Kruus, P., 302, *341*
Krygowski, T. M., 135, *200*
Kubota, N., 328, *345*
Kuhn, S. J., 121, *130*
Kundu, K. K., 260, 309, 329, *339*, *340*, *346*
Kurz, J. L., 158, *200*
Kurz, L. C., 158, *200*

L

Laakso, R., 33, *351*
Lacadie, J. A., 47, *62*
Laidlaw, W. G., 213, 321, *343*, *344*
Laidler, K. J., 41, *64*, 135, *200*, 214, *343*, *346*
Laiken, N., 253, *346*
Laing, T.-M. 85, *129*
Laliberte, L. H., 267, *339*
Lam, K. B., 177, *200*
Lam, L. K. M., 8, 9, 10, 31, *63*, *66*
Lam, S. T., 148, 149, 195, *200*

Lam, S. Y., 325, 334, *346*
Lama, R. F., 290, *346*
Lamb, J. D., 327, *348*
Lambert, J. B., 17, *64*
La Mer, V. K., *207*, *346*
Lamson, D. W., 190, *200*
Lancelot, C. J., 4, 8, 9, 10, 31, *62*, *63*, *64*, *66*
Lanning, J. A., 314, *346*
Lannung, A., 286, *346*
Lantzke, I. R., 329, 335, *340*
Lapinte, C., 189, *198*
Larkin, J. A., 290, 292, *346*
Larsen, J. W., 27, 28, *66*, 97, *131*
Larson, J. W., 212, 295, *338*, *346*
Last, A. M., 143, 144, 145, *196*, *198*
Lateef, A. B., 322, *347*
Lav, C. F., 325, *346*
Laudie, H., 298, *344*
Laughton, P. M., 205, *346*
Lawrence, K. G., 288, 309, 310, *336*, *338*, *341*, *342*
Lazdins, I., 21, 37, *63*
Leduc, P.-A. 217, 265, 267, 269, 270, *342*, *345*, *346*, *349*
Lee, I., 219, 313, 321, *343*, *346*
Lee, J. J., 313, *346*
Lee, M. D., *313*. *346*
Lee, T. P., 272, 275, *347*
Lee, Y., 301, *346*
Le Fevre, P. H., 20, *66*
Leffek, K. T., 26, 45, *64*, 153, *200*
Leffler, J. E., 59, *64*, 71, 77, 81, *130*, 215, 248, *346*
Leibnitz, E., 327, *345*
Lem, B., 3, *67*
Lemberg, H. L., 220, 236, *346*, *350*
Lemon, T. H., 236, *344*
le Noble, W. J., 189, *200*
Lenoir, D., 9, 14, *64*, *65*
Lentz, B. R., 235, *346*
Lepori, L., 253, 298, *339*
Le Roux, L. J., 32, *65*
Letcher, J. W., *338*
Leung, H. W., 181, *197*
Leung, P. S., 264, *350*
Levi, A., 149, *200*
Levine, A. S., 267, 276, *341*, *346*
Levine, S., 209, 236, *346*, *349*
Levy, H. A., 223, 230, *348*, *349*
Lewis, E. S., 154, *200*
Lidwell, O. M., 85, 86, *128*, 151, *196*
Lienhard, G. E., 89, *130*
Lietzke, M. H., 241, *346*
Liggero, S. H., 19, 27, *62*
Lilley, T. H., 219, *340*
Linda, P., *128*
Lindberg, J. J., 325, 326, *345*, *346*
Lindell, E., 333, *351*
Lindenbaum, S., 267, 268, 269, *346*

Linderstrøm-Lang, C. U., 290, *347*
Liotta, C. L., 143, *200*
Lipinski, C. A., 112, *130*
Liu, K.-J., *197*
Liu, K.-T., 19, *62*
Liu, L. J., 182, *198*
Llewellyn, J. A., 26, 45, *64, 65*
Loebenstein, W. L., 229, *342*
Loewenschuss, H., 95, *129*, 153, *199*
Lohman, K. H., 78, 96, *132*
Long, F. A., 272, 278, *344, 347*
Lorimer, J. P., 309, 310, 314, 329, *341*
Lo Surdo, A., 269, 275, *352*
Lowe, B. M., 261, 266, 267, *341, 347*
Lu, B. C. Y., 290, *346*
Lucas, M., 274, *342, 347*
Luck, W. A. P., 231, 232, 233, *339, 347*
Luff, B. B., 260, *341*
Luksha, E., 310, *340*
Lumry, R., 247, 248, *347*
Lyrra, J. P., 323, *351*

M

Macdonald, D. D., 167, *200*, 302, 313, 322, 324, 326, 327, 329, 332, *347*
MacDonald, J. C., 261, *347*
MacGregor, W. S., 134, *200*
MacKellar, W. J., 315, *350*
MacKenzie, A. P., 326, *349*
Mackie, J. D. H., 3, 32, *64*
MacKinnon, M. J., 322, *347*
MacQuarrie, R. A., 88, *131*
Majerski, Z., 17, 25, *64*, 102, *130*
Mak, H. D., 301, *343*
Mak, T. C. W., 227, 228, 229, *347*
Malcolm, G. N., 295, *347*
Malik, S. K., 257, 260, *352*
Mallon, C. B., 115, 116, *130*
Maltamo, S., 324, *351*
Mamantov, A., 114, 115, *130*
Mano, H., 187, *201*
Marcus, R. A., 83, 94, *130*, 157, *197*, *200*
Margerum, D. W., 258, *339*
Margolin, Z., 144, 145, 146, 147, 191, *197*, *200*
Marino, G., 120, *128*
Marsden, P. O., 212, 247, *344*
Marshall, H., 26, *67*
Martin, J. F., 266, *345*
Martinelli, A., 252, *339*
Martti, E., 291, 293, *345*
Maruyama, K., 187, *201*
Marjyamo, M., 252, *348*
Marziano, N. C., 149, *200*
Masterson, W. L., 272, 275, *347*
Mateo, S., 154, *197*
Matesich, M. A., 43, *63, 65*

Mathieson, J. G., 238, 239, 260, 264, *340, 347*
Mathur, H. B., *341*
Matous, J., 291, 295, *347*
Matteoli, E., 252, *339*
Matthews, W. S., 144, 145, 146, 147, 191, *197, 200*
Matts, T. C., 27, 33, *65*
Maugh, T., 259, *347*
Mayer, J. E., 245, *347*
Mazumdar, K., 260, *346*
McCallum, R. J., 144, 145, 146, 191, *200*
McCollum, G. J., 144, 145, 146, 191, *200*
McCrary, T. J. Jr., 16, *65*
McDevit, W. F., 272, *347*
McDonald, M. P., 262, *336*
McDowell, J. M., 265, *347*
McGarvey, J. J., 159, *199*
McGuinness, J. A., 124, *132*
McIver, R. T., 60, *64*
McKee, C., 225, *344*
McKelvey, D. R., 307, 312, 313, *337*
McLennan, D. J., 27, 28, *65*
McMillan, W. G., 245, *347*
McMullan, R. K., 225, 227, 228, 229, *339, 345, 347*
Mecca, T. G., 28, 30, *62*
Megerle, G. H., *201*
Meisenheimer, J., 174, *200*
Membrey, J. R., 331, *338*
Menninga, L., 278, *347*
Meranda, D., 307, *347*
Mikhail, S. Z., 314, *351*
Mikhailov, V. A., 250, *347*
Milakofsky, L., 35, 36, 39, *66*
Millar, E. M., 95, *128*
Miller, A., 190, *197*
Miller, F., 20, *65*
Miller, J., 174, 177, *200*
Miller, L. L., 108, 109, *130*
Miller, S. I., 79, *130*
Miller, S. L., 209, *344*
Millero, F. J., 230, 264, 290, 315, *338, 346, 347*
Millot, F., 167, *202*
Mills, R., 265, *344*
Milne, J. B., 39, *64*
Minasz, R. J., 313, *337*
Minisci, F., 125, 126, *130*
Mintz, E. A., 188, *196*
Mison, P., 14, *65*
Mitchell, A. G., 325, *347*
Mixan, C. E., 17, *64*
Modena, G., 149, *200*
Moelwyn-Hughes, E. A., 206, 207, 208, 212, 250, 256, *343, 347, 348*
Mohanty, R. K., 266, 312, 313, *348, 350*
Moller, E., 85, 86, *128*
Moolel, M., 334, *348*
Moran, G., 306, *338*
Morcom, K. W., 293, 333, 334, *336, 338, 348*

More O'Ferrall, R. A., 85, 89, 92, 95, *130,* 153, 169, *200*
Moreau, C., 335, *348*
Morel, J.-P., 293, 297, 307, 308, 309, 312, 314, 315, 316, 329, *336, 337, 343, 345, 348, 349*
Moriarity, T. C., 147, *196*
Morin, J., 309, 312, 315, 316, *336, 337,* 348
Morokuma, K., 221, *348*
Morris, S. H., 280, *338*
Morrison, T. J., 303, 305, *339*
Morten, D. H., 34, 38, 46, 53, *62, 65*
Moseley, P. G. N., *344*
Mosely, R. B., 8, 32, *67*
Moskowitz, J., 222, *344*
Moss, R. A., 112, 113, 114, 115, 116, *128, 130*
Mowery, P. C., 146, 147, *202*
Mueller, W. A., 135, *196*
Mui, J. Y.-P., 113, *131*
Muir, D. M., 16, *62*
Murakami, S., 290, 307, *348*
Murdoch, J. R., 94, *130,* 158, *200*
Murrill, E., 86, 88, *130*
Murthy, A. S. N., 221, *343*
Murto, J., 39, *65, 348*
Murto, M., 331, 332, 333, *351*
Murto, M. J., 163, 167, *202*
Mylonakis, S. G., 87, 95, *130*

N

Nadas, J. A., 43, *63*
Nakagawa, T., 252, *348*
Nakajima, T., 252, *348*
Naghizadeh, J. N., 60, *62*
Nakanishi, K., *348*
Nakayama, H., 292, 293, *348*
Narten, A. H., 230, *348*
Neal, W. C., Jr., 36, *64*
Nefedov, O. M., 116, *130*
Neggia, P., 311, *348*
Neilson, G. W., 241, *348*
Nelson, D. C., 134, *201*
Nelson, H. D., 249, *348*
Nelson, K. L., 117, *128*
Nelson, N. J., 146, 194, *197*
Nemethy, G., 209, 249, 253, 254, 327, *346, 348, 349*
Neufeld, F. R., 135, *199*
Newman, K. E., 219, *340*
Newsham, D. M. T., 291, *340*
Ng, S. B., 291, *340*
Nicholas, R. D., 107, 110, *131*
Nichols, N. F., 217, 252, *343, 348*
Nichols, R. W., 112, *130*
Nicole, D., *198*
Nielson, W. D., 150, *198*

Niemann, H., 327, *343*
Niemeyer, H. M., 88, *131*
Nishida, S., 110, *130*
Nobilione, J. M., 329, *329,* 330, *350*
Nordlander, J. E., 9, 16, 20, *65*
Norris, A. R., 175, *197*
Novak, D. M., *340*
Novak, J. P., 272, 291, 295, *347*
Nowlan, V., 190, *198*
Noyce, D. S., 111, 112, *129, 130*
Noyes, R. M., 159, *200*
Nudelman, A., 162, *199*
Nychka, N., 125, 126, *131*

O

Oakenfull, D. G., 279, 280, *348*
O'Connell, J. P., 287, *348*
O'Connor, M. E., 263, *350*
O'Donnell, J. P., 147, *201*
Ogg, R., 205, *349*
Ogston, A. C., 78, *130*
Oh, S., 82, 85, *129*
O'Hara, W. F., 247, *344*
Ohara, Y., 267, *351*
Ohno, T., 307, *350*
Okamoto, K., 6, *65,* 102, 103, *130*
Olah, G. A., 118, 119, 120, 121, 127, *130*
Olsen, F. P., 148, 149, *197*
O'Neill, B., 96, *129*
Onsager, L., *348*
Ormondroyd, S., *340*
Orr, J. B., 327, *348*
Orvik, J. A., 176, *200*
Osada, Y., 6, *65*
O'Shaughnessy, D. A., 329, *341*
Ossip, P. S., 39, *63*
Ovenden, P. J., 314, *344*
Overchuk, N., 121, *130*
Owens, P. H., 88, *131*
Owicki, J. C., 220, *350*

P

Packer, K. J., 326, *348*
Padwa, A., 123, 124, *132*
Page, D. I., 233, 234, 324, *342, 348*
Page, M. I., 234, *337*
Pajunen, A., 326, *351*
Palenius, I., 331, *351*
Pall, D. B., 229, *342*
Palm, V. A., 43, *64,* 135, *200*
Palombo, R., 335, *340*
Pánková, M., 183, 185, *200*
Parish, J. H., 16, *62*

Parker, A. J., 2, 65, 134, 136, 137, 138, 140, 141, 148, 160, 161, 162, 174, 177, 182, *196, 197, 198, 200*, 219, 266, 288, 335, *340, 344, 348*
Parker, A. K., 76, *127*
Parlman, R. M., 188, *196*
Passerini, R. C., 149, *200*
Patil, K. J., 298, *345*
Paul, K., *128*
Pauling, L., 209, 223, *348, 349*
Pavez, H. J., 112, *130*
Payne, M. A., 86, 88, *130*
Pearson, R. G., 45, 65, 88, *130*
Pedersen, K. J., 83, 88, *128,* 150, 151, *197*
Pederson, C. H., 182, *201*
Pedley, M., 255, *342*
Pelletier, G. E., 275, *340*
Pelton, A. D., 281, *337*
Perlmutter-Hayman, B., 212, 277, *345, 349*
Peron, J. J., 232, 267, *349*
Perram, J. W., 209, 236, *346, 349*
Perron, G., 260, 266, 294, 301, *337, 340, 341, 345, 349*
Perry, S. G., 279, *339*
Peslak, J., 327, *339*
Peterson, P. E., 9, 20, 31, 45, 47, 48, 55, 65
Peterson, S. W., 223, *349*
Phibbs, M. K., 325, *349*
Philip, P. R., 217, 242, 260, 267, 270, *342, 345, 349*
Philippe, R., 325, 327, *345, 349*
Philpot, J. S. L., 78, *130*
Pick, J., 291, *347*
Picker, P., 217, *349*
Pierrotti, R. A., 251, *349*
Pietra, F., 175, *201*
Piggott, S. P., 325, *339*
Pikal, M. J., 314, *346*
Pinder, K. L., 299, *349*
Pinkston, M. F., 163, *199*
Pitkanen, I. P., 324, 331, 332, *351*
Pitman, I. H., 279, *339*
Pitzer, K. S., 276, *349*
Plumlee, D. S., 328, *343*
Pointud, Y., 308, *349*
Polanyi, M., 72, 122, *129*, 205, *349*
Ponomarova, L. I., 250, *347*
Popkie, H., 236, 265, *345*
Pople, J. A., 60, *64*, 220, 221, 222, *340, 341, 349*
Poranski, C. F., Jr., 175, *199*
Porterhouse, G. A., 219, *340*
Potts, W., 327, *351*
Powles, J. G., 236, *349*
Pratt, K. C., 302, *349*
Price, E., 140, *199*
Price, G. G., 146, *201*
Priesand, M. A., 194, *197*
Prigogine, I., 214, *349*
Pritt, J. R., 6, 7, 13, 14, 17, 40, 59, *62*, 65
Pross, A., 20, 21, 65, 77, 98, 99, 101, 103, 110, *130, 131*
Prue, J. E., 278, *344*
Pruss, G. M., 185, 186, 187, *196*
Przybyla, J. R., 115, *130*
Pudjaatmaka, A. H., 88, *131*
Putz, G. J., 17, *64*

Q

Quadrifoglio, F., 298, *340*
Queen, A., 27, 33, 65, 107, 108, 109, *127*, 256, 258, *349*
Quickenden, M. J., 217, 252, 253, 261, 298, *342, 350, 351*
Quirk, R. P., 147, *196*
Quist, A. S., 209, *342*

R

Raaen, V. F., 23, 24, 28, 65
Raber, D. J., 4, 5, 8, 9, 14, 17, 19, 22, 23, 27, 33, 34, 36, 47, *63, 64*, 65, 77, 97, 98, 107, 110, *131*
Radle, C., 253, 264, *344*
Radom, L., 115, *131*
Rahma, A., 235, *349, 351*
Rajender, S., 247, *347*
Rallo, F., 193, 195, *198*, 328, *350*
Ramanthan, P. S., 245, 247, 270, 271, *349*
Rammler, D. H., 150, *201*
Rand, M. H., 89, *128*
Randall, J. J., 179, *197*
Ranky, W. O., 134, 194, *201*
Rao, C. N. R., 220, 221, *343, 349*
Rapp, M. W., 35, 36, 39, *66*
Rappoport, Z., 36, 39, 43, 65, 77, 108, 109, *131*
Raridon, R. J., 308, *349*
Rasaiah, J. C., 245, *349*
Rasmussen, D. H., 326, *349*
Rat, J. C., 310, *349*
Rath, N. S., 301, 343
Ravenhill, J. R., 234, 252, 262, *342*
Ray, A., 327, *349*
Reddy, T. B., 146, *199*
Reding, F. P., 223, *344*
Rehfeld, S. J., 275, *349*
Reich, I. L., 4, 65
Reichardt, C., 32, 39, 40, 41, 42, 43, 44, *63, 65*, 135, *198, 201, 349*
Reid, D. S., 217, 250, 251, 252, 253, 255, 259, 260, 262, 298, 299, 310, 311, *342, 349*
Remijnse, A. G., 309, *337*
Renard, E., 313, *349*

Renaro, J. A., 334, *349*
Rendall, H. M., 267, *347*
Rice, S. A., 236, *352*
Rickborn, B., 182, *198*
Ridd, J. H., 121, *131*
Ritchie, C. D., 29, 45, 65, 76, 81, 100, 103, *131*, 134, 138, 144, 145, 146, 156 172, *197*, *201*
Rizzo, E., 260, *337*
Robbins, H. M., 28, *66*
Roberts, D. D., 13, 65, 163, *201*
Roberts, J. D., 30, *63*
Robertson, R. E., 23, 26, 45, *64*, 65, 136, *201*, 205, 208, 212, 256, 257, 258, 276, 318, 319, 335, *338*, *345*, *346*, *348*, *349*
Robin, M. B., 50, *62*
Robinson, A. L., 207, 241, *342*
Robinson, B. H., 212, *336*
Robinson, G. C., 99, *132*
Robinson, R. A., 214, 241, *350*
Rochester, C. H., 147, 167, 169, *201*, 253, 309, *350*
Rodante, F., 193, 195, *198*, 328, *350*
Rodewald, R. F., 173, *198*, 328, 333, *343*
Rogers, T. E., 76, *129*
Röll, W., 9, *65*
Romm, R., 162, *199*
Rorabacher, D. B., 315, *350*
Rosenbaum, E. E., 134, *199*
Ross, S. D., 135, *201*
Rosso, J.-C., 291, 298, *339*, *350*
Rothenberg, F., 118, *128*
Rowlinson, J. S., 221, 245, 284, 285, 295, 297, *347*, *350*
Roy, R. N., 308, *350*
Rubenstein, P. A., 88, *131*
Rubienski, G. A., 266, *347*
Rubini, P., 146, *197*
Rudolph, P. J., 147, *196*
Rumney, T. G., 145, 147, 169, *197*, *198*
Runnels, L. K., *348*
Russell, G. A., 124, 125, 126, *128*, *131*
Russell, K. E., 175, *197*
Rutherford, R. J. D., 43, *63*
Ruzziconi, R., 187, *196*

S

Sachs, W. H., 153, *196*
Sada, E., 307, *350*
Sadler, I. H., 113, *131*
Safford, G. J., 264, *350*
Sagatys, D., 86, 88, *130*
Saito, S., 269, *352*
Sakai, H., 219, *345*
Saluja, P. P. S., 218, *339*
Samoilov, O. Ya., 237, *350*
Sandorfy, C., 232, *349*
Sandstrom, J. P., 274, *350*

Sandstrom, W. A., 89, *128*
Sarma, T. S., 266, 267, 312, *348*, *350*
Sarnowski, M., 261, *350*
Sato, Y., 87, 95, *130*
Saunders, M., 14, *65*
Saunders, W. H., Jr., 182, 184, 185, *196*, *201*
Scatchard, G., 135, *201*, 281, 325, *350*
Schaal, R., 144, 167, 169, *199*, *202*
Schadt, F. L., 9, 10, 11, 15, 24, 37, 39, 47, 48, 53, 54, 57, 58, 59, *62*, *66*, 98, *128*
Schaefer, A. D., 184, *202*
Schaleger, L. L., 275, *337*
Scheraga, H. A., 209, 220, 235, 249, 252, 254, *343*, *346*, *348*, *350*
Schichman, S. A., 326, *350*
Schiffer, J., 232, *350*
Schleyer, P. von R., 3, 4, 5, 8, 9, 10, 11, 12, 13, 14, 15, 17, 19, 22, 24, 26, 27, 30, 31, 33, 34, 37, 39, 45, 46, 47, 48, 53, 54, 57, 58, 59, 60, *62*, *63*, *64*, *65*, *66*, 77, 97, 98, 107, 110, *128*, *131*, *135*, *196*
Schneider, G., 295, 298, *341*, *350*
Schneider, H., 254, 334, *338*, *346*, *348*
Schofield, K., 120, *129*
Schrier, E. E., 274, *351*
Schukarev, S. A., 303, *350*
Scorrano, G., 147, 148, 149, *196*, *200*
Scott, C. B., 45, *67*, 78, 79, 96, *131*, *132*
Scott, J. M., 26, *65*
Scott, J. M. W., 136, *201*, 205, 212, *349*
Schubert, W. M., 20, 60, *66*
Schulz, R. A., *62*, *62*
Schurhoff, W., 44, *62*
Sendijarević, V., 17, 25, 26, 27, *64*
Senior, W. A., 232, 233, *350*
Sep, W. J., 43, *66*
Sera, A., 187, *201*
Serratrice, G., 156, *198*
Seyferth, D., 113, *131*
Shafran, R. N., 116, *130*
Shatenshtein, A. I., 111, *131*
Shepler, R. E., 302, *337*
Sheppard, J. G., 258, *336*
Shiner, V. J., Jr., 9, 17, 18, 19, 20, 23, 25, 26, 27, 35, 36, 39, 59, *64*, *66*, 110, *131*
Shinoda, K., 293, *348*
Shipman, L. L., 220, *350*
Shold, D. M., 13, 59, *64*, 81, *129*
Shu, F. R., 315, *350*
Shurupova, L. V., 302, *352*
Sicher, J., 182, 183, *200*, *201*
Siedle, A. R., 190, *199*
Siepmann, T., 39, 41, 42, 44, *63*, 135, *198*
Simmons, E. L., 159, *199*
Simmons, H. E., 116, *131*
Sinnott, M. L., 7, *66*
Sipp, K. A., 9, 20, 31, *65*
Skell, P. S., 112, 114, 115, 116, *131*
Slansky, C. M., 44, *66*

Slaugh, L. H., 123, *132*
Sliwinski, W. F., 13, *66*
Slootmaekers, P. J., 316, *351*
Slutsky, J., 9, *66*
Small, L. E., *196*
Small, R., 146, 147, 159, *199*
Smid, J., 3, 18, *64*, *66*
Smith, F., 298, *344*
Smith, H. T., 217, 244, 252, 298, *343*
Smith, M. D., 327, *347*
Smith, O. H., 298, *344*
Smith, P. J., 182, *201*
Smith, R. D., 116, *131*
Smith, R. W., 293, 333, 334, *338*, *348*
Smith, S., 32, *67*, 179, *202*
Smith, S. G., 33, 37, 46, *66*, 135, *201*
Smitherman, A., 276, *343*
Sneen, R. A., 7, 15, 18, 27, 28, *66*, *67*, 96, 97, 102, 107, *131*
Snell, H., 266, *343*
Sobel, H., 45, *65*
Sobr, J., 291, 295, *349*
Somsen, G., 266, 312, *341*
Songstad, J., 45, *65*
Sorgen, D. K., 302, *337*
Southam, R. M., 16, *62*
Spanswick, J., 125, 126, *131*
Spencer, D., 252, *341*
Starkey, J. D., 146, *201*
Steigman, J., 2, *66*
Stein, A., 297, *350*
Steiner, E. C., 146, 188, *198*, *201*
Steinert, H., 327, *345*
Stern, J. H., 252, 263, 273, 274, 309, 329, 330, *350*
Stern, K. H., 218, *351*
Stevens, C. G., 96, *131*
Stewart, R., 143, 144, 145, 146, 147, 148, 149, 152, 167, 168, 169, 182, *196*, *197*, *198*, *199*, *200*, *201*
Stiles, P. J., 115, *131*
Stillinger, F. H., 220, 222, 235, 236, 250, 251, *338*, *344*, *346*, *349*, *350*, *351*
Stimson, E. R., 274, *351*
Stock, L. M., 117, *128*, *131*
Stocken, L. A., 78, *130*
Stokes, J. M., 143, *199*
Stokes, R. H., 143, *199*, 207, 214, 241, 260, *350*, *351*
Stoller, L., 257, 260, *352*
Stone, J., 248, *343*
Storesund, H. J., 14, *66*
Stoughton, R. W., 241, *346*
Strachan, W. M. J., 257, *346*
Strauss, M. J., 175, *201*
Streitwieser, A., Jr., 2, 9, 10, 16, 26, 45, 54, 58, *66*, 88, *131*, 146, 147, 169, *202*, *205*, *351*
Strich, A., 156, *198*
Strickler, S. J., 96, *131*

Su, T. M., 13, *66*
Subramanian, S., 312, *348*
Sugamori, S. E., 212, 256, 318, 319, 335, *348*, *349*
Suggett, A., 260, 261, 262, 263, *343*, *351*
Suhr, H., 177, *201*
Sunder, S., 267, 312, *348*, *351*
Sunko, D. E., 17, 23, 25, 36, 39, *64*, *66*, 102, *130*
Sutton, J., 191, *200*
Svoboda, M., 185, *200*
Swain, C. G., 8, 32, 39, 44, 45, 47, 53, 54, 56, *67*, 78, 79, 96, *132*
Swanson, J. C., 184, *196*
Swart, E. R., 32, *65*, 258, *336*
Sweeney, W. A., 60, *66*
Sweigart, D. A., 50, *67*
Symonds, J. R., 253, *350*
Symons, E. A., 146, 147, 167, 171, 172, 173, *197*, *201*, 327, *351*
Symons, M. C. R., 18, *63*, *67*, 210, 223, 231, 233, 234, 266, 301, 333, *336*, *338*, *339*, *340*, *341*, *345*, *351*
Szele, I., 36, 39, *66*
Szent-Gyorgi, A., 204, *351*
Szmant, H. H., 134, 193, *201*
Szwarc, M., 18, *67*, 146, 147, *202*

T

Taft, R. W., 136, *197*, 279, *339*
Taher, N. A., 279, *337*
Tait, M. J., 260, 261, *342*, *351*
Takahashi, J., 18, 20, *67*
Takeguchi, N., 252, *352*
Takemura, Y., 6, *65*
Tamaki, K., 267, *351*
Tamura, K., 39, *67*
Tanaka, R., 290, 307, *348*
Tanner, D. D., 125, 126, *132*
Tarhan, H. O., 76, *128*
Tashiro, M., 118, *130*
Taylor, J. W., 23, 28, *63*, *67*
Taylor, R., 76, *129*
Taylor, R. P., 190, *202*
Telkowski, L. A., 14, *65*
Terrier, F., 144, 167, 169, *199*, *202*
Terui, H., 328, *345*
Thain, J. M., 329, 330, 335, *340*
Thomas, R. K., 49, 50, *67*
Thompson, P. T., 241, *342*
Thomson, A., 320, *351*
Thornton, E. K., 82, *132*
Thornton, E. R., 2, 8, 23, 31, 35, *63*, *67*, 82, 89, *132*, 182, *202*
Thuaire, R., 315, *351*
Thyagarajan, B. S., 134, *199*
Ticknor, L. B., 325, *350*
Tiepel, E. W., 287, *351*

Tiilikainen, M., 320, 323, *351*
Timimi, B. A., 212, *351*
Tingoli, M., *196*
Todheide, K., 229, *351*
Tolgyesi, W. S., 121, *132*
Tolmacheva, T. A., 303, *350*
Tomić, M., 36, *67*
Tomkins, R. P. T., 288, 309, *336, 341*
Tomlinson, D. J., 326, *348*
Tommila, E., 41, 44, *67*, 163, 166, 167, *202*, 209, 291, 293, 320, 321, 323, 324, 326, 331, 332, 333, 335, *344, 345, 351*
Tonnet, M. L., 320, *351*
Toporowski, P. M., 326, *340*
Tourigny, G., 36, *63*
Tourky, A. R., 314, *351*
Tranter, R. L., 153, *196*
Treiner, C., 281, 309, *351*
Treloar, N. C., 293, *338*
Trimble, H. M., 327, *351*
Trotman-Dickenson, A. F., 122, *132*
Turner, D. J., 310, *342*
Turner, D. W., 50, *62, 67*
Tutsch, R., 265, *344*

U

Ueda, Y., 259, *344*
Ulrich, P., 190, *200*
Uschold, R. E., 145, 146, 156, *201*
Utaka, M., 165, *199*
Uzan, R., 119, *128*

V

Van de Poel, W., 316, *351*
Van Hook, W. A., 82, *132*
Vanier, N. R., 144, 145, 146, 191, *200*
Vaslow, F., 247, 269, 290, *341, 347, 351*
Veillard, A., 30, *63*, 156, *198*
Venkatasubramanian, N., 164, 165, 166, *196*
Venkoba Rao, G., 164, 165, 166, *196*
Verhoeven, J. W., 43, *66*
Verrall, R. E., 218, 232, 233, *340, 350*
Versmold, H., 265, *344*
Viana, C. A. N., 258, *344*
Vidulich, G. A., 333, *340*
Vierk, A. L., 333, *351*
Villermaux, S., 193, *202*, 310, *349, 352*
Vincent, C. A., 265, *347*
Virtalaine, M. L., 333, *351*
Virtanen, P. O. I., 103, *131*
Vitagliano, V., 260, 261, 298, 313, *337, 340*
Vitullo, V. P., 25, *67*, 87, 95, *130*
Vogel, P., 14, *65*
Vogel, P. C., 265, *344*

Voice, P. J., 309, 310, 329, *341, 342, 345*
Voipo, A., 320, 323, *351*
Vuks, M. F., 302, *352*

W

Waddington, D., 231, 234, 243, 269, 302, *336, 338, 339*
Wadso, I., 217, 252, *343, 345, 348*
Waiss, A. C., Jr., 16, *66*
Wakeham, W. A., 302, *349*
Walden, F. A., 98, 99, *129*
Walker, D. C., 191, *197*
Waller, F. J., 45, 47, 48, 55, *65*
Walling, C., 123, 124, *132*
Walrafen, G. E., 222, 230, 232, 233, 234, 260, *339, 352*
Walsh, T. O., 16, *66*
Wang, J. H., 234, *352*
Warheit, A. C., 76, *129*
Warner, D. T., 261, *352*
Watanabe, K., 49, 50, *67*
Waterman, D. C. A., 95, *129*
Watson, B., 252, 253, 298, 301, *342, 343*
Watts, D. W., 137, 141, *198*, 219, 266, 288, 335, *340*
Watts, H., 281, 291, 292, 301, *343*
Watts, R. O., 265, *343, 352*
Wauchope, R. D., 212, 249, *352*
Weaver, W. M., 76, *132*
Webb, J. G. K., 190, *197*
Wecht, K. W., 234, *339*
Weiner, H., 15, 28, *67*
Weiss, E., 9, *65*
Weitl, F. L., 33, 34, *64*, 110, *129*
Wells, C. F., 292, 296, 309, 310, 329, *352*
Wells, P. R., *352*
Wen, W.-Y., 209, 222, 236, 252, 261, 263, 266, 267, 268, 269, 271, 274, 301, *342, 344, 352*
Wenke, G., 9, *65*
Weres, O., 236, *352*
Westaway, K. C., 25, *67*
Westheimer, F. H., 82, 123, *132*, 155, *202*
Wetlaufer, D. B., 257, *260, 352*
Whalley, E., 224, 323, *337, 338*, 352
Wheeler, J. C., 284, 297, *352*
White, H. F., 223, *344*
Whitehouse, D. R., 314, *344*
Whiting, M. C., 6, 7, 13, 14, 16, 17, 40, 59, *62, 65, 66*, 146, *201*
Whittingham, K. P., 325, *342*
Whittington, S. G., 252, *341*
Wiberg, E., 190, *202*
Wiberg, K. B., 123, *132*, 151, *202*
Wiebbe, R., 286, *346*
Wiegers, K. E., 185, 186, 187, *196*
Wigfield, D. C., 3, *67*

Wiley, P. F., 28, 30, *62*
Wilf, J., 254, *338*
Wilgis, F. P., 25, *67*
Wilhelm, E., 216, *352*
Williams, R. C., 23, *67*
Williamson, A. G., 290, 334, *339*
Willis, C. L., 43, *62*
Willison, M. J., 181, *198*
Wills, R., 136, *199*, 320, *344*
Wilmarth, W. K., 172, *202*
Wilson, D. P., 252, *352*
Wilson, G. T., 224, *352*
Wilson, P. T., 324, *346*
Winey, D. A., 182, *202*
Winick, J., 221, *348*
Winstein, S., 3, 4, 8, 18, 20, 21, 26, 32, 33, 34, 35, 36, 37, 39, 41, 45, 46, 47, 51, 52, 59, 60, *62*, *63*, *65*, *66*, *67*, 79, 99, 109, *129*, *132*, 135, 136, 179, *199*, *201*, *202*, 210, 316, 317, *344*, *352*
Wirth, H. E., 269, 275, *352*
Wold, S., 212, 248, *352*
Wolfe, J. R., Jr., 16, *66*
Wolford, R. K., 333, *352*
Wong, S. M., 154, *202*
Wonkka, R. E., 136, *199*
Wood, D. C., 134, *199*
Wood, M., 219, *340*
Wood, R. H., 243, 255, 260, 267, 269, 275, *336*, *339*, *341*, *346*
Wood, R. M., 262, *336*
Wooley, E. M., 138, *202*
Wootten, M. J., 293, 333, *336*, *338*
Worley, J. D., 233, *352*

Worley, S. D., 50, *63*
Wright, D. J., 103, *131*
Wu, C.-H., 290, 315, *338*, *344*, *347*
Wu, T.-L., 277, *339*
Wu, Y. C., 242, 243, *352*
Wynne-Jones, K. M. A., 89, *128*
Wynne-Jones, W. F. K., 325, *347*
Wyss, H. R., 233, *341*

Y

Yaacobi, M., 254, 255, 303, 305, *338*, *352*
Yamataka, H., 37, *67*
Yates, K., 148, *202*
Yeager, E., 218, *352*
Yee, K. C., 93, *128*
Yoneda, H., 6, *65*
Yoshida, Z., 165, *199*
Young, A. T., 169, *201*
Yukawa, Y., 37, *67*

Z

Zabel, A. W., 146, *197*
Zahler, R. E., 174, *197*
Zana, R., 218, *352*
Zaugg, H. E., 184, *202*
Závada, J., 183, 185, *200*
Zeidler, M. D., 210, 253, *343*, *352*
Zeiss, H. H., 6, 30, *63*
Zollinger, H., 95, *129*, 153, 190, *197*

Cumulative Index to Authors

Allinger, N. L., **13**, 1
Anbar, M., **7**, 115
Arnett, E. M., **13**, 83
Bard, A. J., **13**, 155
Bell, R. P., **4**, 1
Bennett, J. E., **8**, 1
Bentley, T. W., **8**, 151; **14**, 1
Bethell, D., **7**, 153; **10**, 53
Blandamer, M. J., **14**, 203
Brand, J. C. D., **1**, 365
Brinkman, M. R., **10**, 53
Brown, H. C., **1**, 35
Buncel, E., **14**, 133
Cabell-Whiting, P. W., **10**, 129
Cacace, F., **8**, 79
Carter, R. E., **10**, 1
Collins, C. J., **2**, 1
Cornelisse, J., **11**, 225
Crampton, M. R. **7**, 211
de Gunst, G. P., **11**, 225
Eberson, L., **12**, 1
Farnum, D. G., **11**, 123
Fendler, E. J., **8**, 271
Fendler, J. H., **8**, 271; **13**, 279
Ferguson, G., **1**, 203
Fields, E. K., **6**, 1
Fife, T. H., **11**, 1
Fleischmann, M., **10**, 155
Frey, H. M., **4**, 147
Gilbert, B. C., **5**, 53
Gillespie, R. J., **9**, 1
Gold, V., **7**, 259
Greenwood, H. H., **4**, 73
Havinga, E., **11**, 225
Hogeveen, H., **10**, 29, 129
Ireland, J. F., **12**, 131
Johnson, S. L., **5**, 237
Johnstone, R. A. W., **8**, 151
Kohnstam, G., **5**, 121
Kramer, G. M., **11**, 177
Kreevoy, M. M., **6**, 63
Liler, M., **11**, 267

Ledwith, A., **13**, 155
Long, F. A., **1**, 1
Maccoll, A., **3**, 91
McWeeny, R., **4**, 73
Melander, L., **10**, 1
Mile, B., **8**, 1
Miller, S. I., **6**, 185
Modena, G., **9**, 185
More O'Ferrall, R. A., **5**, 331
Neta, P., **12**, 223
Norman, R. O. C., **5**, 53
Nyberg, K., **12**, 1
Olah, G. A., **4**, 305
Parker, A. J., **5**, 173
Peel, T. E., **9**, 1
Perkampus, H. H., **4**, 195
Pittman, C. U., Jr., **4**, 305
Pletcher, D., **10**, 155
Pross, A., **14**, 69
Ramirez, F., **9**, 25
Rappoport, Z., **7**, 1
Reeves, L. W., **3**, 187
Robertson, J. M., **1**, 203
Rosenthal, S. N., **13**, 279
Samuel, D., **3**, 123
Schaleger, L. L., **1**, 1
Scheraga, H. A., **6**, 103
Schleyer, P. von R., **14**, 1
Scorrano, G., **13**, 83
Shatenshtein, A. I., **1**, 156
Shine, H. J., **13**, 155
Silver, B. L., **3**, 123
Simonyi, M., **9**, 127
Stock, L. M., **1**, 35
Symons, M. C. R., **1**, 284
Thomas, A., **8**, 1
Tonellato, U., **9**, 185
Tüdös, F., **9**, 127
Turner, D. W., **4**, 31
Ugi, I., **9**, 25
Ward, B., **8**, 1
Whalley, E., **2**, 93

Williams, J. M., Jr., **6**, 63
Williamson, D. G., **1**, 365
Wilson, H., **14**, 133
Wolf, A. P., **2**, 201
Wyatt, P. A. H., **12**, 131
Zollinger, H., **2**, 163
Zuman, P., **5**, 1

Cumulative Index of Titles

Abstraction, hydrogen atom, from O—H bonds, **9**, 127
Acid solutions, strong, spectroscopic observation of alkylcarbonium ions in, **4**, 305
Acid-base properties of electronically excited states of organic molecules, **12**, 131
Acids, reactions of aliphatic diazo compounds with, **5**, 331
Acids, strong aqueous, protonation and solvation in, **13**, 83
Activation, entropies of, and mechanisms of reactions in solution, **1**, 1
Activation, heat capacities of, and their uses in mechanistic studies, **5**, 121
Activation, volumes of, use for determining reaction mechanisms, **2**, 93
Aliphatic diazo compounds, reactions with acids, **5**, 331
Alkylcarbonium ions, spectroscopic observation in strong acid solutions, **4**, 305
Ambident conjugated systems, alternative protonation sites in, **11**, 267
Ammonia, liquid, isotope exchange reactions of organic compounds in, **1**, 156
Aqueous mixtures, kinetics of organic reactions in water and, **14**, 203
Aromatic photosubstitution, nucleophilic, **11**, 225
Aromatic substitution, a quantitative treatment of directive effects in, **1**, 35
Aromatic substitution reactions, hydrogen isotope effects in **2**, 163
Aromatic systems, planar and non-planar, **1**, 203
Arynes, mechanisms of formation and reactions at high temperatures, **6**, 1
A-S_E2 reactions, developments in the study of, **6**, 63

Base catalysis, general, of ester hydrolysis and related reactions, **5**, 237
Basicity of unsaturated compounds, **4**, 195
Bimolecular substitution reactions in protic and dipolar aprotic solvents, **5**, 173

^{13}C N.M.R. spectroscopy in macromolecular systems of biochemical interest, **13**, 279
Carbene chemistry, structure and mechanism in, **7**, 163
Carbon atoms, energetic, reactions with organic compounds, **3**, 201
Carbon monoxide, reactivity of carbonium ion towards, **10**, 29
Carbonium ions (alkyl), spectroscopic observation in strong acid solutions, **4**, 305
Carbonium ions, gaseous, from the decay of tritiated molecules, **8**, 79
Carbonium ions, photochemistry of, **10**, 129
Carbonium ions, reactivity towards carbon monoxide, **10**, 29
Carbonyl compounds, reversible hydration of, **4**, 1
Catalysis, enzymatic, physical organic model systems and the problem of, **11**, 1
Catalysis, general base and nucleophilic, of ester hydrolysis and related reactions, **5**, 237
Catalysis, micellar, in organic reactions; kinetic and mechanistic implications, **8**, 271
Cation radicals in solution, formation, properties and reactions of, **13**, 155
Cations, vinyl **9**, 185
Charge density—N.M.R. chemical shift correlations in organic ions, **11**, 125
Chemically induced dynamic nuclear spin polarization and its applications, **10**, 53
CIDNP and its applications, **10**, 53
Conformations of polypeptides, calculations of, **6**, 103
Conjugated molecules, reactivity indices, in, **4**, 73

D_2O—H_2O Mixtures, protolytic processes in, 7, 259
Diazo compounds, aliphatic, reactions with acids, 5, 331
Dimethyl sulphoxide, physical organic chemistry of reactions in, 14, 133
Dipolar aprotic and protic solvents, rates of bimolecular substitution reactions in, 5, 173
Directive effects in aromatic substitution, a quantitative treatment of, 1, 35

Electrochemistry, organic structure and mechanism in, 12, 1
Electrode processes, physical parameters for the control of, 10, 155
Electron spin resonance, identification of organic free radicals by, 1, 284
Electron spin resonance studies of short-lived organic radicals, 5, 23
Electronically excited molecules, structure of, 1, 365
Electronically excited states of organic molecules, acid-base properties of, 12, 131
Energetic tritium and carbon atoms, reactions of, with organic compounds, 2, 201
Entropies of activation and mechanisms of reactions in solution, 1, 1
Enzymatic catalysis, physical organic model systems and the problem of, 11, 1
Equilibrium constants, N.M.R. measurements of, as a function of temperature, 3, 187
Ester hydrolysis, general base and nucleophilic catalysis, 5, 237
Exchange reactions, hydrogen isotope, of organic compounds in liquid ammonia, 1, 156
Exchange reactions, oxygen isotope, of organic compounds, 3, 123
Excited molecules, structure of electronically, 1, 365

Force-field methods, calculation of molecular structure and energy by, 13, 1
Free radicals, identification by electron spin resonance, 1, 284
Free radicals and their reactions at low temperature using a rotating cryostat, study of, 8, 1

Gaseous carbonium ions from the decay of tritiated molecules, 8, 79
Gas-phase heterolysis, 3, 91
Gas-phase pyrolysis of small-ring hydrocarbons, 4, 147
General base and nucleophilic catalysis of ester hydrolysis and related reactions, 5, 237

H_2O—D_2O Mixtures, protolytic processes in, 7, 259
Heat capacities of activation and their uses in mechanistic studies, 5, 121
Heterolysis, gas-phase, 3, 91
Hydrated electrons, reactions of, with organic compounds, 7, 115
Hydration, reversible, of carbonyl compounds, 4, 1
Hydrocarbons, small-ring, gas-phase pyrolysis of, 4, 147
Hydrogen atom abstraction from O—H bonds, 9, 127
Hydrogen isotope effects in aromatic substitution reactions, 2, 163
Hydrogen isotope exchange reactions of organic compounds in liquid ammonia, 1, 156
Hydrolysis, ester, and related reactions, general base and nucleophilic catalysis of, 5, 237

Ionization potentials, 4, 31
Ions, organic, charge density—N.M.R. chemical shift correlations, 11, 125
Isomerization, permutational, of pentavalent phosphorus compounds, 9, 25
Isotope effects, hydrogen, in aromatic substitution reactions, 2, 163
Isotope effects, steric, experiments on the nature of, 10, 1
Isotope exchange reactions, hydrogen, of organic compounds in liquid ammonia, 1, 150
Isotope exchange reactions, oxygen, of organic compounds, 3, 123
Isotopes and organic reaction mechanisms, 2, 1

Kinetics, reaction, polarography and, 5, 1
Kinetics of organic reactions in water and aqueous mixtures, 14, 203

Macromolecular systems of biochemical interest, ^{13}C N.M.R. spectroscopy in, 13, 279
Mass spectrometry, mechanism and structure in: a comparison with other chemical processes, 8, 152
Mechanism and structure in carbene chemistry, 7, 153
Mechanism and structure in mass spectrometry: a comparison with other chemical processes, 8, 152

Mechanism and structure in organic electrochemistry, 12, 1
Mechanisms, organic reaction, isotopes and, 2, 1
Mechanisms of reactions in solution, entropies of activation and, 1, 1
Mechanisms of solvolytic reactions, medium effects in the rates and, 14, 133
Mechanistic studies, heat capacities of activation and their use in, 5, 121
Mechanistic applications, the reactivity–selectivity principle, 14, 69
Medium effects on the rates and mechanisms of solvolytic reactions, 14, 1
Meisenheimer complexes, 7, 211
Micellar catalysis in organic reactions: kinetic and mechanistic implications, 8, 271
Molecular structure and energy, calculation of, by force-field methods, 13, 1

N.M.R. chemical shift—charge density correlations, 11, 125
N.M.R. measurements of reaction velocities and equilibrium constants as a function of temperature, 3, 187
N.M.R. Spectroscopy, ^{13}C, in macromolecular systems of biochemical interest, 13, 279
Non-planar and planar aromatic systems, 1, 203
Norbornyl cation: reappraisal of structure, 11, 179
Nuclear magnetic resonance, see N.M.R.
Nucleophilic aromatic photosubstitution, 11, 225
Nucleophilic catalysis of hydrolysis and related reactions, 4, 237
Nucleophilic vinylic substitution, 7, 1

O—H bonds, hydrogen atom abstraction from, 9, 127
Oxygen isotope exchange reactions of organic compounds, 3, 123

Permutational isomerization of pentavalent phosphorus compounds, 9, 25
Phosphorus compounds, pentavalent, turnstile rearrangement and pseudorotation in permutational isomerization, 9, 25
Photochemistry of carbonium ions, 10, 129
Photosubstitution, nucleophilic aromatic, 11, 225
Planar and non-planar aromatic systems, 1, 203
Polarizability, molecular refractivity and, 3, 1
Polarography and reaction kinetics, 5, 1
Polypeptides, calculations of conformations of, 6, 103
Protic and dipolar aprotic solvents, rates of bimilecular substitution reactions in, 5, 173
Protolytic processes in H_2O—D_2O mixtures, 7, 259
Protonation and solvation in strong aqueous acids, 13, 83
Protonation sites in ambident conjugated systems, 11, 267
Pseudorotation in isomerization of pentavalent phosphorus compounds, 9, 25
Pyrolysis, gas-phase, of small-ring hydrocarbons, 4, 147

Radiation techniques, application to the study of organic radicals, 12, 223
Radicals, cation, in solution, formation, properties and reactions of, 13, 155
Radicals, organic, application of radiation techniques, 12, 223
Radicals, organic free, identification by electron spin resonance, 1, 284
Radicals, short-lived organic, electron spin resonance studies of, 5, 53
Rates and mechanisms of solvolytic reactions, medium effects on, 14, 1
Reaction kinetics, polarography and, 5, 1
Reaction mechanisms, use of volumes of activation for determining, 2, 93
Reaction mechanisms in solution, entropies of activation and, 1, 1
Reaction velocities and equilibrium constants, N.M.R. measurements of, as a function of temperature, 3, 187
Reactions of hydrated electrons with organic compounds, 7, 115
Reactions in dimethyl-sulphoxide, physical organic chemistry of, 14, 133
Reactivity indices in conjugated molecules, 4, 73
Reactivity-selectivity principle and its mechanistic applications, 14, 69
Refractivity, molecular, and polarizability, 3, 1
Resonance, electron-spin, identification of organic free radicals, by, 1, 284
Resonance, electron-spin, studies of short-lived organic radicals, 5, 63

Short-lived organic radicals, electron-spin resonance studies of, **5**, 53
Small-ring hydrocarbons, gas-phase pyrolysis of, **4**, 147
Solution, reactions in, entropies of activation and mechanisms, **1**, 1
Solvation and protonation in strong aqueous acids, **13**, 83
Solvents, protic and dipolar aprotic, rates of bimolecular substitution reactions in, **5**, 173
Solvolytic reactions, medium effects on the rates and mechanisms of, **14**, 1
Spectroscopic observation of alkylcarbonium ions in strong acid solutions, **4**, 305
Spectroscopy, ^{13}C N.M.R., in macromolecular systems of biochemical interest, **13**, 279
Stereoselection in elementary steps of organic reactions, **6**, 185
Steric isotope effects, experiments on the nature of, **10**, 1
Structure and mechanism in carbene chemistry, **7**, 153
Structure and mechanism in organic electrochemistry, **12**, 1
Structure of electronically excited molecules, **1**, 365
Substitution, aromatic, a quantitative treatment of directive effects in, **1**, 35
Substitution reactions, aromatic, hydrogen isotope effects in, **2**, 163
Substitution reactions, bimolecular, in protic and dipolar aprotic solvents, **5**, 173
Superacid systems, **9**, 1

Temperature, N.M.R. measurements of reaction velocities and equilibrium constants as a function of, **3**, 187
Tritiated molecules, gaseous carbonium ions from the decay of, **8**, 79
Tritium atoms, energetic, reactions with organic compounds, **2**, 201
Turnstile rearrangement in isomerization of pentavalent phosphorus compounds, **9**, 25

Unsaturated compounds, basicity of, **4**, 195

Vinyl cations, **9**, 185
Volumes of activation, use of, for determining reaction mechanisms, **2**, 93

Water and aqueous mixtures, kinetics of organic reactions in, **14**, 203

QD
476
A4
v.14
1977

JUN 28 1978